Schröder
Rheology of Plastics

Thomas Schröder

Rheology of Plastics

Theory and Practice

HANSER

Print-ISBN: 978-1-56990-387-2
E-Book-ISBN: 978-1-56990-946-1

Bibliographic information of the German National Library:
The German National Library lists this publication in the German National Bibliography; detailed bibliographic
data are available on the Internet at http://dnb.d-nb.de.

© 2025 Carl Hanser Verlag GmbH & Co. KG, Munich
Vilshofener Straße 10 | 81679 Munich | info@hanser.de
www.hanserpublications.com
www.hanser-fachbuch.de
Editor: Dr. Mark Smith
Production Management: le-tex publishing services GmbH, Leipzig
Cover concept: Marc Müller-Bremer, www.rebranding.de, Munich
Cover design: Max Kostopoulos
Cover picture: © MHT Mold & Hotrunner Technology AG
Typesetting: Eberl & Kœsel Studio, Kempten

Preface

Rheology is the science of the flow and deformation behavior of materials. In the field of plastics, rheology plays a crucial role, as these materials undergo melting, flowing, and solidifying processes to take shape in a wide range of applications. Whether in injection molding, extrusion, or 3D printing, the rheological properties of plastics are pivotal for the quality of the final product.

This book is dedicated to the rheology of plastics, a discipline that bridges theoretical foundations and practical applications. Plastics exhibit complex material properties, ranging from viscosity and elasticity to yield stress and relaxation behavior. These properties depend not only on the chemical composition of polymers but also on factors such as temperature, pressure, and deformation rate.

The importance of rheology extends far beyond material science. In industrial practice, it enables the optimization of manufacturing processes, the prevention of defects in final products, and the targeted development of new plastic formulations. For example, studying flow behavior helps in selecting the ideal process parameters for injection molding or improving extrusion processes to achieve more precise and efficient production. In the injection molding process, rheology can be used to design the rheology of distribution systems, such as hot runner systems, and to optimize molded parts (cavities) with regard to pressure loss or material shearing. In extrusion, distributor dies (e.g. fishtail or coat hanger) can be rheologically balanced so that the melt emerges from the dies evenly and the melt is subjected to the lowest possible thermal and mechanical stress.

This book covers both the fundamentals of the rheology of plastics and its practical applications. It aims to provide students, researchers, engineers, and industry professionals with a comprehensive understanding of the subject. Theory and practice go hand in hand: while the physical principles allow a systematic analysis of material properties, the application-oriented approach demonstrates how this knowledge can be leveraged to develop innovative and sustainable solutions.

I invite you to delve into the fascinating world of the rheology of plastics. May this book provide you with both knowledge and inspiration for your work and research.

Thomas Schröder

Foreword

Thomas Schröder's *Rheology of Plastics* is a timely and important addition to the field of polymer technology, offering valuable insights into the rheological behavior of plastics and the practical applications of rheology in processing and process design. With clarity and depth, this book serves as both a theoretical foundation and a practical guide, offering a comprehensive compendium of concepts in plastics rheology. Rooted in the author's extensive teaching experience, this book stands out as an effective pedagogical tool. It is well-suited for instructors seeking a structured path to introduce the core principles of polymer rheology, for students entering the field, and for practicing engineers aiming to deepen their understanding of how rheology impacts processing during product manufacture.

The book begins with a concise presentation of key equations—those that are second nature to seasoned rheologists—before moving into the fundamental principles of rheology. The theoretical groundwork chapters set the stage for an insightful chapter on rheometry, followed by a detailed exploration of constitutive equations. Schröder skillfully ties rheological behavior to variables such as deformation rates and temperature, showing how these relationships are used to model common flow scenarios and link these directly to polymer processing operations. One of the book's great strengths is its unwavering focus on the practical applications of rheology. Complex concepts are consistently anchored in real-world processing challenges—from the design of injection molding hot runner systems to advanced process simulation—ensuring that readers see the relevance of rheology in everyday engineering practice.

Rheology of Plastics is an essential resource for anyone seeking to understand the flow behavior of polymers, bridging the gap between theory and application with clarity, precision, and relevance.

Tim A. Osswald

The Author

Prof. Dr. Thomas Schröder taught the subjects of injection molding, tool technology, simulation technology and rheology at the Darmstadt University of Applied Sciences (h_da) until 2024. After studying mechanical engineering with a specialization in plastics technology at RWTH Aachen University, he completed his doctorate under Prof. Dr. Dr. h.c. Walter Michaeli in the field of gas injection technology. After several years with a well-known plastics processor, he moved to Krupp Corpoplast in Hamburg, where he was responsible for preform manufacturing systems. Following this role, he headed the application technology SPA at the injection molding machine manufacturer Netstal in Switzerland. In 2001, Prof. Schröder was appointed to the Darmstadt University of Applied Sciences. He is a member of the Institute for Plastics Technology Darmstadt ikd and chairman of the Society for the Promotion of Young Engineers GFTN e.V., which is an Institute at the Darmstadt University of Applied Sciences. Prof. Schröder conducts very intensive research in the areas of rheology, injection molding and tool technology as part of funded third-party projects. Furthermore, Prof. Schröder is a member of the Doctoral Center for Sustainability Sciences and thus has the right to award doctorates at the Darmstadt University of Applied Sciences.

Important Rheological Formulas

The *Hagen-Poiseuille* Equations

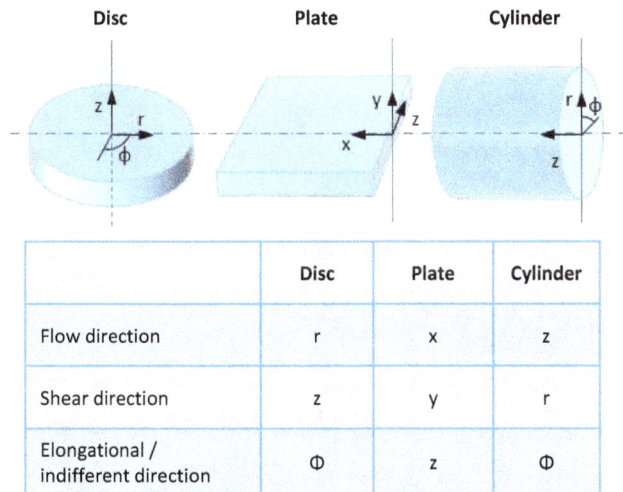

	Disc	Plate	Cylinder
Flow direction	r	x	z
Shear direction	z	y	r
Elongational / indifferent direction	Φ	z	Φ

Figure 1 Geometries for the *Hagen-Poiseuille* Equations

Disk	$\dfrac{\Delta p}{r} = \dfrac{12 \cdot \bar{v}_r \cdot \eta}{H^2}$
Plate	$\dfrac{\Delta p}{x} = \dfrac{12 \cdot \bar{v}_x \cdot \eta}{H^2}$
Cylinder	$\dfrac{\Delta p}{z} = \dfrac{32 \cdot \bar{v}_z \cdot \eta}{D^2}$

With the continuity equation $\dot{V} = \bar{v} \cdot A$ follows:

Disk	$\dfrac{\Delta p}{r} = \dfrac{6 \cdot \dot{V} \cdot \eta}{\pi \cdot R \cdot H^3}$
Plate	$\dfrac{\Delta p}{x} = \dfrac{12 \cdot \dot{V} \cdot \eta}{B \cdot H^3}$
Cylinder	$\dfrac{\Delta p}{z} = \dfrac{128 \cdot \dot{V} \cdot \eta}{\pi \cdot D^4}$

Equations for the Representative Shear Rate

Disk	$\bar{\dot{\gamma}} = e_{rectangle} \cdot \dfrac{6 \cdot \bar{v}_r}{H} = e_{rectangle} \cdot \dfrac{3 \cdot \dot{V}}{\pi \cdot r \cdot H^2}$
Plate	$\bar{\dot{\gamma}} = e_{rectangle} \cdot \dfrac{6 \cdot \bar{v}_x}{H} = e_{rectangle} \cdot \dfrac{6 \cdot \dot{V}}{B \cdot H^2}$
Cylinder	$\bar{\dot{\gamma}} = e_{circle} \cdot \dfrac{8 \cdot \bar{v}_z}{D} = e_{circle} \cdot \dfrac{32 \cdot \dot{V}}{\pi \cdot D^3}$

With $e_{rectangle} = 0.772$ and $e_{circle} = 0.815$

Equations for Calculating the Viscosity

Power-Law Approach (Potency Approach of *Ostwald* and *de-Waele*)

$$\eta = K \cdot a_T{}^n \cdot \dot{\gamma}^{n-1}$$

or

$$\eta = a_T{}^{\frac{1}{m}} \cdot \phi^{-\frac{1}{m}} \cdot \dot{\gamma}^{\frac{1-m}{m}}$$

Carreau-WLF-Approach

$$\eta = \dfrac{a_T \cdot P_1}{(1 + a_T \cdot \dot{\gamma} \cdot P_2)^{P_3}}$$

Cross-WLF-Approach

$$\eta\left(\dot{\gamma}\right) = \frac{\eta_0}{1 + \left(\frac{\eta_0 \dot{\gamma}}{\tau^*}\right)^{1-n}}$$

with

$$\eta_0\left(T\right) = D_1 \cdot e^{\left[\frac{-A_1\left(T-T^*\right)}{A_2+\left(T-T^*\right)}\right]}$$

Equations for Calculating the Temperature Shift Factor a_T

Arrhenius-Approach

$$a_T = e^{\left[\frac{E_0}{R}\left(\frac{1}{T} - \frac{1}{T_{reference}}\right)\right]}$$

William-Landel-Ferry-Approach (WLF-Approach)

$$\log\left(a_T\right) = \frac{8.86 \cdot \left(T_{reference} - T_s\right)}{101.6K + \left(T_{reference} - T_s\right)} - \frac{8.86 \cdot \left(T - T_s\right)}{101.6K + \left(T - T_s\right)}$$

Equation for the Temperature Depending Specific Volume

$$v\left(\vartheta\right) = v\left(\vartheta_0\right) \cdot \left[1 + \alpha \cdot \left(\vartheta - \vartheta_0\right)\right]$$

Contents

1 Introduction

Rheology is a very old discipline and was founded as early as 1930 by E. C. Bingham and M. Reiner in Easton (USA) as an independent discipline. Significant individual contributions were published much earlier, for example, in 1676 by R. Hooke; in 1687, by I. Newton; in 1745, by L. Euler; in 1820, by C. L. M. H. Navier; in 1845, by G. Stokes; in 1847, by J. L. M. Poiseuille; in 1867, by J. C. Maxwell; and in 1908, by L. Prandtl [1]. Approaches from these contributions, such as Newton's law of friction or the law of Hagen–Poiseuille, are also derived in this book [1].

Figure 1.1 Sir Isaac Newton (left) (* January 4, 1643, † March 31, 1727), Robert Hooke (right) (* July 28, 1635, † March 4, 1703) [Source: Wikipedia]

The word rheology is derived from the Greek word *rheos*, which means "to flow". Rheology is therefore the science (*logia*) of the deformation and flow of materials. Flow can be understood as the continuous deformation of a material under the influence of external forces. The task of rheology is to describe, measure, and explain how a solid or a fluid reacts under the influence of external forces and deformations [1] (Figure 1.2).

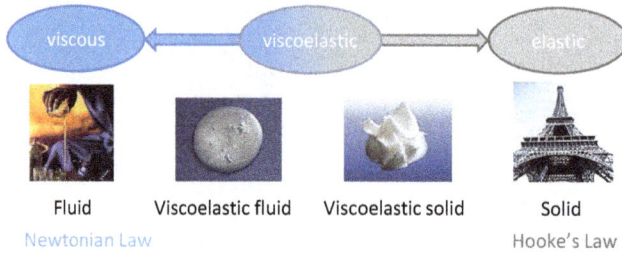

Figure 1.2 Rheology describes the deformation and flow behavior of materials [2]

What Is Rheology?

Rheology is the science of the deformation and flow of materials.

Sooner or later, everyone encounters special rheological phenomena. Whether it is toothpaste and jam in the morning or ketchup in the evening (Figure 1.3), all these substances have their own special flow behavior, which will be discussed in detail in this book.

Figure 1.3 Everyday examples of rheological phenomena

1.1 Why Is Rheology Needed in Plastics Technology?

The rheology of plastics makes it possible, among other things, to describe the flow processes of plastic melts. The approaches and boundary conditions, such as the Hagen–Poiseuille equation or the Navier–Stokes adhesion condition, are explained and derived in the following chapters and then used to calculate flow processes.

With the help of today's computer-aided simulation programs, such as Cadmould®, Moldflow®, Moldex®, Sigmasoft®, or Fluent®, these flow processes can be calculated and graphically displayed with the corresponding boundary conditions (material parameters, processing conditions, etc.). These programs are based on the fundamentals of rheology. These include the rheological material data from rheometry and the mostly empirically determined mathematical approaches, such as material laws and temperature equations of rheology.

If we think of extrusion, rheology is used to describe the complex flow processes in the extruder and thus to design screw geometries. Furthermore, the flow and mixing processes in shearing and mixing parts can be described. The balancing of the melt in the distributor systems—for example, in slot dies or side fed mandrel distributor systems using coat-hanger or fish-tail distributors—is also an important branch of rheology (Figure 1.4). The aim of balancing is to distribute the melt flow in such a way that it has the same velocity at all points at the outlet. This is achieved by balancing the distribution systems in such a way that the pressure loss is the same on all flow paths. These balancing methods will be discussed in detail later.

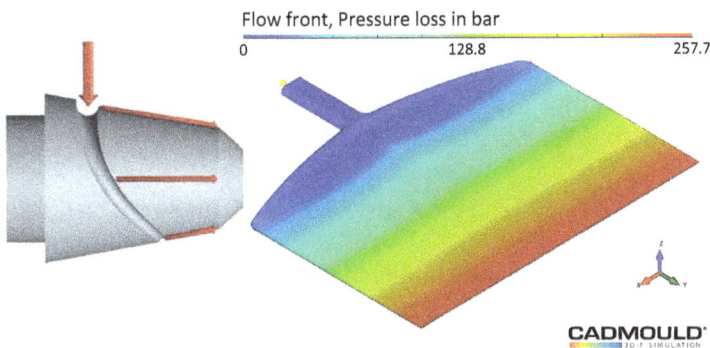

Flow front, Pressure loss in bar

0   128.8   257.7

CADMOULD®

Figure 1.4 Balancing of a side fed mandrel and a slot die using a coat hanger

Alongside extrusion, injection molding is one of the most widely used plastics processing methods. The principles of rheology are also used here in the design of pusher screws, piston accumulators, and shearing and mixing parts. The aim is to be able to describe the flow processes of the plastic in the complete injection molding system, that is, from the granulate to the finished part. A statement about pressure losses, dwell times, temperatures, shear stresses, shear rates, orientations, etc. is possible.

The topic of structural analysis is becoming increasingly important, especially from the point of view of lightweight construction. Simulation programs such as those mentioned above can be used to predict the flow-related orientations of glass fibers in the molded part, for example, and transfer them to a strength calculation program via an interface using mapping. This makes it possible to take the anisotropy of the material into account when designing the component and, as a result, to optimize the component, which is much more realistic and therefore more effective.

In injection molding, however, the focus is on the design of the injection molds including the gating system for the production of plastic molded parts. The design process can be divided into three phases (Figure 1.5). After the mold finding phase, the mold dimensioning follows. In addition to the mechanical and thermal design, this also includes the rheological design. This means that rheological knowledge, i.e., knowledge of the flow behavior of plastic melts, is primarily required at this point [3].

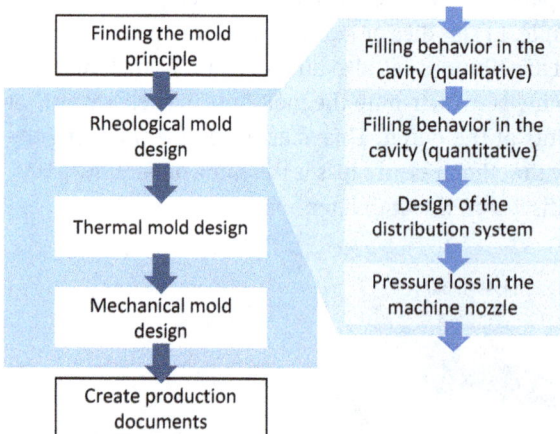

Finding the mold principle

Rheological mold design

Thermal mold design

Mechanical mold design

Create production documents

Filling behavior in the cavity (qualitative)

Filling behavior in the cavity (quantitative)

Design of the distribution system

Pressure loss in the machine nozzle

Figure 1.5 The phases of mold design and the steps of rheological mold design [4]

The rheological mold design is the first design step of the second mold design phase, as hardly any restrictions from the results of the other design steps are to be expected here. It is initially used to determine the position of weld lines and air pockets depending on the type and position of the gates and the wall thickness. Since the material behavior plays a subordinate role for such an analysis and since neither pressures nor velocities of the melt are required here, this design step is designated as qualitative in Figure 1.5. The filling pattern method can be used here, for example. This method will be discussed in detail later.

Once the flow paths of the melt in the mold cavity have been determined by this analysis, a quantitative analysis can be carried out. This requires knowledge of the material behavior, that is, the viscosity function and thermal material values. In this step of the design, the pressure required to fill the mold cavity is calculated. This depends not only on the material behavior but also on the process parameters. The optimum injection rates and melt and wall temperatures are therefore also determined here. Furthermore, the limit values (pressure, temperature, shear rate, shear stress, etc.) of the process parameters can be defined; if exceeded, they lead to material damage due to excessive shear rates or excessive frictional heating.

The first two rheological design steps result in the positions of gates and the required melt flows as well as the required melt temperature. These are decisive boundary conditions for the distribution system to be designed in the third step. The position of the gates determines the rough dimensions and possible distributor variants. In most cases, the designer must dimension the diameters of the distributor channel using the same simulation calculation as in step two. The results here are also additional pressure losses in the distributor system as well as temperature and shear stress on the material.

After this dimensioning, the pressure required to fill the mold is determined. This must be provided by the machine. In addition, the pressure loss occurring in the machine nozzle itself must also be taken into account. Depending on the amount of pressure loss, it must be checked whether the machine pressure (injection pressure) is sufficient.

When calculating pressure losses in the machine nozzle and in the melt distribution system, it is also crucial to take into account the so-called inlet or extensional pressure losses. These always occur when there is a change in the flow channel cross section (cross-sectional jump). For example, the diameter of a machine nozzle always changes from the screw diameter to a diameter that is smaller than that of the sprue bushing. This results in additional pressure losses, which must be added to the shear pressure losses. This topic will also be discussed in detail later.

Computer-aided simulation programs usually neglect these pressure losses because the rheological material data is either not available or inadequate. This can lead to incorrect calculations, especially when designing hot runner systems. Depending on the complexity of the distributor system, the calculated pressure losses are often far below the actual pressure losses.

The rheological design is followed by the thermal design. Some restrictions from the rheological calculation must already be taken into account here. For example, the melt temperature is specified and the temperature at the mold cavity wall is also fixed within narrow limits.

The laws of rheology are also used to predict the flow and filling behavior for molds with several mold cavities or large molds with multiple injection points. Rheological balancing is particularly important for multi-cavity (Figure 1.6) and family molds. The aim of balancing is to complete the filling of all mold cavities at the same time.

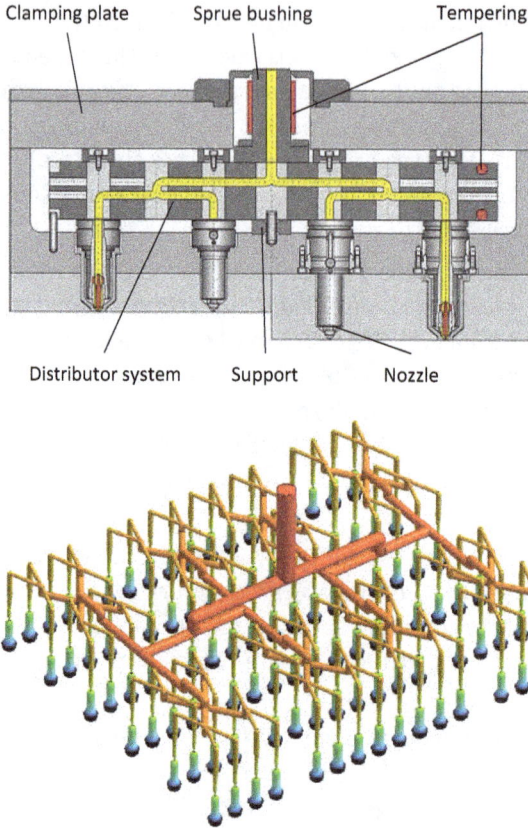

Figure 1.6 Multi-cavity mold with hot runner distributor system [5, 6]

For multi-cavity molds, two options for rheological balancing are available:

- Natural rheological balancing
- Mathematical rheological balancing

Family molds (see Figure 1.7), in which the cavities are not identical, are always mathematically rheologically balanced, while natural rheological and mathematical rheological balancing can be used for multi-cavity molds with identical cavities. As a rule, natural rheological balancing is preferred for these systems due to the independence of the operating point and the simpler design. This topic will also be discussed in more detail later.

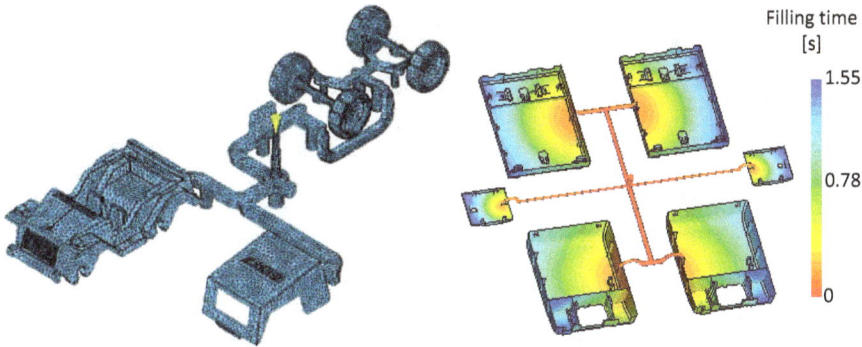

Figure 1.7 Examples of family molds [7]

1.2 Computer-Aided Simulation Programs for the Design of Injection Molds

Since the empirical determination of the optimum mold design—for example, using test molds—is quite time-consuming and cost-intensive, computer-aided simulation programs are increasingly being used in the mold finding phase. These computer-aided design (CAD) programs are able, for example, to predict the flow processes in the mold quite accurately (Figure 1.8). To make this possible, however, a number of mathematical approaches are required to describe the flow behavior of the plastic melt. Furthermore, data on the characteristic flow properties of the plastic melt is re-quired to calculate the flow processes. This is where rheology comes into play. The typical flow properties of the plastic melts are measured and recorded with the help of various measuring devices.

Based on these rheological principles, the computer-aided simulation programs can calculate and graphically display the filling behavior (isovels), the temperature field (isotherms), the filling pressure (isobars), orientations, shrinkage and warpage, and other flow-dependent variables.

For more complex mold geometries (bumpers, dashboards, etc.), a computer-aided rheological design of the mold is usually carried out at an early stage. This allows the position and number of injection points to be defined and varied. Furthermore, the CAD programs make it possible to predict the weld lines and warpage. By varying the injection points, the complex molded part can be optimized in advance. This reduces the costs and time required for development, mold production, and subsequent sam-pling. Special processes such as cascade injection molding or Dynamic Feed® can now also be simulated.

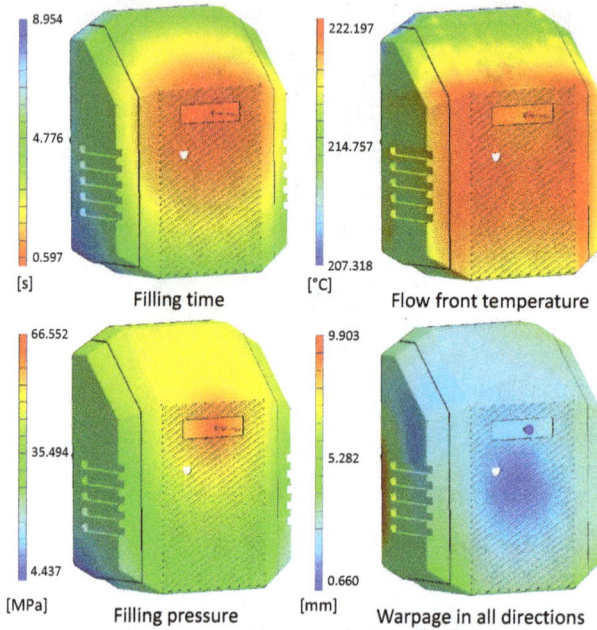

Figure 1.8 Results of a computer-aided simulation calculation

Figure 1.9 shows a typical example of a simulation of a bumper mold. In this case, pressure-controlled cascade injection molding (Dynamic Feed®) was simulated.

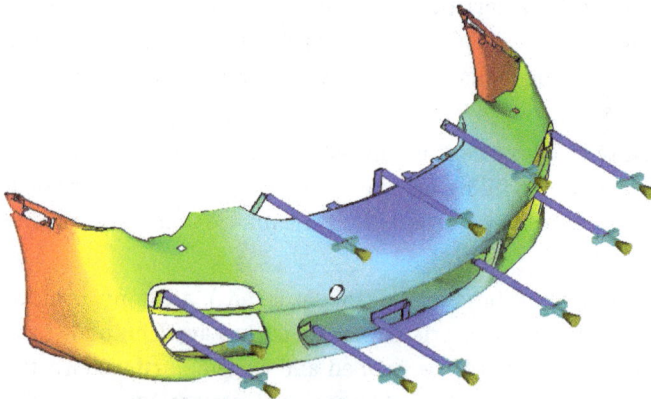

Figure 1.9 Computer-aided simulation of a bumper

Furthermore, the quality of the injection-molded parts is decisively defined by the part formation process in the injection mold. This includes the flow processes of the melt in the mold, in the injection and the holding pressure phase (see Figure 1.10 and Table 1.1).

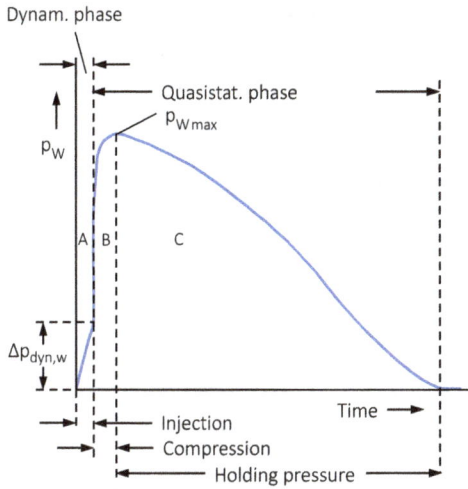

Figure 1.10
Cavity pressure with the process phases

Table 1.1 Molded Part Quality Depending on the Process Phases [8]

	Injection phase	Compression phase	Holding pressure phase
Variables with an influence	▪ Injection rate ▪ Melt and mold temperature ▪ Melt viscosity	▪ Switchover to holding pressure ▪ Setting of the pressure limitation	▪ Holding pressure level and duration ▪ Mold temperature ▪ Deformation of the mold ▪ Stability of the clamping unit ▪ Clamping force
Material parameters with an influence	▪ Melt viscosity ▪ Molecular degradation ▪ Crystallinity ▪ Orientation in the surface layer	▪ Crystallinity ▪ Anisotropies	▪ Crystallinity ▪ Orientation inside the molded part ▪ Shrinkage
Properties of molded parts affected	▪ Surface quality	▪ Degree of molding ▪ Flash formation ▪ Weight	▪ Weight ▪ Dimensional accuracy ▪ Blowhole ▪ Sink marks ▪ Relaxation ▪ Demolding behavior

All key quality characteristics, such as weight, dimensional accuracy, surface quality, etc., are primarily determined in these three-part formation phases. In this respect, knowledge of the pressure curve in the process phases is of great importance.

Computer-aided simulation programs are also very helpful here. These programs can be used to make statements about the part formation process (pressure, flow front rate, shear, shear stress, temperature, etc.). Since the part formation process, as Figure 1.10 shows, is directly related to the quality of the molded part, it is possible to make statements about possible weak points or part defects in advance. However, this generally requires a high level of specialist knowledge, as the interrelationships are usually quite complex.

References

[1] Pahl, M.; Gleißle, W.; Laun, H.-M.: *Praktische Rheologie der Kunststoffe und Elastomere*. VDI-Verlag GmbH, Düsseldorf, 1995

[2] Mezger, T. G.: *The Rheology Handbook*. Vincentz Network, Hanover, 2016

[3] Hopmann, C.; Menges, G.; Michaeli, W.; Mohren, P.: *Spritzgießwerkzeuge Auslegung, Bau, Anwendung*. Hanser, Munich, 2018

[4] Lichius, U.; Schmidt, L.: *Rechnergestütztes Konstruieren von Spritzgießwerkzeugen: systematisches Entwickeln von Betriebsmitteln, Aufbau und Funktion von Spritzgießwerkzeugen*. Vogel, Würzburg, 1986

[5] N. N.: Ewikon Heißkanaltechnik, Technical document, 1992

[6] N. N.: MHT Mold & Hotrunner Technology AG, *http://www.mht-ag.de*

[7] N. N.: CADMOULD 3D-F User Manual

[8] Johannaber, F.; Michaeli, W.: *Handbuch Spritzgießen*. Hanser, Munich, 2004

2 Rheological Phenomena

In rheology, a distinction is made between three basic rheological properties:

- Viscosity
- Plasticity
- Elasticity

All real materials have these basic rheological properties. These vary and depend on the stress level, stress duration, temperature, etc. [1].

The basic properties of elasticity, plasticity, and viscous material behavior can be clearly demonstrated with spheres that have these properties. To do this, select a steel ball, a putty ball, and a drop of water and drop them from a not too great height h onto a clean steel plate (Figure 2.1). The steel ball bounces up, comes to rest at some point, and remains undeformed. In this stress range, the steel ball corresponds to a purely elastic solid. The putty ball deforms plastically on impact and retains this deformation at rest. This ball exhibits plastic-elastic solid behavior. After impact, the water droplet flows apart until the interfacial tension is reached and forms a film. Water is therefore a viscous fluid. If you now take a silicone rubber ball and drop it onto the plate like the other balls, the ball bounces several times like an elastic solid. In the end, it remains in place and melts on the surface after a long time. Silicone rubber is a viscoelastic fluid [1].

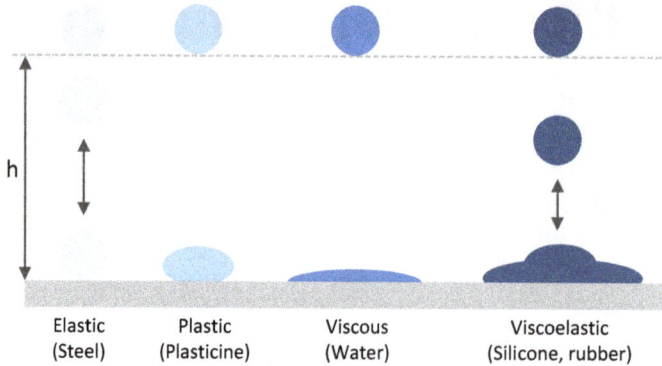

Figure 2.1 Examples of rheological properties [1]

> Plastic melts have viscoelastic flow behavior.

In general, a distinction must be made between low-viscosity and high-viscosity media. While air has a very low viscosity and flows almost frictionlessly, plastic, for example, has a very high viscosity in the processing region. The flow resistance of plastic is correspondingly high. This is noticeable during processing: Due to the high viscosity (toughness), high pressure is required to inject the plastic into a mold, for example.

> Flow resistance serves as a measure of viscosity. The unit of viscosity is Pa·s.

Table 2.1 shows the viscosity values of some materials under ambient conditions. It also shows that the plastic melt has a much higher viscosity under processing conditions (temperature, pressure, etc.) than water, for example.

Table 2.1 Viscosity Values of Some Materials

Material	Viscosity η [Pa·s]	Consistency
Air	10^{-5}	Gaseous
Water	10^{-3}	Low-viscosity fluid
Glycerin	1	Fluid
Polymer melt*	10^1 to 10^6	High-viscosity fluid
Glass	10^{21}	Solid-like

* at processing temperature

2.1 Shear-Thinning Viscosity

For Newtonian media, the viscosity η is purely a material quantity and only depends on the pressure and temperature. For shear-thinning viscous media, η is still influenced by the deformation (shear) rate $\dot{\gamma}$ and the time t.

The following example is intended to illustrate the difference between a Newtonian and a shear-thinning viscous medium. Two glass tubes with a capillary of the same geometry are used for this purpose (Figure 2.2). The left tube is filled with a Newtonian fluid; the other one is filled with a shear-thinning viscous polymer solution to the same level.

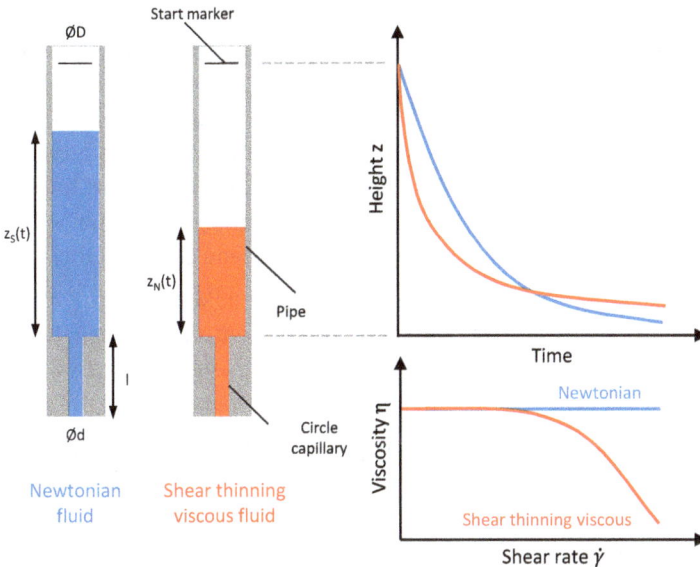

Figure 2.2 Flow phenomena in a Newtonian fluid and a polymer solution [2, 3]

The zero viscosities η_0 of the two media are first measured using a falling ball viscometer (see Chapter 6). The experiment shows that the ball sinks at the same rate in both tubes, which initially indicates an identical zero viscosity η_0 ($\eta_{0N} = \eta_{0S}$). If the two media are now allowed to escape through the capillaries, the shear-thinning viscous medium initially flows faster. After a certain time, both menisci move at the same velocity.

This behavior can be explained as follows: At the beginning of the experiment, the column of fluid is large. The weight of the column generates a high pressure in the capillaries, which initially results in a high flow velocity and thus a high shear rate $\dot{\gamma}$. For the shear-thinning viscous medium, the viscosity η decreases with increasing shear rate $\dot{\gamma}$ (flow rate). As a result, this fluid initially flows faster than the Newtonian fluid and the tube empties correspondingly faster. Due to the decreasing pressure, the fluid flows more slowly through the capillary as the filling level decreases. The shear-thinning viscosity of the fluid means that with decreasing shear rate $\dot{\gamma}$ the viscosity η increases again until the so-called zero viscosity η_0 is reached. A further reduction in the shear rate does not lead to a further increase in η. From a certain filling level, the velocity (shear rate $\dot{\gamma}$) is so low that η_0 is reached. The subsidence of the menisci is no longer dependent on the shear rate $\dot{\gamma}$ [1].

Shear-Thinning Viscous Flow Behavior of Plastics

Plastics are so-called macromolecules (polymers) that have the shape of a chain. These thread-like macromolecules are submicroscopically small and have a hydrody-namic diameter of 5 to 50 nanometers in their disordered resting state. The length of the chains is determined by the number of monomer units contained in the chain. In the case of polyethylene, this can be up to 10^4 units. The most important parameter that reflects the chain length is the molecular weight [4]. The longer the chain, the greater the molecular weight. If, for example, we take a polyethylene with a molecu-lar weight corresponding to 100,000 g/mol, the filamentous molecular chains have a length of approximately 1000 nm and a diameter of approximately 0.5 nm when stretched. This results in a length-to-diameter ratio of 2000. If you compare this to a spaghetti with a diameter of 1 mm, this spaghetti would be 2 m long.

At rest, the macromolecules seek the state of greatest possible entropy and are there-fore disordered, in the shape of a ball. The molecular chains are intertwined. Initially, a lot of energy is required to move this polymer ball, that is, to make it flow.

Due to the flow process (wall adhesion, maximum flow rate in the center of the flow channel, and laminar layer flow), the individual fluid layers move in relation to each other. As a result of this process, shear stresses are transferred between the layers due to internal friction, which is referred to as dissipation.

These shear stresses act on the individual polymer chains of the plastic melt and cause the disordered chains to align in the direction of flow. The more the molecular chains are stretched (i.e., aligned), the less energy is required to allow them to slide past each other. For this reason, the viscosity decreases as the shear rate increases. As a result, the shear stress does not display a linear relationship, but is an increasing, concave-down curve. This behavior of a fluid is referred to as shear-thinning viscosity (Figure 2.3). If the shear rate increases even further, the viscosity no longer decreases from a certain point, but runs towards a horizontal plateau. This plateau value is re-ferred to as infinite viscosity η_∞.

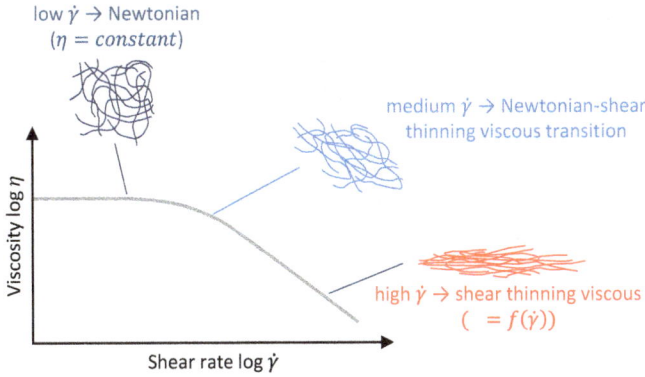

low $\dot{\gamma} \rightarrow$ Newtonian
$(\eta = constant)$

medium $\dot{\gamma} \rightarrow$ Newtonian-shear
thinning viscous transition

high $\dot{\gamma} \rightarrow$ shear thinning viscous
$(= f(\dot{\gamma}))$

Viscosity log η

Shear rate log $\dot{\gamma}$

Figure 2.3 Shear-thinning viscous behavior of plastic melts

> Plastic melts exhibit a shear-thinning viscous behavior. The following applies:
> $\eta = f(\dot{\gamma})$ and $\tau = f(\dot{\gamma})$ (The shear stress τ does not display a linear relationship, but is an increasing, concave-down curve.)

The shear-thinning viscous flow behavior is a function of the polymer type. Depending on the structure of the plastic, the viscosity of the plastic changes differently as a function of the shear rate. This is illustrated in Figure 2.4. There are plastics, such as polycarbonate (PC), which exhibit almost Newtonian flow behavior over a large shear rate range. Later we will see that for a PC, for example, the value m (flow exponent), which describes the shear-thinning viscous flow behavior, is almost $m = 1$. Newtonian flow behavior is present for $m = 1$. In contrast, the ABS in Figure 2.4 shows a distinctly shear-thinning viscous flow behavior.

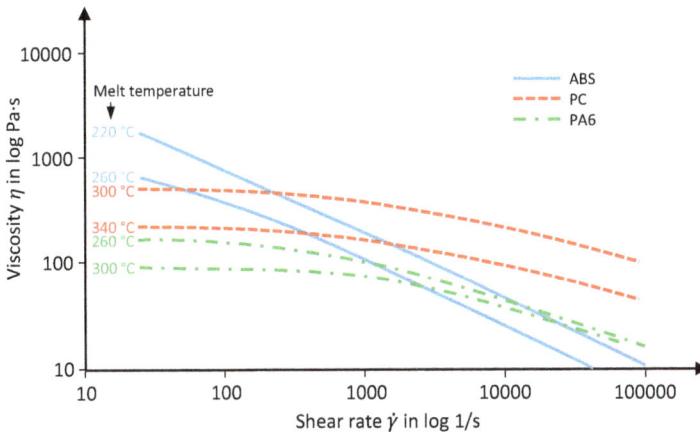

Figure 2.4 Shear-thinning viscous flow behavior for different types of plastics

If the molecular orientation becomes very large, the energy required for the flow increases again. This means that the viscosity increases again at very high shear rates. However, these shear rates are so high that they are of no significance in practice.

The polymer molecule is thus brought out of its most energetically favorable position by the flow process. However, it strives to return to this disordered state, as this is the state of greatest possible entropy. For this reason, restoring forces that counteract the thrust forces and are in equilibrium with them are formed. These restoring forces are greater

- The stronger the molecular orientation/alignment is

- The higher the temperature is—a high temperature means a high activation energy and therefore great restoring forces

A further phenomenon can be observed with shear-thinning viscous fluids. Under shear stress, the fluid attempts to deflect in a direction perpendicular to the direction of stress, that is, the shear direction (see the Weissenberg effect). This results in additional stresses, the so-called normal stresses.

2.2 Dilatancy

Dilatancy was discovered by Osborne Reynolds in 1885. While the viscosity of media with shear-thinning viscous behavior decreases with increasing stress (shear), the viscosity of dilatant media increases with increasing shear. A starch/H_2O slurry in a laboratory beaker can serve as an example (Figure 2.5). A rod standing in it falls over due to its own weight if it is slightly inclined. If the rod is pulled up quickly, the slurry solidifies to such an extent that it is possible to lift the beaker. A similar experiment was carried out by students at a university. These students filled a large basin with a dilatant fluid. Due to the dilatant behavior of the fluid, the students are able to walk over the fluid from one side to the other without sinking. Finally, a student remains standing in the middle of the pool. As he stands still and does not move, the viscosity of the medium decreases and he sinks in.

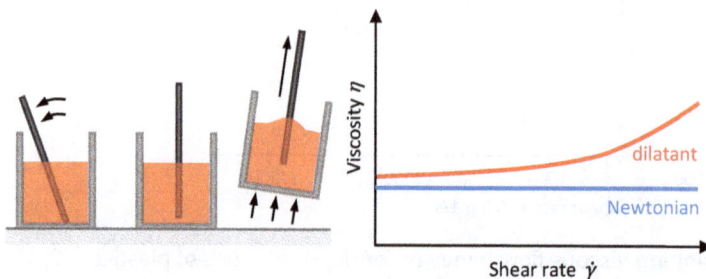

Figure 2.5 Stirring test of a dilatant fluid (starch/H_2O slurry) [1]

The phenomenon of an increase in viscosity when the shear rate is increased is called dilatancy. It mainly occurs with highly concentrated suspensions [1].

> For both shear-thinning viscosity and dilatancy, the viscosity is a function of the shear rate.
>
> $\eta = f(\dot{\gamma})$

Example of Dilatancy

If there is a dilatant medium between two discs, this can be used to transmit force. If one disk rotates while the other disk is stationary, the medium is sheared. As a result, the viscosity of the medium between the disks increases and the power transmission to the second disk increases. The function is similar to a clutch.

The US manufacturer Dow Corning has developed a bouncing putty with dilatant behavior from a silicone polymer. This putty can be kneaded normally. When subjected to sudden mechanical stress, the compound behaves in a completely different way. If the putty is thrown onto the floor as a ball, it bounces back like a rubber ball. If a piece is hit very quickly with a hammer, it shatters into many small sharp-edged pieces, almost like ceramic. Sharp edges and smooth fracture surfaces are also formed when it is torn. No technical applications are known.

A material with similar properties has been developed as protective clothing. This makes it possible to produce the so-called "active protectors", which protect the person from injuries such as impact in sensitive areas. Pads are filled with dilatant media. This has the advantage that the wearer's freedom of movement is not restricted. If the wearer of the protective clothing hits the ground in a motorcycle accident, for example, or is hit by a bullet, the material in the protective padding hardens suddenly, as in the previous case, and protects the wearer from injury.

The use of these dilatant fluids is being tested in conjunction with Kevlar® fabrics in the manufacture of bullet- and stab-resistant protective vests. The fabric provided with a dilatant fluid has an extremely high resistance to penetration by objects due to the properties of the fluid. Tests have shown that even metal arrowheads fired from heavy hunting bows are unable to penetrate a fabric just a few millimeters thick.

2.3 Thixotropy and Rheopexy

Everyone knows the phenomenon of thixotropy from their own experience. If you want to enjoy ketchup, it will not flow out of the bottle on its own. The flow resistance of ketchup is so great that it will not flow out of the bottle even if you turn it upside down. What do you do? You shake the bottle, which reduces the flow resistance and causes it to flow out of the bottle. If you put the ketchup back on the table, the viscosity increases again and the shaking procedure will start again the next time you use it. The decrease in viscosity as a result of continuous stress and the increase (complete or partial) after the end of the stress (hysteresis) is called thixotropy (Figure 2.6). If a yoghurt is stirred, biological superstructures are destroyed and the viscosity decreases. As this process is irreversible, the thixotropy is irreversible, in contrast to the case of ketchup [1].

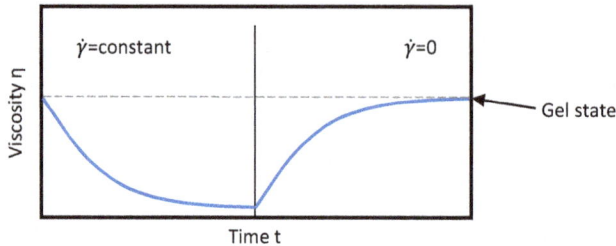

Figure 2.6 Viscosity behavior of thixotropic materials

Paints and varnishes can also be thixotropic. The paint should not run off the roller or brush before application but should adhere to it. If the paint is applied with a roller, the viscosity of the paint decreases, and it can be applied thinly and evenly. As soon as the roller leaves the paint and no more energy is applied to the paint, the viscosity of the paint increases again, and the paint no longer runs. There is no undesirable dripping (nosing).

The properties of thixotropy can also be used specifically in the field of cosmetics. Nail polish, for example, must meet different rheological requirements. Before applying it to the brush, the bottle is shaken so that the brush can be sufficiently wetted. It should be thick enough to adhere to the brush, but thin enough to be transferred from the brush to the fingernail. After application, the nail polish should flow well so that no brush strokes remain visible. Finally, the polish should set again as quickly as possible.

The increase in viscosity as a result of mechanical stress and the decrease (complete or partial) after the end of the stress is called rheopexy (Figure 2.7). This is the inverse phenomenon to thixotropy. Rheopexy is much rarer than thixotropy. Chemical reactions are excluded [1].

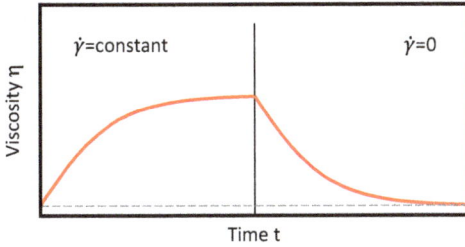

Figure 2.7
Viscosity behavior of materials showing rheopexy

Examples of Rheopexy in Fluids

In the case of rheopexy, an example of the behavior can be shown with dry sand:

- If you fill a balloon with dry sand and tie it in a knot, it is initially soft and elastic. If you throw this filled balloon onto the floor, the grains of sand "cling" together and the balloon feels stable. The balloon envelope ensures that this condition remains reasonably stable.

- If you fill a container with dry sand and insert a stick into the sand, you can easily pull the stick out of the sand again. Now, if you tap the container, the sand compacts and the grains of sand "cling" together. If you then pull on the stick, you shear the grains of sand even more and so they "cling" even more. You can use the rod to lift the entire container filled with sand. After some time, the grip between the grains of sand relaxes again and the stick can be removed from the container without any problems [5].

2.4 Bingham Behavior

Everyone knows this phenomenon: Toothpaste does not come out of an open tube under the effect of gravity alone. Only when slight pressure is applied to overcome the Bingham-behavior limiting stress, plastic flow occurs (Figure 2.8).

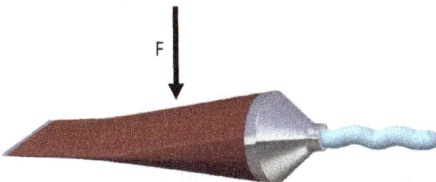

Figure 2.8
Plastic material behavior of toothpaste

The same behavior can be observed with compacted bulk solids in silos. If, for example, the Bingham limiting stress is not exceeded at all points in agitated tanks with plastic-viscous media, stagnation zones without a mixing effect are created [1].

The flow behavior of the Bingham fluid is also characterized by the Bingham limiting stress τ_0, which must first be overcome. Once this limiting shear stress has been overcome, the medium in a Bingham model generally flows in a Newtonian manner.

Examples of Bingham fluids are ketchup, toothpaste, yeast dough, and certain wall paints, but also blood [1]. Technically used suspensions such as electro- and magneto-rheological fluids and concrete can also be described using a Bingham model. These fluids only enter a flow state when a certain shear stress is reached.

However, many materials also exhibit elastic material behavior in practice. In rheology, these materials are also described by the Bingham model:

$$\tau = \tau_0 + \eta \frac{dv}{dy} \tag{2.1}$$

where
v is the flow velocity
y is the coordinate perpendicular to the flow

This means that dv/dy is the velocity gradient (i.e., the change in velocity).

In addition to the Bingham model, other models, such as those of Casson or Herschel-Bulkley, can also be used to describe the yield point. These models are characterized by the fact that they do not use Newton's law of friction to describe the flow behavior of the materials, but rather, for example, the power approach.

The flow behavior of highly filled thermoplastics can also be described using the Herschel-Bulkley model under certain circumstances. In order to initiate the flow process, a lot of energy must first be applied (for example, through the injection pressure). As soon as the flow process takes place, the viscosity decreases and the flow resistance drops. The energy required for the flow process decreases.

Equation 2.2 can therefore describe the flow behavior of a highly filled plastic melt. Here too, a yield point τ_0 must first be overcome before the melt exhibits shear-thinning flow behavior, which is described using the power approach. We will see later that, depending on the filler content, the melt can exhibit pronounced shear-thinning viscous flow behavior.

The equation for describing the yield point according to Herschel-Bulkley is as follows:

$$\tau = \tau_0 + b \cdot \dot{\gamma}^p = \tau_0 + K \cdot \dot{\gamma}^n \tag{2.2}$$

In these models, τ_0 is the Bingham-behavior limiting stress. In the Herschel-Bulkley equation (Equation 2.2), b or K describes the consistency of the fluid and n is the viscosity exponent of the fluid, where $p = n$.

Figure 2.9 shows the measurement of the shear stress as a function of a linearly specified shear rate for ketchup. The limiting shear stress that must be overcome for the ketchup to flow is 47 Pa in this case. The Herschel-Bulkley model was used here to describe the yield point.

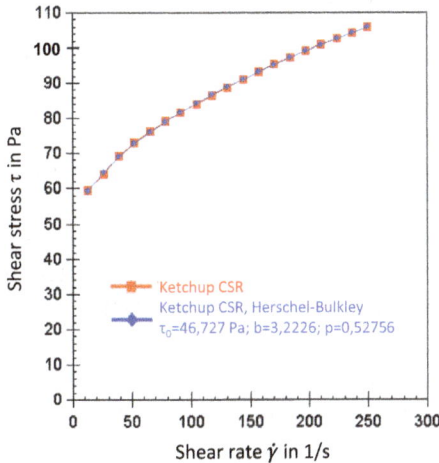

Figure 2.9
Measurement of the yield point of ketchup [6, 7]

Similar effects can be used for wall paints. If you dip the roller into the paint bucket and move it towards the wall, the paint will adhere well to the roller. If you press the roller with paint against the wall and then roll it to apply the paint, the shear forces become high and the paint changes into a flowing state. As soon as the paint adheres to the wall and is not in contact with the roller, it returns to its original state and no longer melts on the wall. This prevents dripping.

Figure 2.10 summarizes the behavior (flow curves) of various media under the effect of shear.

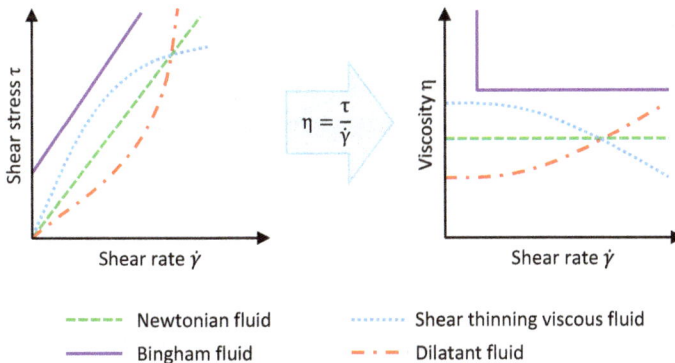

Figure 2.10 Flow curves of different fluids

2.5 Normal Stresses

2.5.1 Origin, Definition, and Characterization

Until now, it was assumed that shear deformation only generates shear stresses in the material. If the media have a viscoelastic flow behavior, as is the case with plastics, for example, normal stresses are generated in addition to the shear stresses. Figure 2.11 shows an example of how these normal stresses arise in a solid element with visco-elastic properties. Due to the flow process, shear stresses τ act on this volume element, deforming it. This deformation leads to a compression of the volume element perpendicular to the direction of flow and to an extension of the volume element in the direction of flow. This results in a compressive stress ($b' < b$) perpendicular to the direction of flow and a tensile stress ($a' > a$) in the direction of deformation (direction of flow) of the volume element. These stresses, which act in a direction perpendicular to the shear stresses and are present on all surfaces of the volume element in visco-elastic media, are referred to as normal stresses N.

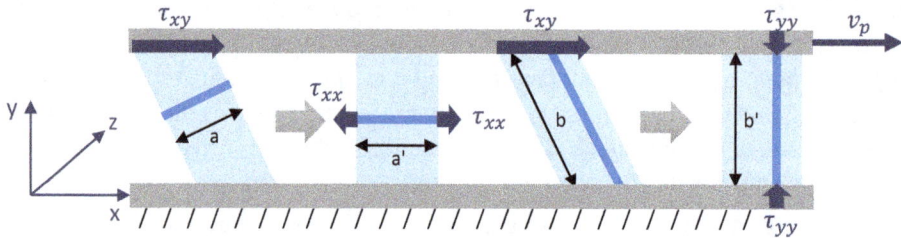

Figure 2.11 Development of normal stress using the example of the deformation of a parallelogram [1]

In practice, the so-called normal stress differences N_1 and N_2 are formed in order to eliminate the height-dependent pressure component. Here, τ_{xx} corresponds to the normal stress in the direction of flow, τ_{yy} to the normal stress perpendicular to the direction of flow (i.e., in the shear direction), and τ_{zz} to the normal stress in the indifferent direction.

$$N_1 = \tau_{xx} - \tau_{yy}$$
$$N_2 = \tau_{yy} - \tau_{zz}$$

(2.3)

The detailed description of the normal stresses and their effects on the flow process, as well as the measurement of the normal stress differences and the normal stress coefficients, are discussed in detail in Chapter 6.

2.5.2 Viscoelastic and Normal Stress Effects

The occurrence of normal stresses can lead to effects that are not known from or contradict the behavior of Newtonian fluids. Some of these effects will be presented below. For the most part, only simple explanations are given, and less emphasis is placed on the analytical formulation.

2.5.2.1 Weissenberg Effect

Karl Weissenberg (1893–1976) studied viscoelastic effects in detail. If a viscoelastic fluid is sheared under certain conditions (high velocity and not too high temperatures), it exhibits the Weissenberg effect (rod-climbing effect) when stirred: If a rotating rod is placed in a large vessel filled with a Newtonian fluid, a drum is formed at suitably high velocities (Figure 2.12). This lowering of the fluid level in the direction of the axis of rotation is caused by the radial forces at work. If a viscoelastic fluid is used instead of a Newtonian fluid, normal stresses are formed—in particular, a tensile stress in the direction of flow. If this tensile stress is imagined as a tensioned rubber ring, a resulting force effect in the direction of the axis of rotation becomes clear. If this force effect is greater than the radial acceleration acting in the opposite direction, material accumulates and the fluid level rises in the direction of the axis of rotation (rotating rod). This climbing effect is known as the Weissenberg effect [1, 4].

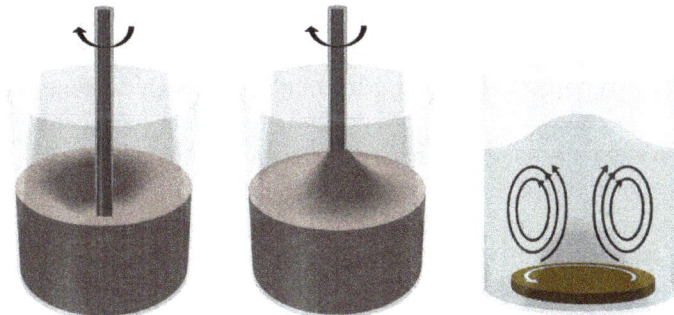

Figure 2.12 Stirring test; left: Newtonian fluid; center and right: viscoelastic fluid [1]

Here too, the dominance of radial acceleration or normal tension determines the quality of the flow. If the normal stress dominates, a flow towards the center is triggered near the rotating part. In the case of a rotating disk located at the bottom, this results in a flow rising in the middle, which leads to an increase in the fluid surface in the middle of the container as a result of the momentum exchange during the deflection to the upper cross flow (swelling effect). In contrast, in a Newtonian fluid, the classic formation of a vortex can be observed, with the fluid level rising towards the outside.

Furthermore, the pressure and flow conditions in such an apparatus change. With a Newtonian fluid, the rotating rod in the middle of the container leads to a pressure gradient from the edge to the center. With a viscoelastic fluid, these conditions are reversed. As a result, the flow directions in a bypass are also different, as Figure 2.13 illustrates.

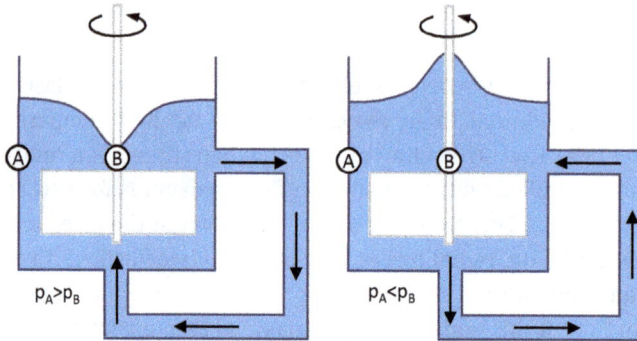

Figure 2.13 Pressure and flow conditions of a Newtonian (left) and a viscoelastic (right) fluid in a stirred vessel [8]

In principle, this behavior could be used in extrusion to promote polymer melting. The example in Figure 2.14 shows a so-called elastic melt extruder developed by B. Maxwell and A. J. Scalora [9, 10]. A rapidly rotating disk (rotor) causes the sheared viscoelastic fluid to be drawn into the shear gap, creating a pressure maximum in the middle of the shear gap. This results in the fluid being pumped through the centrally located die. As the normal stresses of the viscoelastic fluid are responsible for this effect, this extruder is also known as a normal stress extruder. As a result, the efficiency is essentially determined by the viscoelastic behavior of the fluid. The conveying behavior was further improved by H. G. Fritz [11, 12] by introducing spiral grooves in the disk. Despite this, the elastic melt extruder did not gain any significance in extrusion.

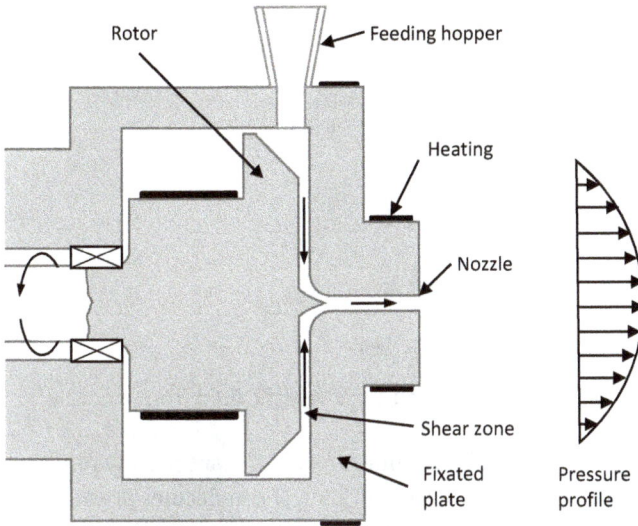

Figure 2.14 Design of the elastic melt extruder by Maxwell and Scalora [8]

2.5.2.2 Die-Swelling Effect

If you look at the free jet of a fluid as it emerges from a circular, vertical pipe, you will quickly notice a difference between a Newtonian medium and a medium with visco-elastic properties. In the case of a Newtonian medium, gravity causes the fluid to accelerate and the velocity to increase with increasing distance from the outlet. For reasons of mass conservation, the increase in velocity in incompressible media must be absorbed by reducing the cross section, resulting in a constricting jet. Even with a viscoelastic fluid, gravity also acts in the manner described above. In addition, the shear deformation near the outlet causes a tensile stress in the direction of flow and a compressive stress (normal stress) transverse to it. This leads to a widening of the jet in relation to the nozzle diameter. This effect is also referred to as strand or extrudate sills and the resulting strand shape as onion formation [8].

For a polymer melt, this behavior can be explained as follows. If a viscoelastic volume element (polymer chain) flows through a nozzle with a very small diameter d_1, the fluid element is accelerated by the decreasing flow cross section in the nozzle (Figure 2.15). As a result, forces act in the direction of flow and they not only align the viscoelastic volume element but also stretch it in the direction of flow. As soon as the volume element leaves the nozzle, it tries to relax again as it "remembers" its previous shape. The extent to which the volume element can "remember" the elastic stresses applied depends on various factors. A very important factor is time. The more time passes before the relaxation process can take place, the less the volume element "remembers" the applied stresses. The relaxation time of plastics is discussed in more detail in Chapter 6.

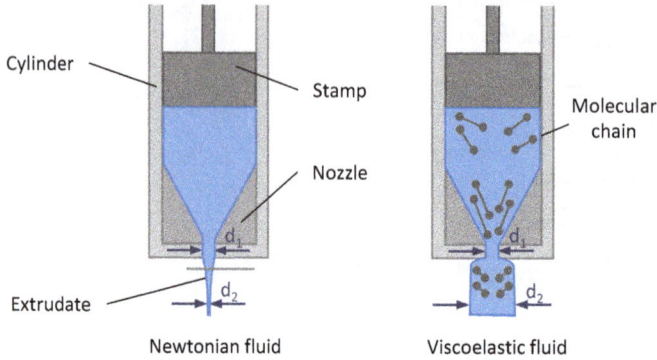

Figure 2.15 Outflow behavior of a Newtonian and a viscoelastic fluid from a nozzle

However, the abovementioned memory effect alone is not sufficient to fully describe the strand extension. As already explained in Section 2.5.1, the molecules are also oriented and stretched by the shear flow in the pipe. The resulting normal stresses, which are compensated for in the pipe by the boundary condition on the pipe wall, lead to strand extension after leaving the pipe. As the shear rate in the pipe increases, the orientation and the normal stress effects are stronger and therefore the strand extension is greater.

In summary, the following statements can be made about strand widening: The effect, also known as die swelling (or "extrudate swelling"; see Figure 2.16), is stronger,

- The shorter the nozzle is
- The higher the flow rate is
- The smaller the nozzle diameter is

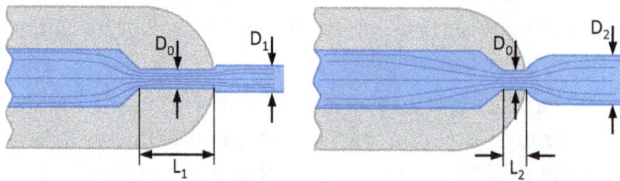

Figure 2.16 Schematic representation of extrudate swelling [13]

Chapter 6 describes in detail how such strand-widening effects can be measured and how the first normal stress difference can be determined from this, for example.

During extrusion, the strand extension may have to be taken into account. If, for example, the target is a rectangular profile, the extrudate will have outwardly curved side surfaces with this flow geometry in the extrusion die (Figure 2.17). The strand extension is greater in the middle of the side surfaces than in the corners, as the velocity gradients and therefore the elastic deformations are smaller in the corners. One measure would be to modify the nozzle geometry appropriately. To date, the optimum nozzle cross section has mostly been determined experimentally during processing. One of the reasons for this is that the necessary material parameters (normal stresses, storage and loss modulus, etc.) for a prediction using simulation (Fluent®, among others) are generally not available.

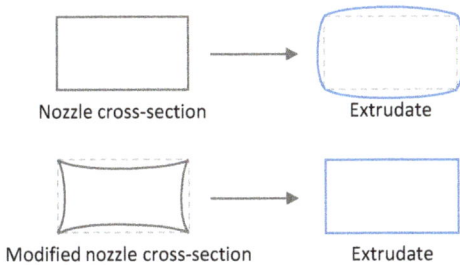

Nozzle cross-section   Extrudate

Modified nozzle cross-section   Extrudate

Figure 2.17
Effect of the die cross section on the extrudate [13]

References

[1] Pahl, M.; Gleißle, W.; Laun, H.-M.: *Praktische Rheologie der Kunststoffe und Elastomere*. VDI-Verlag GmbH, Düsseldorf, 1995

[2] Bonten, C.: *Plastics Technology: Introduction and Basics*. Hanser, Munich, 2014

[3] Werner, F. C.: *Über die Turbulenz in gerührten newtonschen und nicht-newtonschen Fluiden*. Herbert Utz Verlag, Munich, 1997

[4] Menges, G.; Haberstroh, E.; Michaeli, W.; Schmachtenberg, E.: *Menges Werkstoffkunde Kunststoffe*. Hanser, Munich, 2011

[5] N. N.: *http://www.chemie.de/lexikon/Rheopexie.html* (accessed March 26, 2025)

[6] Mezger, T. G.: *Applied Rheology: With Joe Flow on the Rheology Road*. Anton Paar GmbH, Graz, 2014

[7] Mezger, T. G.: *The Rheology Handbook*. Vincentz Network, Hanover, 2016

[8] Giesekus, H.: *Phenomenological Rheology: An Introduction*. Springer, Berlin · Heidelberg, 1994

[9] Maxwell, B.; Scalora, A. J.: *Moderne Kunststoffe*, 1959, 37, 107, Oct.

[10] Maxwell, B.; Scalora, A. J.: Patent Elastic Melt Extruder, US3301933A

[11] Fritz, H. G.: *Kunststofftechnik*, 1968, 6, 430

[12] Fritz, H. G.: PhD thesis, Universität Stuttgart (1971)

[13] Michaeli, W.: *Extrusionswerkzeuge für Kunststoffe und Kautschuk: Bauarten, Gestaltung und Berechnung*. Hanser, Munich · Vienna, 1991

3 Rheological Models

Figure 3.1 shows the behavior of different media under stress. The stress is a shear stress that jumps to a certain value at a time t_0 and remains constant until a time t_1. At t_1, the shear stress suddenly returns to zero.

- A Newtonian fluid deforms linearly as long as the load is maintained. If the load is removed, the Newtonian medium remains with the deformation.

- The ideal elastic solid follows the load abruptly with a deformation. As soon as the stress is removed, the deformation of the ideal elastic solid returns to its original state.

- The viscoelastic fluid initially follows the load abruptly with a deformation. It then retards (creeps) as long as the stress is maintained. The deformation increases degressively (concave-down curve). As soon as the stress is removed, the deformation partially returns to a smaller value. It then retards again, that is, the deformation partially decreases again [1–5].

Plastic melts also exhibit this last behavior, as they have viscoelastic properties. Simple basic bodies that can be used to describe this behavior are the spring and the damper. These basic bodies are described below.

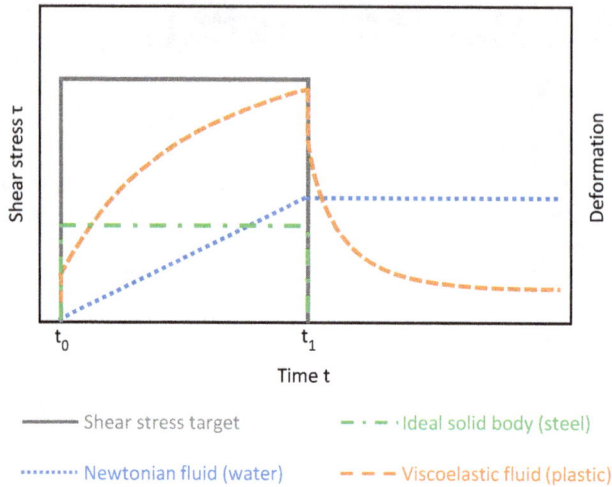

Figure 3.1 Shear stress and deformation as a function of time for an ideal solid and Newtonian and viscoelastic fluids under the same shear stress

3.1 The Ideal Elastic Model

A Hooke's solid is defined as an ideal elastic material in which the entire deformation occurs without delay when a force is applied and there is proportionality between force and deformation. After unloading, the stresses and deformations return to their initial value. An ideal spring exhibits such properties. It therefore serves as a substitute model for a Hooke's solid (Figure 3.2).

Figure 3.2
The spring as a model for a Hooke's solid

As described above, the ideal elastic material (Hooke's solid) reacts with a rectangular function of the deformation, which begins and ends at the same time as the shear stress jump.

Hooke's Solid

In an ideal elastic Hooke's solid, a shear stress that oscillates in phase with the shear occurs. The stress amplitude $\hat{\tau}$ is the product of the shear modulus G_H and the shear amplitude $\hat{\gamma}$. The following applies:

$$\tau = G_H \gamma = G_H \hat{\gamma} \sin \omega t = \hat{\tau} \sin \omega t \tag{3.1}$$

The elastic spring periodically stores the volume-related energy W_{el}, which is proportional to G_H and the square of the shear amplitude.

$$W_{el} = \int\limits_{\gamma=0}^{\hat{\gamma}} \tau d\gamma = G_H \frac{\hat{\gamma}^2}{2} \tag{3.2}$$

One parameter that can be used to describe the elastic behavior of plastics is the so-called storage modulus, which is explained in Section 6.5.

3.2 The Ideal Viscous Model (Newtonian Fluid)

An ideal viscous model or Newtonian fluid is a material in which there is a proportionality between the acting stress and the resulting deformation rate. In the case of gradual loading, the deformation changes linearly with the loading time. After unloading, the deformation achieved (i.e., shear) is maintained (ideal viscous behavior).

Such properties are exhibited, for example, by a damper that consists of a piece of pipe filled with oil in which a fitted piston moves (Figure 3.3). It serves as a substitute model for the Newtonian fluid.

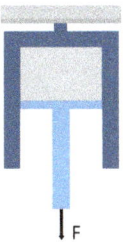

F

Figure 3.3
The damper as a model for ideal viscous behavior

A Newtonian fluid during shear is characterized by the dynamic shear viscosity η and by the extensional viscosity η_E or Trouton viscosity in extension. The following relationships apply to the shear stress and the normal (strain) stress σ with the shear rate $\dot{\gamma}$ and the strain rate $\dot{\varepsilon}$:

$$\tau = \eta \cdot \dot{\gamma} \tag{3.3}$$

and

$$\sigma = \eta_E \cdot \dot{\varepsilon}$$

The unit for the two parameters η and η_E is Pa·s.

In contrast, the viscosity related to the density ρ of the fluid, the so-called kinematic viscosity v, is:

$$v = \frac{\eta}{\rho} \tag{3.4}$$

and its unit is m²/s.

The following approach is often used for incompressible Newtonian fluids to describe the extensional viscosity:

$$\eta_E = 3 \cdot \eta_0 \tag{3.5}$$

The shear viscosity at very low shear rates (zero viscosity) is denoted by η_0. The fact that this relationship only applies to a good approximation is discussed in Chapter 6.

Newtonian Fluid

For the ideal viscous Newtonian fluid, the instantaneous shear stress is proportional to the shear rate, which is ahead of the shear in phase by $\frac{\pi}{2}$. The resulting shear stress therefore also leads the shear by $\frac{\pi}{2}$. The shear stress amplitude is proportional to the viscosity η_N, the shear amplitude $\hat{\gamma}$, and the angular frequency ω:

$$\tau = \eta_N \dot{\gamma} = \eta_N \omega \hat{\gamma} \sin\left(\omega t + \frac{\pi}{2}\right) = \hat{\tau} \sin\left(\omega t + \frac{\pi}{2}\right) = \hat{\tau} \cos \omega t \tag{3.6}$$

For each oscillation cycle, a volume-related energy W_{dis}—proportional to the Newtonian viscosity η_N, the shear amplitude, and the angular frequency—is dissipated:

$$W_{dis} = \int_{\overline{\omega t}=0}^{2\pi} \tau d\gamma = \pi \eta_N \omega \hat{\gamma}^2 \tag{3.7}$$

One parameter that can be used to describe the viscous behavior of plastics is the loss modulus. This parameter is discussed in more detail in Section 6.5.

3.3 The Viscoelastic Model

The viscoelastic material initially reacts to the stress with a spontaneous deformation. The deformation then increases degressively (concave-down curve), that is, the material creeps (retards). When the shear stress returns to a smaller value, the deformation returns to a smaller value spontaneously. The material then retards, that is, there is a recovery phase. Viscoelastic models are therefore a combination of spring and damper.

To describe this deformation behavior, springs and dampers are thus connected in parallel and/or in series in different models. Two models that can also be combined to characterize the viscoelastic properties of polymers are those of Maxwell (series connection) and Kelvin–Voigt (parallel connection); see Figure 3.4.

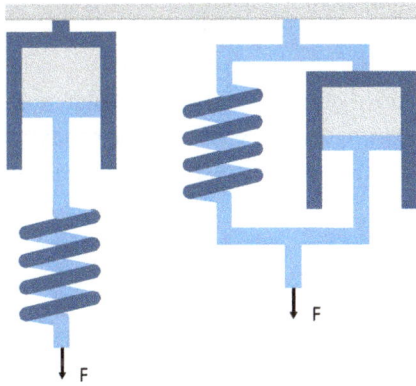

Figure 3.4
Maxwell model (left) and Kelvin–Voigt model (right)

3.3.1 The Kelvin–Voigt Model

In a viscoelastic material, shown here for the Kelvin–Voigt model, there is both an elastic contribution to the stress (spring) in phase and a viscous contribution (damper) out of phase. The following applies:

$$
\begin{aligned}
\tau &= G_H \hat{\gamma} \sin \omega t + \eta_N \omega \hat{\gamma} \cos \omega t \\
&= \hat{\gamma} \left[G_H \sin \omega t + \eta_N \omega \cos \omega t \right] \\
&= \hat{\gamma} |G^*| \sin (\omega t + \delta)
\end{aligned}
\tag{3.8}
$$

The total stress is proportional to the shear amplitude and can be summarized as a sine function that leads the shear in phase by the angle δ_V:

$$
\delta_V = \arctan \frac{\omega \eta_N}{G_H}
\tag{3.9}
$$

The index V is used here to identify the Voigt model, as the phase angle depends on the selected switching of the spring and damper elements.

The stress amplitude is the product of the shear amplitude and the magnitude of a complex modulus G^*, which in the Voigt model becomes:

$$|G^*|_V = \sqrt{G_H^2 + (\omega \eta_N)^2} \qquad\qquad\qquad (3.10)$$

The magnitude of the complex modulus also depends on the model and, like the phase angle, changes with the oscillation frequency.

> Plastics behave in a viscoelastic way: This applies to both the solid and the molten state and means that they initially deform spontaneously like in a purely elastic model (spring) when stress is applied. Subsequently, the deformation increases as a function of time while the stress remains constant. This behavior corresponds to that of the fluid and is called *retardation*. If a deformation is imposed on the plastic, the stress decreases over time. This behavior is called *relaxation*.

Figure 3.5 shows an example of a series connection of Maxwell and Kelvin–Voigt models, the so-called Burgers model. With this series connection, the viscoelastic behavior of plastics can be represented very well. It corresponds, in the stress–extension curve, to Figure 3.1, at the beginning of the chapter.

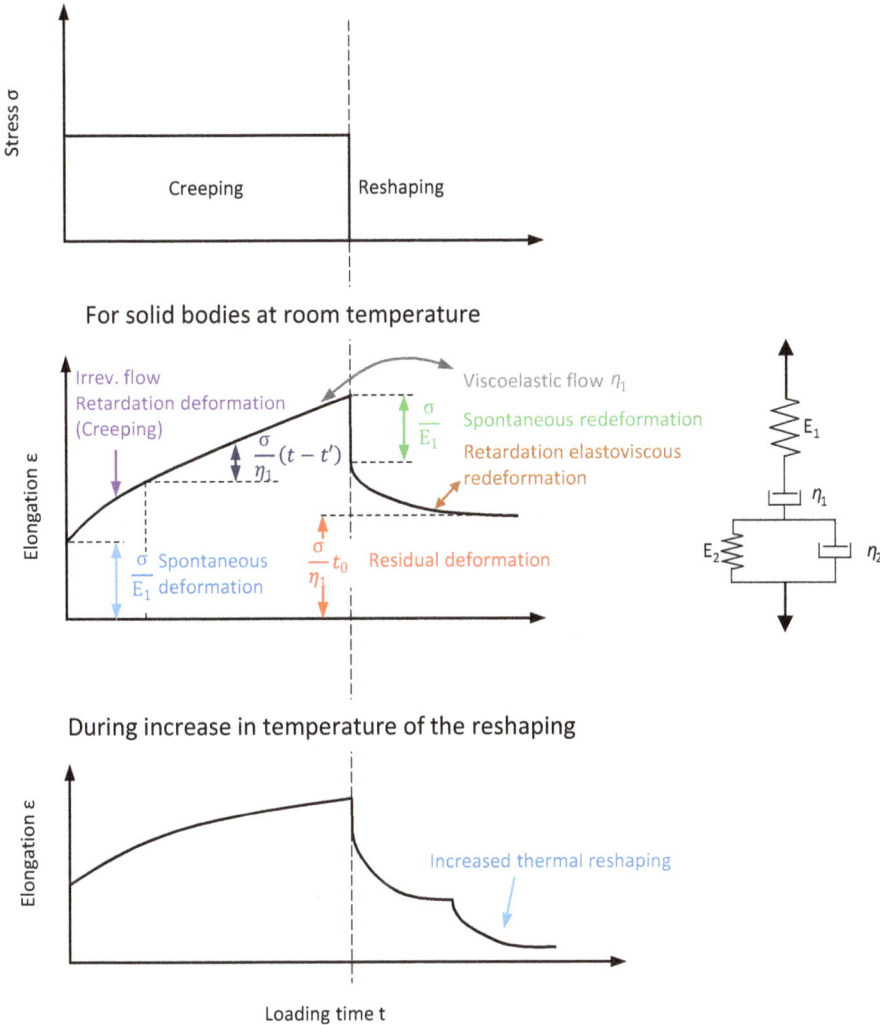

Figure 3.5 Series connection of a Kelvin–Voigt and a Maxwell model [6]

3.3.2 General Viscoelastic Material

In the case of oscillatory shear, the linear behavior of any viscoelastic material, whether solid or fluid, can be described by a parallel connection of a spring and a damper with frequency-dependent values of the spring modulus and the damper viscosity. This is not the Kelvin–Voigt model because its model parameters are frequency-independent by definition. As shown in Figure 3.6 on the left, the elastic component is now characterized by the storage modulus $G'(\omega)$ and the viscous component by the dynamic viscosity $\eta'(\omega)$.

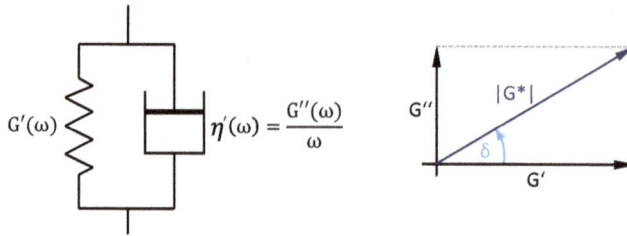

Figure 3.6 The general viscoelastic material

The latter can also be represented as the quotient of a loss modulus $G''(\omega)$ and the angular frequency:

$$\eta'(\omega) = \frac{G''(\omega)}{\omega} \tag{3.11}$$

For a fixed frequency, the model parameters are constant, and the shear stress can be calculated in the same way:

$$\begin{aligned} \tau &= \hat{\gamma}\left[G'(\omega)\sin\omega t + G''(\omega)\cos\omega t\right] \\ &= \hat{\gamma}\left|G^*\right|(\omega t)|\sin\left[\omega t + \delta(\omega)\right] \end{aligned} \tag{3.12}$$

The portion of the shear stress determined by G' oscillates in phase, while the portion determined by G'' oscillates out of phase in shear. The relationship between the components G' and G'' and the magnitude of the complex modulus G^* can be illustrated by the diagram in Figure 3.6 (on the right).

The calculation rules for the magnitude and phase angle can also be read from:

$$\left|G^*\right| = \left|G'^2 + \left|G''^2\right|\right|^{0.5} \tag{3.13}$$

$$\delta = \arctan\frac{G''}{G'} \tag{3.14}$$

References

[1] Baur, E.; Drummer, D.; Osswald, T. A.; Rudolph, N.: *Saechtling Kunststoff Taschenbuch*. Hanser, Munich, 2013

[2] Bonten, C.: *Plastics Technology: Introduction and Basics*. Hanser, Munich, 2014

[3] Braun, D.; Becker, G.; Carlowitz, B.: *Die Kunststoffe: Chemie, Physik, Technologie – Kunststoff-Handbuch 1*. Hanser, Munich, 1990

[4] Ferry, J. D.: *Viscoelastic Properties of Polymers*. John Wiley & Sons, Toronto, 1980

[5] Baur, E.; Harsch, G.; Moneke, M.: *Werkstoff-Führer Kunststoffe: Eigenschaften, Prüfungen, Kennwerte*. Hanser, Munich, 2019

[6] Menges, G.; Haberstroh, E.; Michaeli, W.; Schmachtenberg, E.: *Menges Werkstoffkunde Kunststoffe*. Hanser, Munich, 2011

4 The Shear Test and the Derivation of Newton's Law of Friction (Material Law)

As already described in Chapter 1, rheology can be considered as the science of the deformation and the flow of materials. Deformation refers to the relative displacement of material elements in relation to each other [1].

The deformations are caused by forces or stresses (shear stress, tensile stress, etc.). Depending on the type of stress, a distinction is made between

- Shear
- Strain (or extension)
- Bending
- Others

The shear test is described below and Newton's law of friction (material law) is derived from the shear test. Furthermore, the term shear rate is explained using different types of flow.

4.1 The Shear Test

To define the variables

- Shear stress τ
- Shear γ
- Shear rate $\dot{\gamma}$

a solid element is considered between two parallel plates (Figure 4.1). The lower plate is firmly clamped. The upper plate is loaded by the tangential force F_T. The plates have the contact area $A = b \cdot l$ with the solid element of width b and length l. The distance between the plates is h.

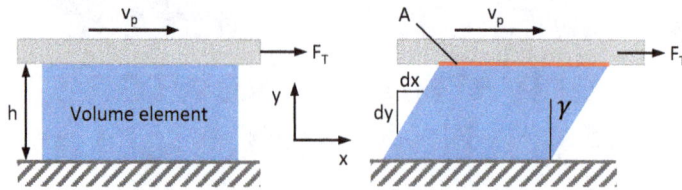

Figure 4.1 Deformation of a volume element between two plates [1]

Shear Stress

If no other forces are acting, a constant shear stress is created in the volume element over the height h:

$$\tau = \frac{F_T}{A} = \text{constant} \tag{4.1}$$

For solids, the following relationship exists between the shear stress τ and the shear γ (Hooke's law):

$$\tau = G \cdot \gamma \tag{4.2}$$

where
G is the shear modulus

It is independent of γ for an ideal elastic solid.

If the volume element consists of a fluid and the contact condition is fulfilled, the fluid adheres to the upper and lower plates. In our case, the upper plate is now pushed by the tangential force F_T at a velocity v_p (see Figure 4.2) to the right. y is the coordinate over the gap height h. This means that the following relationships apply to the velocity v_x, that is, the velocity curve of the fluid between the two plates

$$v(y = h) = v_{max} = v_p$$
$$v(y = 0) = 0 \tag{4.3}$$
$$v(y) = v_p \cdot \frac{y}{h}$$

If a drag flow is considered, i.e., the fluid adheres to a moving plate while the second plate is at rest, the following statement can be made: For homogeneous media, the flow velocity increases linearly with height in the case of plane layer flow between two parallel plates.

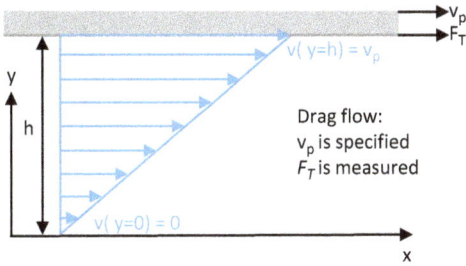

Drag flow:
v_p is specified
F_T is measured

Figure 4.2
Velocity profile between two parallel
plates (drag flow)

The following proportionality applies to drag flow:

$$\frac{F_T}{A} \propto \frac{v_p}{h} \tag{4.4}$$

with

$$v(y) = v_p \cdot \frac{y}{h} \Rightarrow \frac{v(y)}{y} = \frac{v_p}{h}$$

Thus, Equation 4.4 becomes:

$$\frac{F_T}{A} \propto \frac{v(y)}{y} = \frac{dv}{dy}$$

and

$$\frac{F_T}{A} = \tau \Rightarrow \tau \propto \frac{dv}{dy}$$

The shear stress causes a deformation of the volume element (Figure 4.3). The deflection is referred to as shear dy.

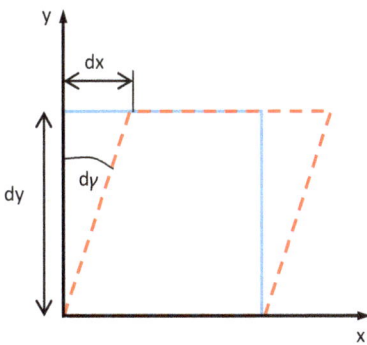

Figure 4.3
Shear deformation of a volume element

From Figure 4.3 it follows, for the angle dy, the trigonometric function:

$$\tan dy = \frac{dx}{dy} \tag{4.5}$$

It still applies:

$$dv = \frac{dx}{dt} \Rightarrow dx = dv \cdot dt$$

Introducing it in Equation 4.5, the result is

$$\tan dy = \frac{dv \cdot dt}{dy}$$

The following applies for small angles dy: $\tan dy$ approaches dy. It follows:

$$dy = \frac{dv \cdot dt}{dy} \Rightarrow \frac{dy}{dt} = \frac{dv}{dy} \tag{4.6}$$

This results in the proportional relationship:

$$\tau \propto \frac{dv}{dy} = \frac{dy}{dt} \tag{4.7}$$

which results in the proportional relationship between the shear stress and the shear rate:

$$\tau \propto \frac{dy}{dt} \Rightarrow \tau \propto \dot{y}$$

Furthermore, using the proportionality factor η, this results in Newton's law of friction:

$$\tau = \eta \cdot \dot{y} \tag{4.8}$$

From Equation 4.6 it can be deduced that the shear rate is the velocity gradient (change in velocity) over the flow cross section. The following applies:

$$\dot{y} = \frac{dv}{dy} \tag{4.9}$$

Since the velocity of a pure drag flow is always linear regardless of the medium across the channel cross section, the shear rate (like the shear stress) must be constant.

Velocity and Shear Rate in a Pressurized Flow

In the following, the term shear rate will be discussed in more detail. In general, the gradient of the flow velocity dv/dy is referred to as the shear rate $\dot{y}$. In other words, the shear rate is nothing other than the change in velocity over the channel cross section (y direction). In the following, we will not consider a drag flow, but a pressure flow. The fluid, which is located between two plates at rest, adheres to the two plates again (Figure 4.4). Assuming Newtonian flow behavior, a parabolic velocity profile is formed between the two plates. The maximum velocity $v(y = 0) = v_{max}$ is present in the middle of the flow channel, while the velocity at the top and bottom of the channel wall is zero due to the wall adhesion $v(y = h/2) = 0$. The velocity, and therefore the shear rate, changes between these points. This change is maximum at the channel wall, which means that the maximum shear rate is $\dot{\gamma} = \dot{\gamma}_{max}$, while the shear rate at the center of the channel is $\dot{\gamma} = 0\ s^{-1}$. Between these points, the shear rate must be linear as a derivative of the quadratic parabola for a Newtonian fluid.

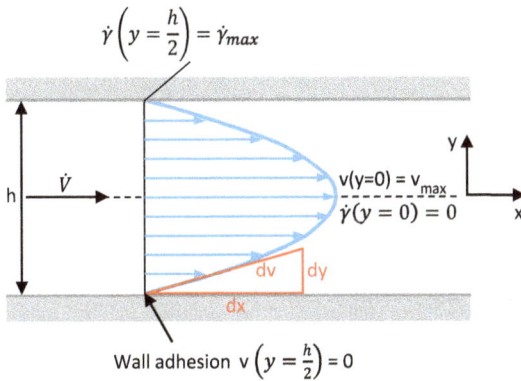

$$\dot{\gamma}\left(y = \frac{h}{2}\right) = \dot{\gamma}_{max}$$

$$v(y=0) = v_{max}$$
$$\dot{\gamma}(y = 0) = 0$$

Wall adhesion $v\left(y = \frac{h}{2}\right) = 0$

Figure 4.4

Velocity profile of a pressure flow

Newtonian Fluids

Fluids are called "Newtonian fluids" if:

- They behave according to Newton's law of friction $\tau = \eta \cdot \dot{\gamma}$ and the following applies: $\eta \neq f(\dot{\gamma})$
- The shear changes linearly with time when the load changes abruptly
- The shear γ is completely maintained after unloading

> Another way of writing Newton's law of friction is:
>
> $$\eta = \frac{\tau}{\dot{\gamma}}$$

For Newtonian fluids ($\eta \neq f(\dot{\gamma})$, $\eta = $ constant), the viscosity corresponds to the slope of the function $\tau = \eta \cdot \dot{\gamma}$; see Figure 4.5.

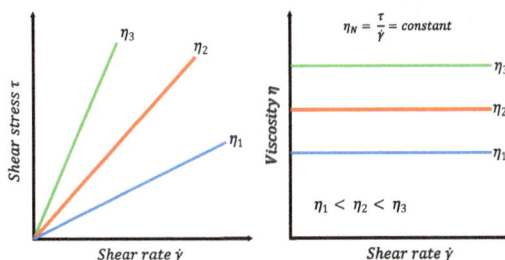

Figure 4.5

Flow curves of a Newtonian fluid

Non-Newtonian Fluids

All fluids that do not behave as just described are called "non-Newtonian fluids".

For non-Newtonian fluids, for example, shear-thinning viscous fluids, the relationships that apply are $\eta = f(\dot{\gamma})$ and $\tau = f(\dot{\gamma})$. Plastic melts are not Newtonian, but shear-thinning viscous. This means that the viscosity decreases with increasing shear rate and the increase in shear stress is degressive (concave-down curve; see Figure 4.6).

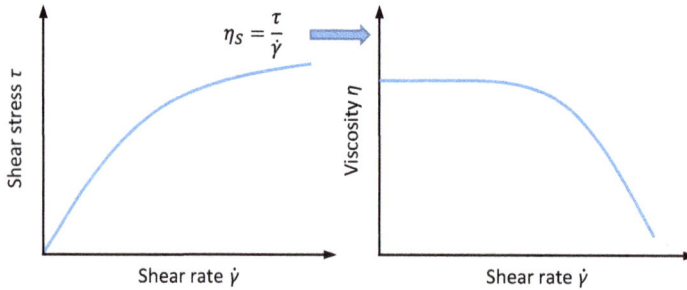

Figure 4.6 Flow curves of a shear-thinning viscous fluid

> The graphical representation of $\tau = f(\dot{\gamma})$ and $\eta = f(\dot{\gamma})$ is called flow curves. $\tau = f(\dot{\gamma})$ and $\eta = f(\dot{\gamma})$ are material laws.

4.2 Important Rheological Material Laws

In the following, the rheological properties of solids and liquids will be compared. In general, Hooke's law is used to characterize the behavior of solid bodies under load. This law describes the relationship between the stresses, the deformations, and the respective moduli of the material. The situation is similar for ideal fluids. The behavior of the ideal fluid under load (stress) is described by Newton's law of friction. The deformation speed and the viscosity of the medium play a role here.

To characterize the rheological behavior of a purely elastic solid, the compression modulus K and the transverse strain (Poisson) number v are sufficient:

- The shear modulus G in shear

- The modulus of elasticity E when stretched

For purely viscous media

- The shear viscosity η in shear

- The strain or extensional viscosity λ

are to be determined.

In general, the material laws listed in Table 4.1 apply to solids and fluids.

Table 4.1 Material Laws for Solids and Fluids

Type of stress	Material laws: solids		Material laws: fluids	
Tensile stress	Hooke's law	$\sigma = E \cdot \varepsilon$	Newton's law of friction	$\sigma = \lambda \cdot \dot{\varepsilon}$
Shear stress		$\tau = G \cdot \gamma$		$\tau = \eta \cdot \dot{\gamma}$
Link	$E = 3 \cdot G \, (v = 0.5)$		$\lambda = 3 \cdot \eta \, (\dot{\varepsilon} \text{ or } \dot{\gamma} \to 0)$	
σ = Tensile/normal stress τ = Shear stress	E = Modulus of elasticity G = Shear modulus λ = Strain or Troutonian viscosity η = Shear or Newtonian viscosity v = Transverse strain number (Poisson's ratio)		ε = Strain (or extension) γ = Shear $\dot{\varepsilon}$ = Strain rate $\dot{\gamma}$ = Shear rate, velocity gradient	

References

[1] Pahl, M.; Gleißle, W.; Laun, H.-M.: *Praktische Rheologie der Kunststoffe und Elastomere*. VDI-Verlag GmbH, Düsseldorf, 1995

5 Flow Types

A basic distinction is made between two types of flow in the flow processes of plastic melts. In addition to the **shear flow**, the **extensional flow** must also be taken into account, especially with viscoelastic fluids.

In the following, these two types of flow will be considered for a flow in an injection mold. In this case, the temperature of the mold is controlled and is usually significantly lower than that of the incoming melt.

5.1 Shear Flow

If a laminar layered flow is considered in the gap, the fluid is subjected to shear stress by the flow process. Assuming the wall adhesion ($v_{wall} = 0$) according to Stokes and the maximum flow rate ($v_{middle} = v_{max}$) in the center of the channel (see light blue curve in Figure 5.1), a velocity gradient is formed in the flow channel. This velocity gradient is referred to as the shear rate. The shear rate (blue curve) between the melt layers causes a shearing of the volume elements of the melt in the flow channel. The shear is greatest near the flow channel wall, as this is where the maximum shear rate (velocity gradient) is present. As the shear leads to friction (dissipation) of the polymer chains against each other, heat is generated by this process. Most heat is generated in the area of the highest shear rate (near the edge) (see red curve in Figure 5.1).

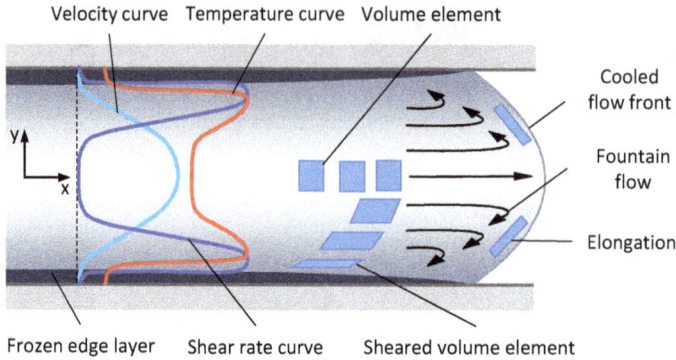

Figure 5.1 Flow process of a plastic melt in the gap

If we consider the flow processes of a plastic melt at the flow front during the flow phase, there is an additional stress on the polymer.

5.2 Extensional Flow

Due to the contact with the cold air in the injection mold, the melt cools at the flow front and a "cooled" skin forms on the surface. In the flow processes of plastic melts, a fountain flow is assumed at the flow front. This means that the melt particles at the flow front are deflected outwards from the center of the flow channel towards the mold wall. This flow process, in conjunction with the cooled melt at the flow front, leads to an alignment and extension of the volume element.

5.3 Orientations

Both the extensional flow and the shear flow effects have an impact on the molecular orientation in the molded part. In the case of transparent molded parts, this can be made visible using polarized light. In the case of plastics with glass fibers or other fillers, the flow effects also affect the orientation of the fillers. As a result, the strength of such a molded part can be decisively defined by the flow processes. These processes will be described in more detail below and illustrated using examples. For this purpose, the alignment of glass fibers in the polymer is considered.

Alignments of glass fibers in the molded part are the result of velocity gradients. Since the layer thickness of the orientations on the surface due to the extensional flow at the flow front is marginal and has almost no influence on the mechanical properties of the molded part, it is neglected in the following. The orientations of the boundary layer near the wall and the middle layer are considered.

The shear rate curve (velocity gradient) behind the flow front causes a high alignment of the glass fibers in the boundary layer. Where the shear rate is at its maximum, the orientation in the molded part is also at its maximum. This orientation caused by the shear rate always takes place in the direction of flow of the melt. Some examples of this are shown below.

5.3.1 Example of Orientations: Molded Panel with Lateral Injection Point

The first molded part is a plate with a lateral injection point. As shown in Figure 5.2, the orientations (blue lines) in the boundary layer are all in the direction of flow, starting from the injection point.

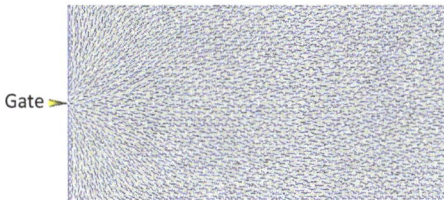

Gate

Figure 5.2
Orientation in the boundary layer in a panel molding with lateral injection point

The orientations in the middle layer depend on the flow in the molded part, that is, the filling behavior of the plastic melt in the mold, and are again caused by differences in velocity in the plastic melt. If concentric flow circles are formed starting from the injection point, the orientation of the middle layer corresponds to these circles in the molded part. This can be seen in the molded part (plate) in Figure 5.3 (red lines). Starting from the injection point, concentric circles are initially formed. The orientations follow these concentric circles until the flow fronts are almost parallel. As there are now no more velocity differences, the orientations remain in the previously introduced position, namely transverse to the direction of flow.

Gate

Figure 5.3
Orientation in the middle layer in a panel molding with lateral injection point

5.3.2 Example of Orientations: Molded Part with Bar Sprue and Lateral Film Gate

Another molded part is a Campus shrinkage plate (test specimen in accordance with ISO type D2 mold according to ISO 294-4), which is injected with a bar sprue. Starting from the bar sprue, the melt is distributed via a distributor to the film gate. The film gate is the connection to the molded part. As shown in Figure 5.4, the orientations in the boundary layer are all in the direction of flow.

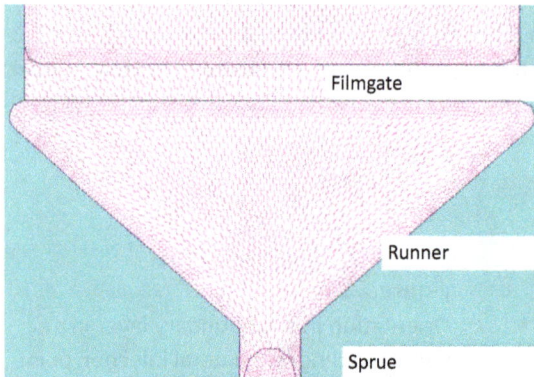

Filmgate

Runner

Sprue

Figure 5.4 Orientation in the boundary layer in a panel molding with bar sprue, distributor channel, and film gate

The orientation in the middle layer will be considered below. Starting from the bar sprue, the molded part in the multi-cavity is also filled with concentric circles (Figure 5.5). As a result, the orientations in the middle layer again follow these concentric circles. Now the melt flow front reaches the film gate. As the film gate has a smaller wall thickness than the distributor channel, the flow cross section is also reduced. An increase in velocity in the direction of flow is the result of the cross-sectional jump. This changes the direction of orientation. The forces acting in the direction of flow lead to an alignment of the fillers in the direction of flow. Downstream of the cross-sectional jump, the velocity decreases to a greater or lesser extent depending on the wall thickness of the molded part. However, this only has a marginal effect on the orientations. As a result, almost all orientations in the molded part are in the direction of flow. This is also desirable, as the plates are used to investigate the shrinkage of the plastic in the direction of flow and transverse to it. Furthermore, to investigate the anisotropic properties of the plastic, tension rods are milled out of this plate in and transverse to the direction of flow and used for mechanical tests (tensile test, among others).

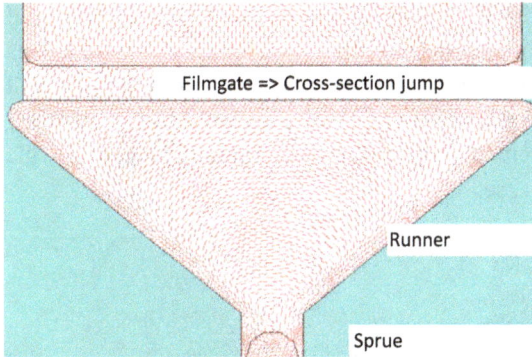

Figure 5.5 Orientation in the middle layer in a panel molding with bar sprue, distributor channel, and film gate

5.3.3 Example of Orientations: Molded Part with Central Gating

The following molded part is a container that in reality is under an internal pressure of up to 30 bar. To improve the mechanical properties under pressure, a 30 % of glass fibers was added to the plastic. If we look at the simple boiler formula for an initial calculation, it quickly becomes clear that the tangential stresses σ_T are twice as high as the axial stresses σ_A (Equation 5.1). As a result, an orientation of the glass fibers transverse to the direction of flow, that is, tangential in the circumferential direction of the molded part, would be desirable. Figure 5.6 shows the orientation of the fillers in the boundary layer. Again, the orientations follow the direction of flow of the melt. Thus, all orientations in the boundary layer are in the axial and not in the tangential container direction. This would not be optimal for the load cases.

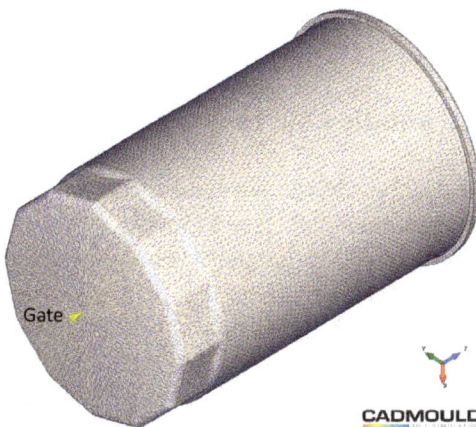

Figure 5.6
Orientation in the boundary layer in a pressure vessel with central gating in the base

Looking at the geometry, according to theory, the fillers in the middle layer would spread across the direction of flow due to the concentric melt fronts in the base. This is also illustrated by Figure 5.7. As there are initially no flow changes in the side wall and therefore no forces acting on the glass fibers that could lead to a change in orientation, the orientations of the layer remain and are therefore initially tangential. This would be advantageous for the load case.

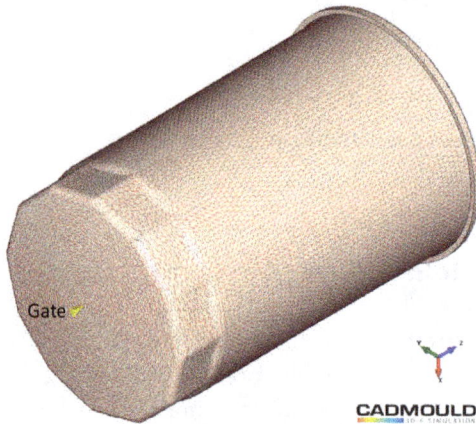

Figure 5.7

Orientation in the middle layer in a pressure vessel with central gating in the base

However, the melt front now reaches a cross-sectional jump in the side wall. As in the example of the plate with a film section, the flow rate of the melt will increase at this point in the direction of flow. This results in a new alignment of the fillers in the direction of flow. As a result, all fibers from this cross-sectional jump onwards are now in the axial direction and no longer correspond to the boundary conditions of the load case:

$$\sigma_T = 2 \cdot \sigma_A \tag{5.1}$$

5.3.4 Orientations: Boundary and Middle Layers

Since two main areas of orientation can be distinguished in the molded part, the percentage of the respective layer (boundary layer and middle layer) also plays a role in the analysis (e.g., mechanical properties). It can be stated that the percentage share of orientations in the boundary layer is predominant and is 60 % to 70 % of the cross section of the molded part. This is illustrated by Figure 5.8. In this case, the proportion of the boundary layer oriented in the direction of flow is approximately 70 %.

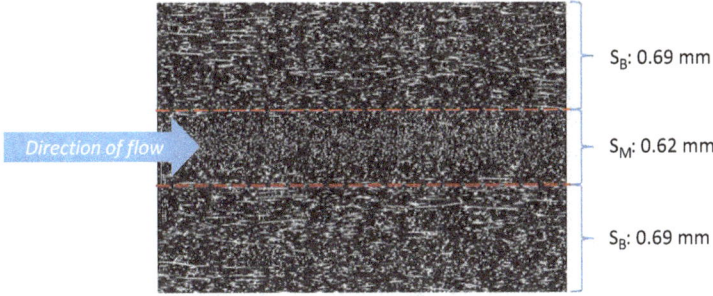

S_B: 0.69 mm

S_M: 0.62 mm

S_B: 0.69 mm

Direction of flow

Figure 5.8 Measured orientations (computer tomography, CT) in a molded part viewed over the flow cross section (wall thickness: 2 mm, injection rate: 100 mm/s, PBT GF30)

This means that the boundary layer has the greater influence on the orientations and thus on the mechanical properties of the molded part. However, this proportion depends, among other things, on the flow behavior—that is, the shear-thinning viscous flow behavior of the plastic melt. The shear-thinning viscosity of the plastic melt is described in the material laws (Carreau, Cross, Ostwald and de Waele) by parameters $(m, n, C = P_3)$ (see Chapter 8); see also Figure 5.9.

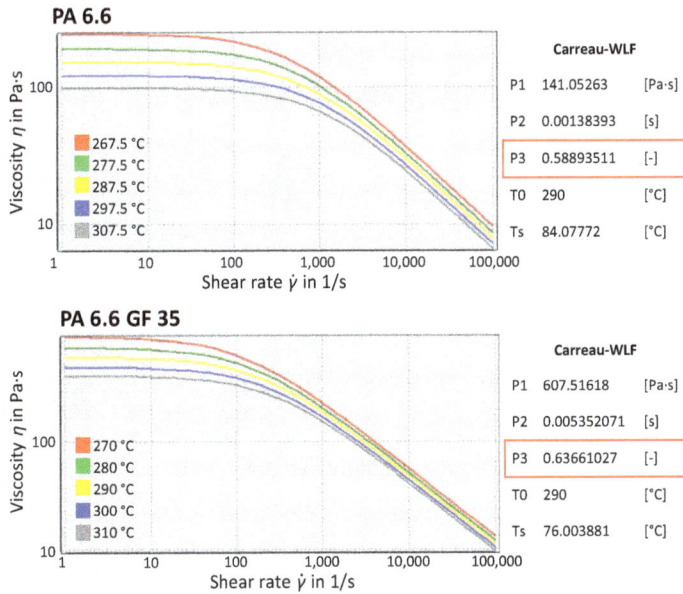

PA 6.6

Viscosity η in Pa·s

267.5 °C
277.5 °C
287.5 °C
297.5 °C
307.5 °C

Shear rate $\dot{\gamma}$ in 1/s

Carreau-WLF		
P1	141.05263	[Pa·s]
P2	0.00138393	[s]
P3	0.58893511	[-]
T0	290	[°C]
Ts	84.07772	[°C]

PA 6.6 GF 35

Viscosity η in Pa·s

270 °C
280 °C
290 °C
300 °C
310 °C

Shear rate $\dot{\gamma}$ in 1/s

Carreau-WLF		
P1	607.51618	[Pa·s]
P2	0.005352071	[s]
P3	0.63661027	[-]
T0	290	[°C]
Ts	76.003881	[°C]

Figure 5.9 Shear-thinning viscous flow behavior and material parameters (Carreau approach) for PA 6.6 and PA 6.6 GF35

The shape of the velocity profile of the plastic melt and thus the shear rate profile changes depending on the shear-thinning viscosity flow behavior of the melt (see Chapter 9, Figures 9.8 and 9.9). The more shear-thinning viscous a plastic melt flows (here, P_3 becomes larger), the "blunter" (flatter) the profile of the velocity curve becomes. The shear rate profile changes accordingly. The maximum shifts to the outside, and in the middle, the shear rate no longer changes over a wide range in the case of highly shear-thinning viscous materials, such as highly filled plastics (Figure 5.10). The influence of the middle layer on the mechanical properties increases. In contrast, the area of the oriented boundary layer increases with low shear-thinning viscosity, as shown in Figure 5.10 in the bottom illustration.

Strong Shear Thinning Viscosity

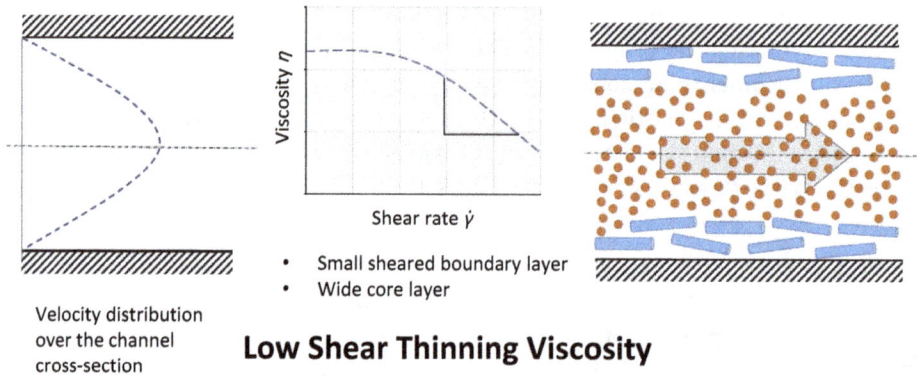

- Small sheared boundary layer
- Wide core layer

Velocity distribution over the channel cross-section

Low Shear Thinning Viscosity

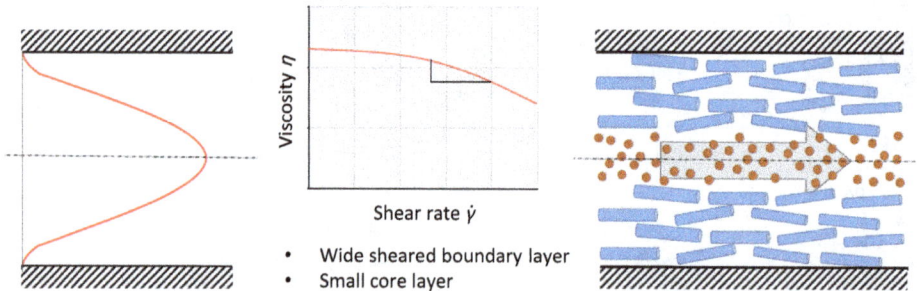

- Wide sheared boundary layer
- Small core layer

Figure 5.10 Influence of the shear-thinning viscous flow behavior on the proportion of orientations in the middle and boundary layers

Computer-aided simulation programs can predict these orientations quite well with the implementation of models (Folgar–Tucker , RSC, etc.), as Figure 5.11 illustrates. This allows the anisotropic properties of the plastics to be taken into account when designing the component.

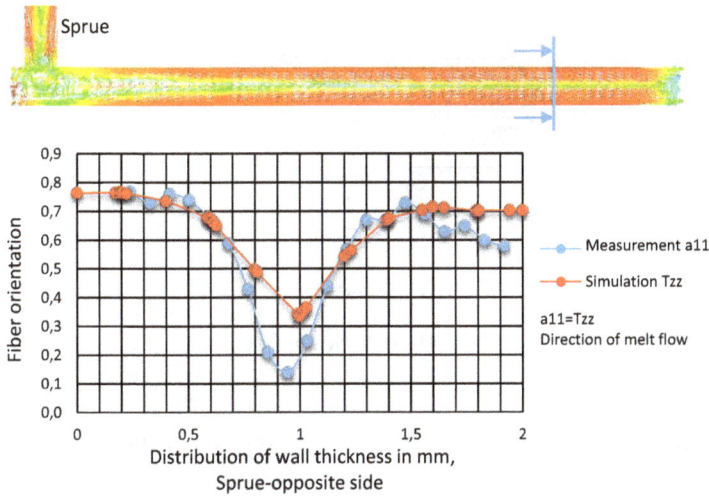

Figure 5.11 Calculated orientations in a molded part viewed in the flow direction and across the flow cross section (wall thickness: 2 mm, injection rate: 100 mm/s, PBT GF30)

5.4 Influence of Extensional Flows

At cross-sectional transitions in nozzles, not only the shear flows but also the extensional flows—the associated extensional pressure losses and the orientations introduced—must be taken into account. Such extensional flows can occur, for example, in hot runner systems during injection molding or in extrusion dies. The volume element of the melt is sheared as it enters the cross section and, due to its viscoelastic properties, is elastically deformed and thus viscously stretched (Figure 5.12). Part of this extension is caused by reversible molecular orientations. This part of the extensional deformation can partially relax time-dependently while flowing through the nozzle with a constant cross section. A pure shear flow is present in the nozzle itself. The elastic extension (molecular orientations) that has not yet dissipated when leaving the capillary, together with the elastic shear in front of the die, leads to a swelling of the extrudate (see die swelling and normal stress effects in Section 2.5.2).

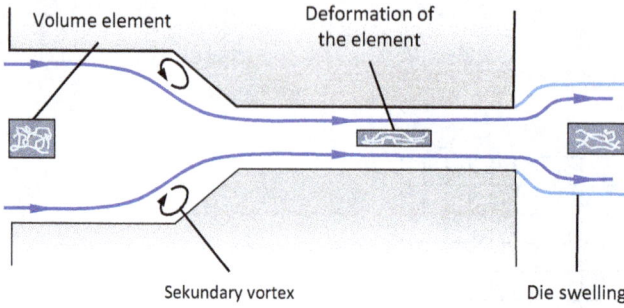

Figure 5.12 Deformation of a cylindrical fluid element when flowing through a nozzle [1]

The higher the extrusion rate and the shorter the capillary, the higher the strand extension.

The extensional deformation in the inlet and the energy of dissipating vortices (secondary flows) that may occur cause an inlet pressure loss.

Stretching, that is, extension of the plastic melt, can also be observed during extrusion. The alignment (orientation) of the polymer chains can be used to increase the mechanical strength and improve the barrier properties in tubular film and hollow solid extrusion (Figure 5.13). The effects of extensional rheology are also used in the production of tapes and shrink films.

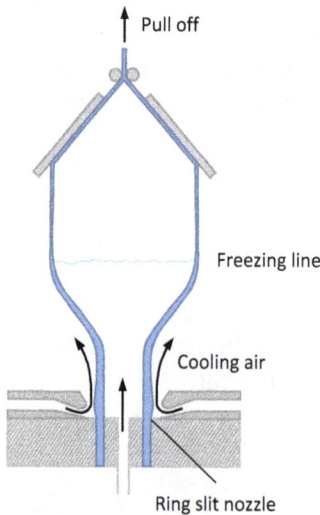

Figure 5.13
Stretching of a tubular film [2]

While extensional flows are generally neglected in the computer-aided simulation of flow processes in a plastic melt, they are taken into account in the simulation of blowing processes, for example.

In practice, extensional rheological parameters are usually determined using shear rheology.

5.5 Trouton Approach for Determining the Extensional Viscosity

For low shear rates, the Trouton approach is often used to determine the extensional viscosity:

$$\eta_D = 3 \cdot \eta_0 \tag{5.2}$$

where
η_0 is zero viscosity
η_D is the extensional viscosity

Comparison of Steady-State Extensional and Shear Viscosities

Figure 5.14 shows a comparison of the steady-state extensional and shear viscosities of a PE-LD melt at the same deformation rate values. Starting from the constant value of zero viscosity η_0, the shear viscosity function (blue curve) only shows a decrease with increasing shear rate. According to the Trouton relationship, the constant extensional viscosity η_D has three times the value of the zero viscosity in shear at low extensional rates. However, it then increases with increasing extensional rate and only falls again after passing through a maximum.

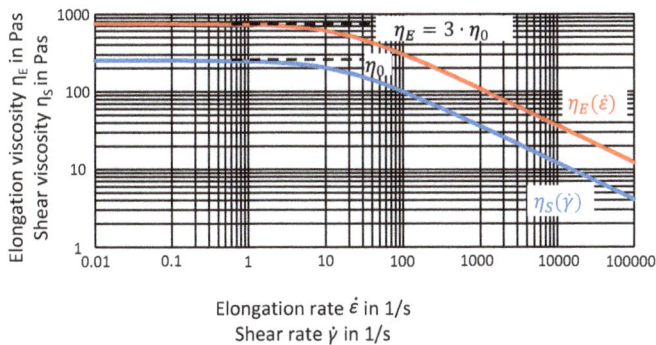

Figure 5.14 Qualitative progression of shear and extensional viscosity as a function of deformation rates

Outside the Newtonian range, the ratio of extensional to shear viscosity may not be constant but may assume values greater than three. For this reason, the Trouton ratio can only be used to a limited extent to determine the extensional viscosity. It is better to determine the true extensional viscosity using suitable viscometers (Chapter 6).

References

[1] Menges, G.; Michaeli, W.; Mohren, P.: *Spritzgießwerkzeuge: Auslegung, Bau, Anwendung*. Hanser, Munich, 2018

[2] Limper, A.: *Process Engineering in Thermoplastic Extrusion*. Hanser, Munich, 2012

Rheometry: Viscometry and Material Data Determination

Viscometers and rheometers are used to determine material data of different media (Figure 6.1). In contrast to a viscometer, which can only measure the viscosity of a fluid in a limited range of conditions, a rheometer is able to measure the viscosity and elasticity of non-Newtonian materials over a wide range of conditions. Key properties that can be measured include viscoelasticity, flow behavior, thixotropy, extensional viscosity, creep behavior and creep recovery, as well as parameters relevant to processing such as extrudate thresholds and melt fracture.

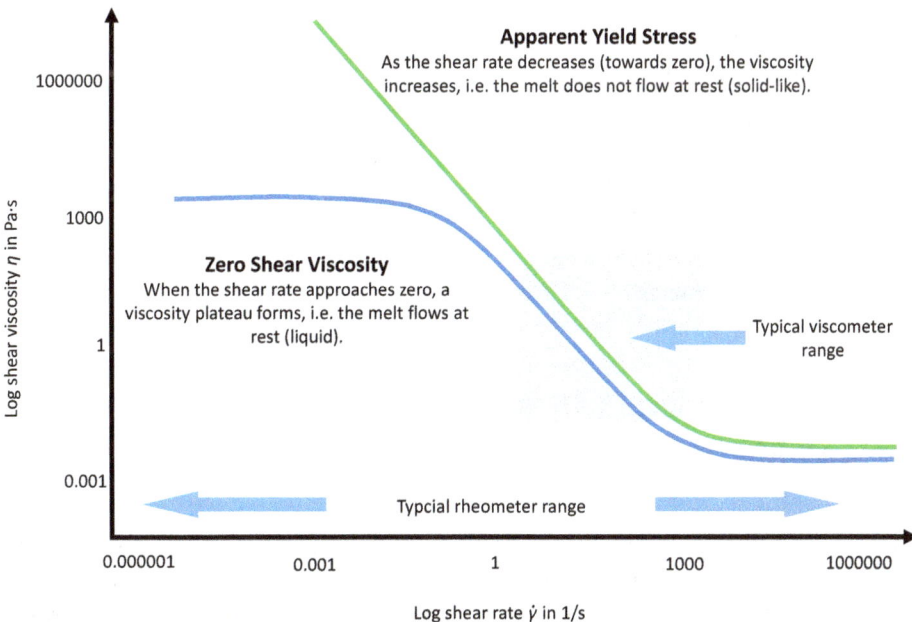

Figure 6.1 Applications of rheometers and viscometers

The viscosity of low-viscosity fluids, such as engine oils or plasticizers, can be determined easily. Falling viscometers can be sufficient for this purpose. If we want to measure the viscosity of a plastic melt, these measurements should be carried out with a rotational/oscillation rheometer or a capillary rheometer, as we also want to map the viscoelastic behavior of the plastic melts for higher shear rates. With the help of rheometry, flow curves such as $\tau = f\,(\dot{\gamma})$ and $\eta = f\,(\dot{\gamma})$ can be recorded and plotted. The oscillation reflects the *viscous (loss modulus)* and *elastic components (storage modulus)* of the plastic. Furthermore, the parameters for the material laws (Ostwald and de Waele power approach, Carreau approach, Cross approach, etc.) can be determined for plastics.

As a result, different rheometers and viscometers are used to determine material data and flow curves depending on the area of application and use. The best-known measuring devices are

- Falling viscometers
    - Falling ball viscometer
    - Rolling ball viscometer
- Viscosity balance
- Rotational and oscillation rheometers
    - Plate–plate rheometer
    - Cone–plate rheometer
    - Coaxial cylinder systems
        - Couette system
        - Searle system
- Capillary rheometers
    - Low-pressure capillary rheometer (MVR, MFI)
    - High-pressure capillary rheometer (HPCR)
    - Inline melting rheometer
- Extensional rheometer

6.1 Ranges of Application of the Different Viscometer and Rheometer Types

The falling ball viscometer and the viscosity balance are mainly used for low-viscosity Newtonian media. They can be used to determine the viscosity of engine oil, for example. The falling velocity of the internal steel ball shows which oil has the lowest resistance. The faster the ball falls down through the medium, the lower the viscosity of the engine oil.

Rotational rheometers are preferred for measuring the material data of fluids such as plasticizers, casting resins, adhesive solutions, paints, and the like. For plastic melts, rotational rheometers are used to determine the zero viscosity (viscosity at very low shear rates). Viscosities at higher shear rates can only be determined with low accuracy due to the increased heating of the measured material caused by frictional heat. The reason for this is that the same measuring material is always sheared during a measurement.

Rotational rheometers are also used to measure normal stresses in plastic melts. In particular, cone–plate rheometers are used for this purpose. Under shear stress, the material being measured tries to push the cone upwards as it is drawn into the measuring gap due to the special behavior of the viscoelastic fluid. With cone–plate rheometers, the normal stress is determined by measuring the force required to hold the cone on the plate. In plate–plate and cone–plate rheometers, the pressure dependence of the viscosity is measured. For this purpose, holes are provided on the base to enable pressure measurement and thus the determination of the normal stresses.

The high-pressure capillary rheometer is suitable for determining material data in a high shear rate range (Figure 6.2). This means that the material data can be determined in the shear rate range in which the most important processing steps of plastic melts take place. For this reason, the capillary rheometer is of the greatest importance for practical rheology. In the capillary rheometer, the material to be measured is injected into the open air. This means that the shearing of the measured material has no effect on the measurement result.

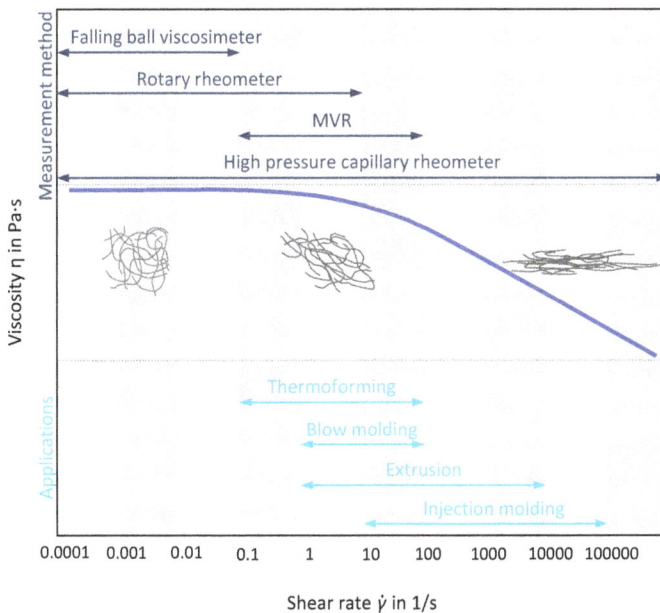

Figure 6.2 Shear rate ranges of different manufacturing processes and measuring methods

6.2 Prerequisite for Determining the Material Data

One of the prerequisites for determining material data using the abovementioned viscometers is the assumption of a laminar layered flow of the fluid. The fact that this assumption is also correct for the high shear rates in the capillary rheometer for plastic melts will be briefly explained. The boundary between a laminar layered flow and a turbulent flow is defined by the Reynolds number. If a Reynolds number is smaller than 2300, the flow is laminar. If the value is greater than 2300, turbulent flow must be assumed.

The Reynolds number is calculated as follows:

$$Re = \frac{\bar{v} \cdot D}{v} = \frac{\bar{v} \cdot D \cdot \rho}{\eta} \tag{6.1}$$

where

$\bar{v}$ is the average flow rate

D is the pipe diameter

v is the kinematic viscosity

ρ is the density

η is the dynamic toughness, i.e., viscosity

> The Reynolds number thus describes the ratio:
>
> $$Re = \frac{\text{Inertial forces}}{\text{Viscous forces(toughness)}}$$

Some extreme values for plastic melts (high flow front rate, large flow channel diameter, low viscosity, etc.) are assumed for estimating the Reynolds number. See Table 6.1; these values result in a Reynolds number of 2.

Table 6.1 Extreme Values for Estimating the Reynolds Number

Quantity	Symbol	Unit	Value
Average flow rate	$\bar{v}$	m/s	10
Channel diameter	D	mm	10
Viscosity	η	Pa · s	50
Density	ρ	g/cm^3	1
Reynolds number	Re	[−]	2

> A laminar layer flow can always be assumed for plastic melts, as the viscous forces predominate due to the high viscosity.

One reason for the low Reynolds number is the relatively high viscosity of plastic melts compared to other media such as water or air. Table 6.2 provides an overview of the viscosities of different materials.

Table 6.2 Viscosities of Different Materials

Material	Viscosity η [Pa·s]	Consistency
Air	10^{-5}	Gaseous
Water	10^{-3}	Low viscosity
Glycerin	1	Fluid
Polymer melt*	10^1 to 10^6	Viscous
Glass	10^{21}	Solid-like

* at processing temperature

6.3 Falling Viscometers

Falling or rolling viscometers are based on the measurement of the sinking or rolling rate of a measuring solid in the fluid under investigation. The sinking process is influenced by the geometry of the measuring solid or the measuring vessel, the difference in density between the measuring solid and the fluid, and the viscosity of the fluid. Therefore, the density of the fluid must be known in order to derive the viscosity. The flow in the falling viscometer is inhomogeneous and difficult to describe, particularly due to the triggered backflow of the fluid. Calibration is therefore always advisable. These disadvantages are offset by the very simple design of the device and its ability to measure at high temperatures and pressures.

6.3.1 Determining the Viscosity with Falling Viscometers Using Stokes's Law

In falling ball viscometers, the viscosity is determined using Stokes's law. The procedure is described below as an example. There is a ball in a container, which falls into a fluid at an initial velocity of $v = 0\,\text{mm/s}$. Initially, it is accelerated downwards due to the force of gravity. The acceleration decreases the further the ball sinks, until it is moving through the fluid at a constant velocity. The force of gravity F_G is then compensated by the frictional force F_R. Therefore, the following equilibrium of forces applies in constant sinking:

$$F_G = F_A + F_R \tag{6.2}$$

where

F_A is the buoyant force

and

$$F_G - F_A = m_{eff} \cdot g = (\rho_k - \rho_{Fl}) \frac{4}{3} \pi R_K^3 \cdot g \tag{6.3}$$

where

R_K is the radius of the sphere

g is the downward acceleration due to gravity

ρ_k is the density of the solid

ρ_{Fl} is the density of the fluid

Gravity is reduced by the buoyancy of the sphere, which is expressed in Equation 6.3 by the effective mass m_{eff}. If $\rho_k > \rho_{Fl}$, the ball sinks; if $\rho_k < \rho_{Fl}$, it rises (Figure 6.3).

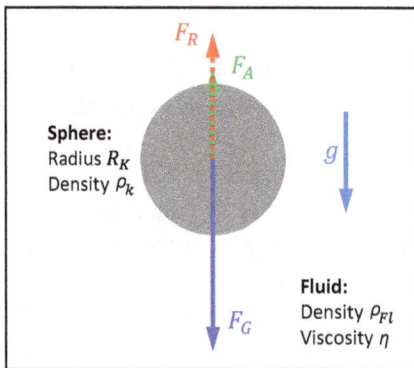

Figure 6.3
Forces on a ball sinking in fluid

By experimenting with spheres of different radii and in different fluids, the following law was discovered, which is also known as Stokes's law:

$$F_R = 6 \cdot \pi \cdot \eta \cdot R_K \cdot v_0 \tag{6.4}$$

where
η is the viscosity of the fluid

It becomes clear that the frictional force is not only dependent on the viscosity of the fluid, but also on the radius of the ball.

It is thus possible to determine the velocity v_0 with the help of the equilibrium of forces (Equation 6.3):

$$v_0 = \frac{2}{9} \cdot g \cdot \frac{R_K^2}{\eta} \cdot (\rho_k - \rho_{Fl}) \tag{6.5}$$

With known velocity and density, it is now possible to determine the viscosity of the fluid, as is the case in a falling ball viscometer:

$$\eta = \frac{2}{9} \cdot g \cdot \frac{R_K^2}{v_0} \cdot (\rho_k - \rho_{Fl}) \tag{6.6}$$

However, Stokes's law only applies to small radii. Osen's approximation formula must be used for larger radii.

6.3.2 Falling Ball Viscometer

The falling ball viscometers are often used because of their simplicity. They are preferably used to measure the viscosity of low-viscosity, Newtonian fluids (oil, plasticizers, etc.), as it is difficult to determine the actual shear rate. In the case of non-Newtonian media, they are used in the range of low shear rates ($\dot{\gamma}$ values) to check the consistency of production.

As described in Section 6.3.1, the viscosity η can be determined quite accurately from the sinking velocity of a ball without great effort. The measuring principle is as follows (Figure 6.4): First, the steady-state sinking velocity of a measuring solid (ball, cylinder, needle) is determined in a reservoir of the measuring fluid. After the ball is launched, it is accelerated to the steady-state velocity. The time of the falling movement is then determined over a defined measuring distance and the rate of descent is derived from this. The viscosity can then be derived using the Stokes equation (see Section 6.3.1) for known densities [1].

- Advantages:
  - Low price, simple assembly
  - Large measuring and examination area
- Disadvantages:
  - Complicated flow shape requires calibration for more accurate measurements
  - Density of the measured material must be known
  - Only steady-state shear viscosity can be determined

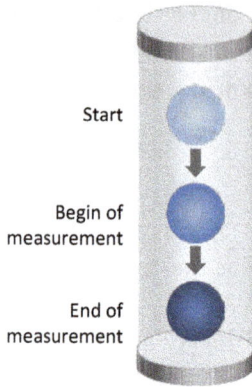

Figure 6.4
Principle of the falling ball viscometer

6.3.3 Rolling Ball Viscometer

In the rolling ball viscometer, the ball does not fall freely in a measuring cylinder but rolls along an inclined measuring cylinder (Figure 6.5). By tilting the measuring cylinder, the load can be changed quickly and easily by gravity. This extends the measuring range. If, for example, a ball falls far too quickly for a measurement, the measured time can be extended with a slight inclination in the rolling ball viscometer, thus increasing the accuracy. The disadvantage is the complex flow form during the rolling process. The measuring principle corresponds to that of the falling ball viscometer. Instead of the sinking velocity, the rolling velocity is determined [1].

- Advantages:
  - Low price
  - Uncomplicated setup, operation, measurement, and cleaning
- Disadvantages:
  - Complicated flow shape requires calibration for more accurate measurements
  - Only steady-state shear viscosity can be determined
  - The density must be known

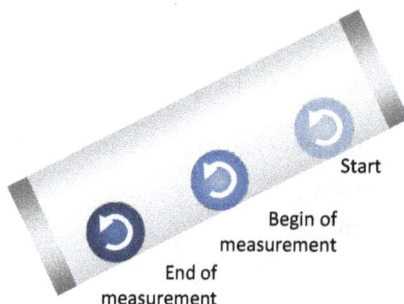

Figure 6.5
Principle of the rolling ball viscometer

6.4 Viscosity Balance

For measurements in the falling ball viscometer, the density of the fluid to be measured must be known. Furthermore, the fluids usually must be transparent. The so-called viscosity balance avoids these restrictions. Here, a ball in the fluid to be measured is suspended on one side of a balance (Figure 6.6). First, the ball is balanced in the fluid on the other side of the balance, using weights. This corresponds to determining the density of the fluid according to Archimedes's law of buoyancy. Further weights are then added or removed, accelerating the ball. When the steady-state velocity is reached, the rate of ascent or descent is determined by measuring the time and distance [1].

- Advantages:
  - Low price
  - Easy handling, even with opaque samples
  - Density does not have to be known
- Disadvantages:
  - Calibration required for more accurate measurements
  - No determination of dynamic variables

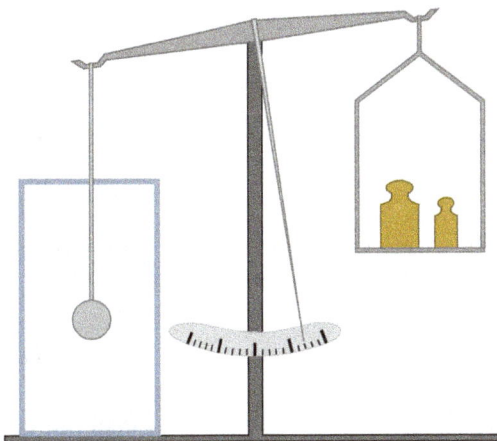

Figure 6.6
Structure of the viscosity balance

6.5 Rotational and Oscillation Rheometers

In the largest number of rotational rheometers, a measuring solid rotates in a fluid sample, or the sample is limited by a rotating wall. This is generally referred to as a rotational rheometer. The rotation creates a drag flow in the fluid. The measured variables are the rotational speed and the torque. These measured variables correlate with the rheometric variables shear rate and shear stress. The relationship depends on the type of measuring geometry. Three of the most common geometries are presented in the following sections and their basic relationships for calculating the flow curves are given [1].

6.5.1 Plate–Plate Rheometer

In the plate–plate system, a measuring fluid is sheared between parallel plates, one of which rotates (Figure 6.7). The measuring range of the shear rate is approximately $10^{-2}\,s^{-1}$ to $10^{2}\,s^{-1}$.

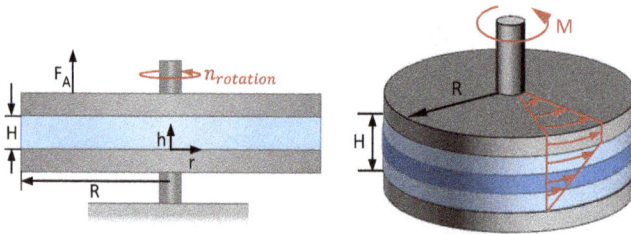

Figure 6.7 Plate–plate rheometer

In contrast to the cone–plate system, the shear rate is not the same throughout the gap but increases from the center ($r = 0$) to the edge ($r = R$). The result for the shear rate $\dot{\gamma}$ is

$$\dot{\gamma}(r) = \frac{2 \cdot \pi \cdot r \cdot n_{rotation}}{H} \tag{6.7}$$

As can be seen from Equation 6.7, the shear rate increases linearly with the radius and reaches its maximum at the outer edge. The following applies:

$$\dot{\gamma}_{edge} = \frac{2 \cdot \pi \cdot R \cdot n_{rotation}}{H} \tag{6.8}$$

The shearing velocity can be varied by changing the rotational speed $n_{rotation}$ or the gap height H.

The following applies to the shear stress τ with Newtonian behavior:

$$\tau = \frac{2 \cdot M}{\pi \cdot r^3} \tag{6.9}$$

If the non-Newtonian behavior is unknown, a correction must be made. In this case, the local derivative is read from a plot of the measured torque M versus the characteristic shear rate: $dM/d\dot{\gamma}_c$. The corrected shear stress can then be determined as follows:

$$\tau_{corr} = \frac{\tau}{4} \cdot \left(3 + \frac{dM}{d\dot{\gamma}_c}\right) \tag{6.10}$$

- Advantages:
  - Shear rates can be adjusted by changing the measuring gap height and the rotational speed
  - Good measurability of normal stress differences and dynamic variables
- Disadvantages:
  - Leakage of measuring material from the measuring gap (edge failure) is possible
  - Precise calibration and measuring gap control are required
  - Inhomogeneous shear rate distribution

Up to a limited shear rate, the shear stress and, thus, also the viscosity of a plastic melt can be determined by means of rotation, as Figure 6.8 shows. The diagram displays the measurement of the so-called zero viscosity (viscosity at very low shear rates) for EVAL™ using a plate–plate rheometer. These values can, for example, be superimposed very well with the measurement of the high-pressure capillary rheometer in order to be able to depict the complete measurement curve for a wide shear rate range.

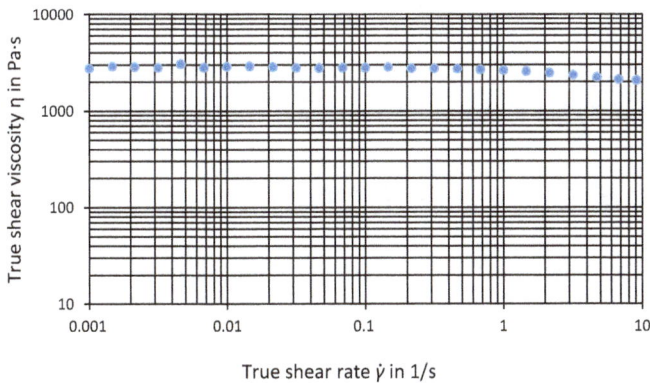

Figure 6.8 True shear viscosity as a function of the true shear rate for EVAL

6.5.2 Cone–Plate Rheometer

The measurements obtained using the cone–plate rheometer correspond to those of the plate–plate rheometer. In the same measuring device, only the plate is replaced by a cone, which has the advantage of a homogeneous shear rate. The angle α is approximately 1° to 2°. Control is achieved via rotational speed (controlled shear rate, CSR) or torque (controlled shear stress, CSS). The rotational speed or torque is measured accordingly. The normal stresses can be derived using force transducers on the drive shaft or on the underside of the cone. Due to the narrow gap in the middle (distance of approximately 40 μm), only a limited maximum particle size is permitted when testing dispersions (Figure 6.9). Therefore, filled plastics (suspensions) should be measured with the plate–plate rheometer, as the fillers could distort the measurement result. The maximum permissible particle size is described in the measuring system standards ISO 3219 and DIN 53019. The ISO standard recommends using a cone angle of 1° and describes cones with angles greater than 4° as nonstandard.

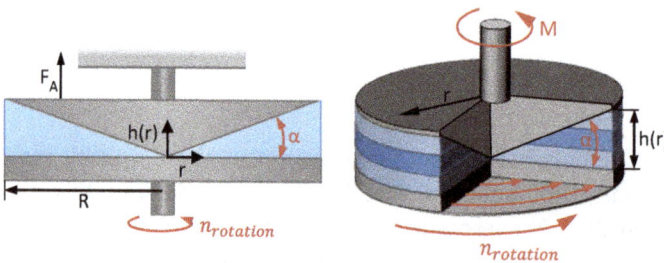

Figure 6.9 Shear gap of a cone–plate rheometer

The circumferential velocity on the cone surface increases towards the outside. At the same time, the vertical gap width increases due to the cone shape. As a result, the shear rate in the vertical direction remains constant over the radius. In order for the vertical shear gradient to dominate, the angle of inclination of the cone must be correspondingly small (flat cone tip). The shear rate in the gap is then calculated as follows:

$$\dot{\gamma} = \frac{2 \cdot \pi \cdot n_{\text{rotation}}}{\alpha} \cdot \left(1 - \alpha^2 + \frac{\alpha^2}{3}\right) \approx \frac{2 \cdot \pi \cdot n_{\text{rotation}}}{\alpha} \tag{6.11}$$

The angle of inclination of the cone tip α must be given in radians [rad] in the calculations.

The shear stress results from the torque M according to:

$$\tau = \frac{3 \cdot M}{2 \cdot \pi \cdot r^3} \tag{6.12}$$

- Advantages:
  - Homogeneous shear rate distribution
  - Good measurability of normal stress differences and dynamic variables
- Disadvantages:
  - Leakage of measuring material from the measuring gap (edge failure) is possible

- Precise calibration and measuring gap control are required
- Filled measuring materials can alter the measurement result if the fillers accumulate under the tip of the cone

6.5.3 Normal Stresses and Viscoelastic Behavior

Due to their extensional molecular structure, high-molecular-weight fluids, such as plastic melts and solutions, have the property of exhibiting elastic effects that are actually only known from elastic solids. This elasticity of fluids manifests itself in many flow phenomena that are unusual for normal fluids. This behavior of plastics and plastic melts is referred to as viscoelastic and can be described in simplified terms using spring/damper models (Maxwell, Kelvin–Voigt, etc.).

Until now, it was assumed that only shear stresses are generated in the material by shear deformation. In the case of viscoelastic materials, namely polymers in particular, normal stress components are also generated during shear. These act in addition to the hydrostatic pressure and are usually different for each spatial direction. Figure 6.10 shows this situation using the example of a parallelogram. If the parallelogram is transformed into the shape of a rectangle by a force acting at the top and to the right, the material layers are stretched or subjected to tensile stress in the direction of the deformation force ($a' > a$) and compressed or subjected to compressive stress in the direction of the deformation gradient ($b' < b$).

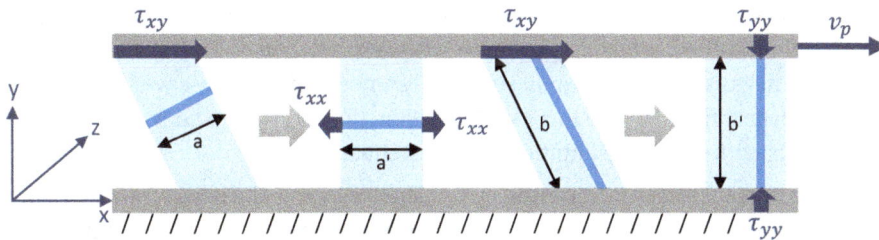

Figure 6.10 Normal stress components using the example of a parallelogram [1]

Normal stresses thus correspond to changes in length, which are referred to as extensions and compressions (tensile and compressive stresses) in relation to the initial length. When viscoelastic materials are moved or deformed, not only one-dimensional forces or stresses occur in the direction of deformation. A three-dimensional deformation state is always present, which can be described with a (3×3) tensor. This tensor contains the three normal stresses τ_{yy}, τ_{xx}, and τ_{zz} (Figure 6.11). By definition, x denotes the direction of movement and y the direction of the gradient, while z represents the indifferent direction.

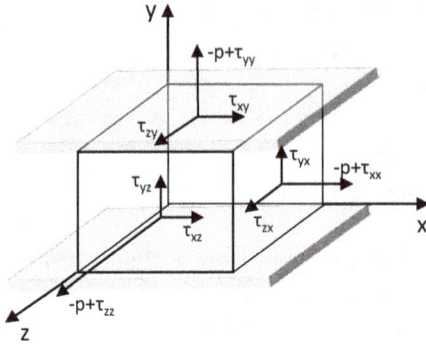

Figure 6.11
Stress states in a viscoelastic fluid
with steady-state shear flow [2]

Since normal stresses contain a height-dependent pressure component, the so-called normal stress differences are formed. This means that the result no longer depends on the ambient pressure. A distinction is made between the first normal stress difference N_1 (difference between the normal stresses in the direction of flow and in the direction of the velocity gradient) and the second normal stress difference N_2 (difference between the normal stresses in the direction of the velocity gradient and in an indifferent direction).

The following applies:

- First normal stress difference: $N_1 = \tau_{xx} - \tau_{yy}$

- Second normal stress difference: $N_2 = \tau_{yy} - \tau_{zz}$

With most materials, shear causes tensile stress in the flow direction and compressive stress in the gradient direction. This means that the fluid element is pulled in length in the direction of flow, while it is compressed perpendicularly. This leads to a positive first or a negative second normal stress difference. N_1 is usually several orders of magnitude greater than N_2. The increase in the functions $N_1 = f(\dot{\gamma})$ and $N_2 = f(\dot{\gamma})$ is constant in the logarithmic representation for small shear rates.

Dividing the normal stress differences by the square of the shear rate results in the normal stress coefficients Ψ_1 and Ψ_2, whose curve is similar to a viscosity function (Figure 6.12). The first normal stress coefficient is considerably greater than the second normal stress coefficient.

The following applies:

- First normal stress coefficient: $\Psi_1 = \frac{N_1}{\dot{\gamma}^2}$

- Second normal stress coefficient: $\Psi_2 = \frac{N_2}{\dot{\gamma}^2}$

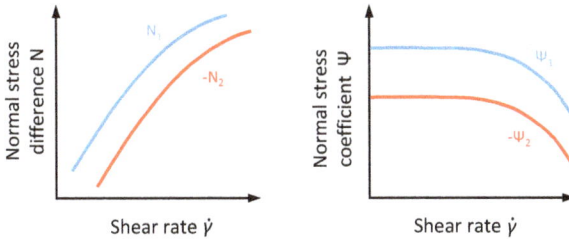

Figure 6.12 Schematic diagram of normal stress differences or coefficients as a function of the shear rate

6.5.4 Measurement of the Normal Stresses of Fluids Using Rotational Rheometry

Rotational rheometers can be used, for example, to determine normal stresses or normal stress differences. In contrast to purely viscous materials, which may follow an outward movement due to the rotational movement of the rheometer and the resulting centrifugal force, viscoelastic media behave differently. The viscoelastic behavior causes the fluid in the rotation test to move inwards rather than outwards like a Newtonian fluid. In elastic fluids, the shear rate in the shear direction—in the cone–plate rheometer, in the circumferential direction—creates a tensile stress. Due to the curvature of this tensile stress, the resulting stresses are in the radial direction. These radial components add up in the direction of the center of the shear gap (support effect). The pressure $p(r)$ increases linearly with $\ln(R/r)$ as the radius decreases. This results in a buoyant force F_A in the axial direction in the rheometer. The normal stress differences and normal stress coefficients can be calculated from the axial forces F_A [2].

Among rotational rheometers, cone–plate systems are particularly suitable for measuring forces or pressures with regard to the derivation of normal stress differences. There are different ways of measuring the normal stress differences with these devices [1].

The first very simple way to determine the first normal stress difference N_1 is to measure the buoyant force F_A using a sensor. The variables shown in Figure 6.13 are determined as follows:

The shear rate in the rheometer is calculated from

$$\dot{\gamma} = \frac{dw_u(r)}{dh(r)} = \frac{w_u}{h(r)} = \frac{r\,\Omega}{r\,\tan\alpha} = \frac{\Omega}{\alpha} \tag{6.13}$$

The shear stress results from

$$\tau = \frac{3}{2 \cdot \pi \cdot R^3} \cdot M_D \tag{6.14}$$

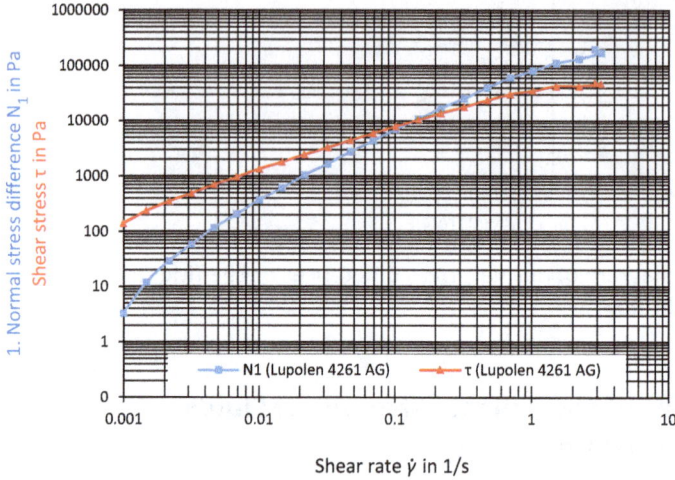

Figure 6.13 Shear stress and first normal stress difference for a Lupolen 4261 (PE-HD)

The elasticity of the medium creates an overpressure in the shear gap, which causes the cone and plate to be pushed apart with the axial force. The axial force F_A results from the integration of the pressure $p(r)$ over the shear gap area:

$$F_A\left(\dot{\gamma}\right) = \int_0^R p\left(r\right) \cdot 2 \cdot \pi \cdot r \cdot dr \tag{6.15}$$

The relationship between the buoyant force and the first normal stress difference can be formulated as follows:

$$F_A\left(\dot{\gamma}\right) = \frac{\pi \cdot R^2}{2} N_1\left(\dot{\gamma}\right) \tag{6.16}$$

The relationship between the first normal stress difference and the integral normal force is obtained by integrating the normal stresses over the radius:

$$N_1\left(\dot{\gamma}\right) = \frac{2 \cdot F_A\left(\dot{\gamma}\right)}{\pi \cdot R^2} \tag{6.17}$$

The specified relationship applies if the radial acceleration of the fluid, which can influence the measurement result, is disregarded. However, depending on the angular frequency and the radius, a correction of the first normal stress difference may be necessary. This correction can be carried out as follows:

$$N_1\left(\dot{\gamma}\right) = \frac{2 \cdot F_A\left(\dot{\gamma}\right)}{\pi \cdot R^2} - 0.15 \cdot \rho \cdot \omega^2 \cdot R^2 \tag{6.18}$$

Here, ρ is the density of the material under investigation and ω is the angular velocity. In addition to the negative pressure, a secondary flow can be triggered by the radial acceleration, which can also contribute to the measured normal force. In most cases, however, the correction specified above is sufficient [1].

The diagram in Figure 6.13 shows the measurement of the shear stress and the first normal stress difference as a function of the shear rate using a cone–plate rheometer. The viscoelastic fluid (here, a PE-HD) pushes the upper plate upwards, creating a buoyant force F_A. This force is measured using a very sensitive sensor in the rheometer and the first normal stress difference N_1 is determined.

It is clear that with this material, which is a very-long-chain blow-molded material, even at very low shear rates of $\dot{\gamma} = 0.2$ s^{-1}, the normal stress effects outweigh the shear stress effects. This means that the elastic flow effects are superimposed on the viscous flow effects. This point is discussed in more detail in the chapter on multilayer flows (Chapter 16).

It is also possible to measure normal stresses using rotational rheometers by directly recording pressures or forces in the gap, provided that this sensor technology is available on the rheometer. Figure 6.14 (left) shows the pressure curve in the gap of the rheometer. It is clear that the pressure increases from the outside to the inside when measuring a viscoelastic fluid.

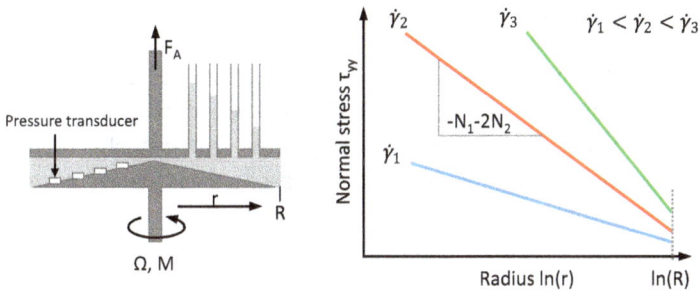

Figure 6.14 Measurement and representation of the radial pressure curve in a cone–plate system

As the right-hand diagram in Figure 6.14 shows, the normal stress τ_{yy} results from the measured pressure. This shows a linear function over the natural logarithm of the radius (Figure 6.15). The increase corresponds exactly to the term $-(N_1 + 2N_2)$. As the shear rate increases, both the gradient and the normal stress τ_{yy} increase. The following applies:

$$p\,(r, \dot{\gamma}) = N_2\,(\dot{\gamma}) - [(N_1 + 2N_2)\,(\dot{\gamma})] \cdot \ln\left(\frac{R}{r}\right) \tag{6.19}$$

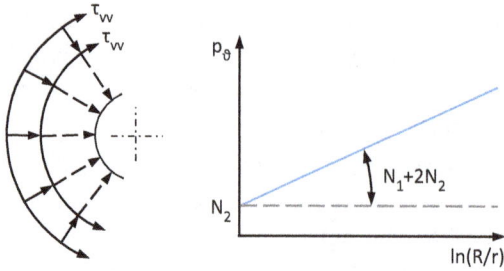

Figure 6.15
Normal stress as a function of the radius

The pressure $p(r)$, which increases linearly over $\ln(R/r)$, intersects the ordinate at $r = R$ or $\ln(R/r) = 0$, which corresponds to the second normal stress difference N_2. Consequently, neglecting the surface tension, the measured or extrapolated pressure value at the outer edge ($r = R$) can be equated to the second normal stress difference N_2. Finally, the first normal stress difference N_1 can be determined from the slope of the straight line [1].

One way of calculating the first normal stress difference is by measuring the shear stress after a shear jump $\Delta\gamma$ (Lodge–Meissner relationship):

$$N_1 = \tau_{xy} \cdot \Delta\gamma \tag{6.20}$$

In principle, it is also possible to measure the normal stress differences with other rotating measuring geometries. However, this usually causes problems with regard to recording the forces or determining the individual variables. In a plate–plate system, for example, the difference $N_1 - N_2$ can be derived by measuring the normal force, but not the two individual contributions N_1 and N_2. In a coaxial cylinder system, it is possible to derive N_1, but this requires a measurement of the pressures acting radially on the cylinders. What is still possible for the stationary cylinder is extremely complicated for the rotating cylinder. For this reason, cone–plate geometry is always recommended for normal stress measurement in rotating systems.

6.5.5 Measurement of the Viscoelastic Properties of Fluids Using Oscillation Rheometry

Measurements in oscillatory shear are being used increasingly in rheometry. Plate–plate or cone–plate rheometers, for example, can be employed for this measurement. In this case, the measuring element (cone or plate) does not rotate but oscillates.

In these vibration measurements, the sample is subjected to a sinusoidal shear with amplitude $\hat{\gamma}$ and the angular frequency ω, where f is the frequency of the oscillation: $\gamma = \hat{\gamma} \sin \omega t$. The torque $M(t)$ is measured, which can be used to calculate the shear stress.

The following applies:

$$\tau = \frac{2 \cdot M\,(t)}{\pi \cdot r^3} \tag{6.21}$$

$$M\,(t) = M_0 \sin\,(\omega t - \delta) \tag{6.22}$$

For an ideal elastic solid, the phase shift is $\delta = 0$, that is, the response occurs immediately with the excitation. This can also be seen in Figure 6.16. The shear stress reacts immediately to the deflection in this ideal elastic solid.

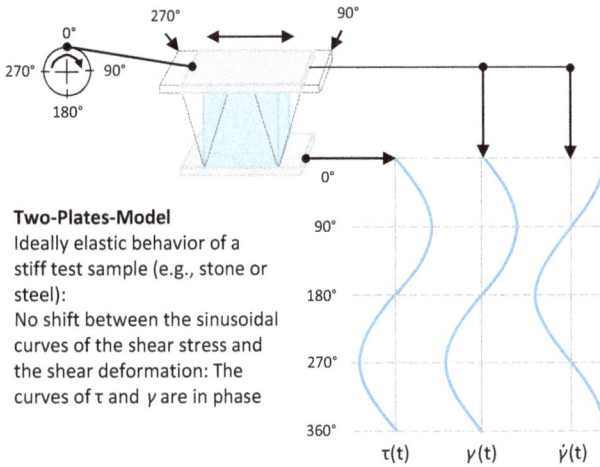

Two-Plates-Model
Ideally elastic behavior of a stiff test sample (e.g., stone or steel):
No shift between the sinusoidal curves of the shear stress and the shear deformation: The curves of τ and γ are in phase

Figure 6.16 Shear rate and shear stress curve of an ideal elastic solid in oscillatory shear [3, 4]

While the frequency f, as the reciprocal of the period of the sinusoidal oscillation, has the unit hertz [Hz], the unit of the angular frequency is s^{-1}. The shear rate $\dot{\gamma}$ is obtained from the derivative with respect to time:

$$\dot{\gamma} = \frac{d\gamma}{dt} = \omega \cdot \hat{\gamma} \cos\,(\omega \cdot t) = \hat{\gamma} \sin\left(\omega \cdot t + \frac{\pi}{2}\right) \tag{6.23}$$

It oscillates at the same frequency as the shear but is 90° ahead of it in phase (see Figure 6.16). The shear rate amplitude $\hat{\dot{\gamma}}$ is given by the product of the shear amplitude and the angular frequency. The greater the phase shift δ, the more the viscous components of the medium predominate.

The behavior of an ideal viscous medium is shown in Figure 6.17. The phase shift between the excitation and the response is 90°, that is, the shear and the shear stress are 90° out of phase.

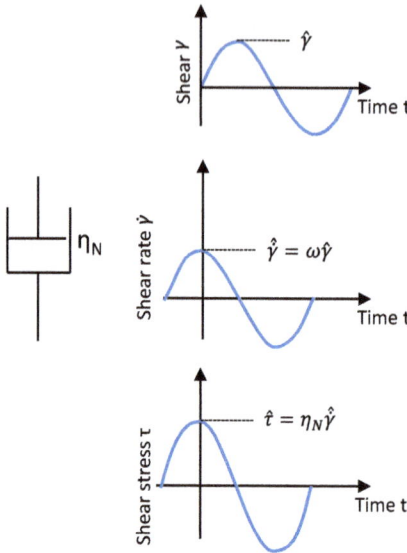

Figure 6.17
Shear rate and shear stress curve
for an ideal viscous medium

Plastics have a viscoelastic behavior, which can lead to a phase shift δ, with $0° \leq δ \leq 90°$, between the shear deformation (sample excitation) and the resulting moment or shear stress (sample response); see Figure 6.18.

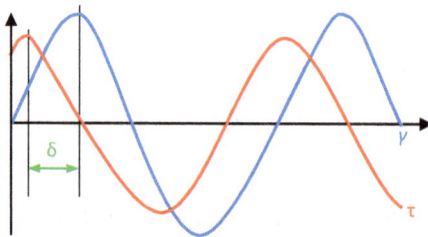

Default   Constant frequency and constant amplitude

Result   Most measurement samples show viscoelastic behavior with the phase shift δ between the sine curves of the measurement specification (e.g. deformation) and the measurement result (then: shear stress)

It applies: 0° (ideal-elastic) ≤ δ ≤ 90° (ideal-viscous)

Figure 6.18 Shear stress curve with a given deformation for viscoelastic behavior with phase shift [3, 4]

This measurement is started with the so-called amplitude sweep. This means that the deflection (amplitude) of the measuring solid (plate or cone) is gradually increased and the storage modulus G' and the loss modulus G'' are recorded. The aim of this measurement is to determine the so-called linear viscoelastic (LVE) range; see Figure 6.19 and Figure 6.20. As long as the storage and the loss moduli are linear and constant, the LVE range is reached. The permissible amplitude for the frequency sweep is thus determined.

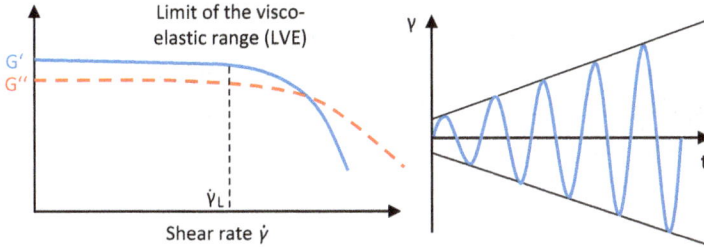

Figure 6.19 Amplitude sweep to determine the linear viscoelastic range

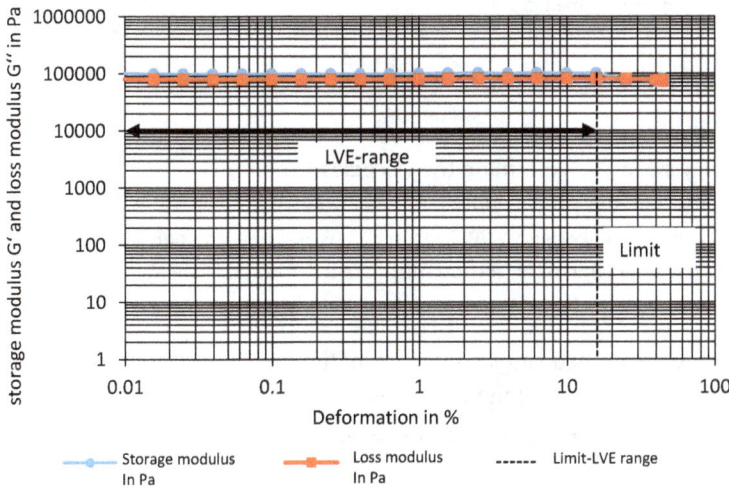

Figure 6.20 Amplitude sweep to determine the linear viscoelastic range for a Lupolen 4261 (PE-HD)

For small deformations, G' and G'' are constant and the sample structure is undisturbed, which corresponds to the LVE range. As soon as the moduli decrease, the structure is disturbed, that is, the end of the LVE range is reached. The plateau value of G' in the LVE range describes the stiffness of the sample at rest; the plateau value of G'' is a measure of the viscosity of the unsheared sample. The ratio of one modulus to the other allows a statement to be made about the consistency of the material: If the storage modulus is higher than the loss modulus, the sample has the character of a viscoelastic solid with predominantly solid properties. Conversely, if $G'' > G'$ in the LVE range, the sample has the properties of a viscoelastic fluid. The further apart the modules are, the more the properties tend towards pure fluid or pure solid.

The amplitude test can also be used to determine the yield point. Two particular points are suitable for this, namely the end of the LVE range and the intersection of the curves G' and G''.

This is followed by the so-called frequency sweep. This means that for a given amplitude, which is in the LVE range, the angular frequency of the measuring solid is gradually increased. The phase shift between the excitation (angular frequency) and the response (torque) can be employed to determine the elastic components G', using Hooke's law, and the viscous components G'', using Newton's law of friction (Figure 6.21).

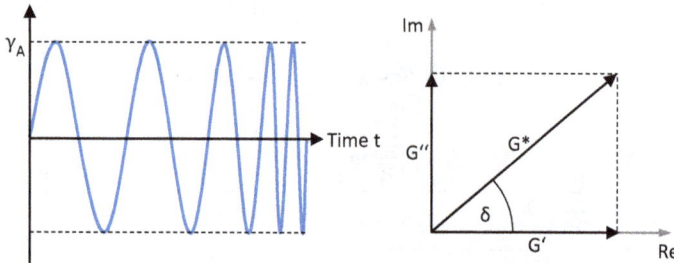

Figure 6.21 Frequency sweep to determine the viscoelastic behavior via loss and storage moduli and phase shift
G' [Pa]: storage modulus, elastic component; G'' [Pa]: loss modulus, viscous component
Physical meaning: G' for the stored and G'' for the lost (dissipated) deformation energy
$\tan\delta$ [-] = G''/G' loss factor or damping factor as a ratio of the viscous and elastic components

The sample response to an applied sinusoidal deformation can have different characteristics and depends on its deformation properties:

Elastic (Hooke's law), describes the linear elastic behavior of solids:

$$\tau(t) = G' \cdot \gamma(t) = G' \cdot \gamma_0 \sin(\omega t) \tag{6.24}$$

Viscous (Newton's law):

$$\tau(t) = G'' \cdot \gamma(t) = \eta \cdot \dot{\gamma}(t) = \eta \cdot \omega \cdot \gamma_0 \cdot \cos(\omega t) \tag{6.25}$$

Viscoelastic:

$$\tau(t) = G' \cdot \gamma_0 \sin(\omega t) + G'' \cdot \gamma_0 \cdot \cos(\omega t) \tag{6.26}$$

The loss factor (or the phase angle) $\tan\delta$ is calculated as follows:

$$\tan\delta = \frac{G''}{G'} \tag{6.27}$$

Figure 6.22 shows the storage and the loss moduli as well as the loss factor for a plastic. It can be seen that at low frequencies, the viscous properties predominate in this case. If the frequency is now increased, the two curves intersect, and the elastic properties of the plastic predominate at higher frequencies. The intersection of the two curves is called the crossover point (COP) and allows a statement to be made about the molecular weight and the molecular weight distribution of the solid measured (Figure 6.23).

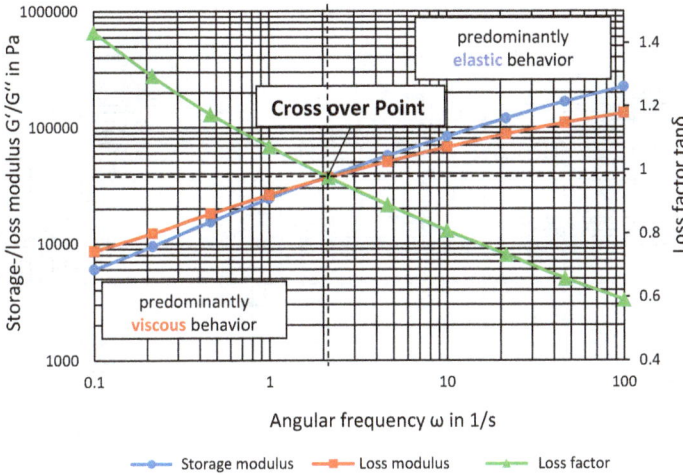

Figure 6.22 Storage and loss moduli and loss factor for a plastic

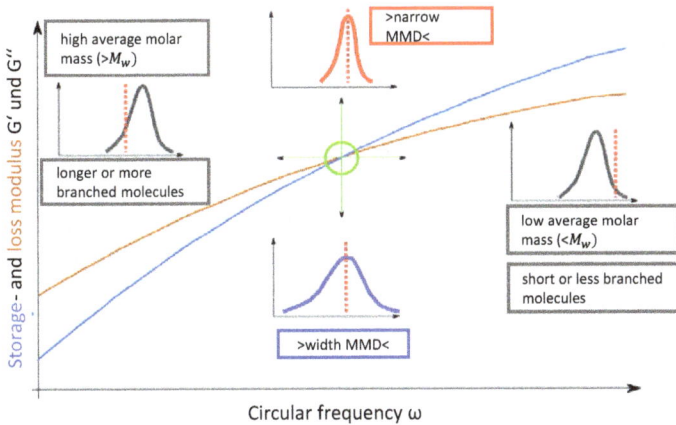

Figure 6.23 Displacement of the crossover point as a function of molecular weight and molecular weight distribution

With a narrow molecular weight distribution, the COP shifts upwards, while a wide distribution of the molecular weight causes a downward shift. If it is a plastic with long polymer chains or branched molecular chains, the COP shifts to the left. If the polymer chains are shorter or less branched, the COP shifts to the right.

It should be noted that this separation between the viscous and elastic properties can only be distinguished so clearly because it is a double-logarithmic plot (Figure 6.24). If the measurement results are plotted double-linearly, the elastic properties of the plastic melt predominate, especially at high frequencies, as shown in Figure 6.25.

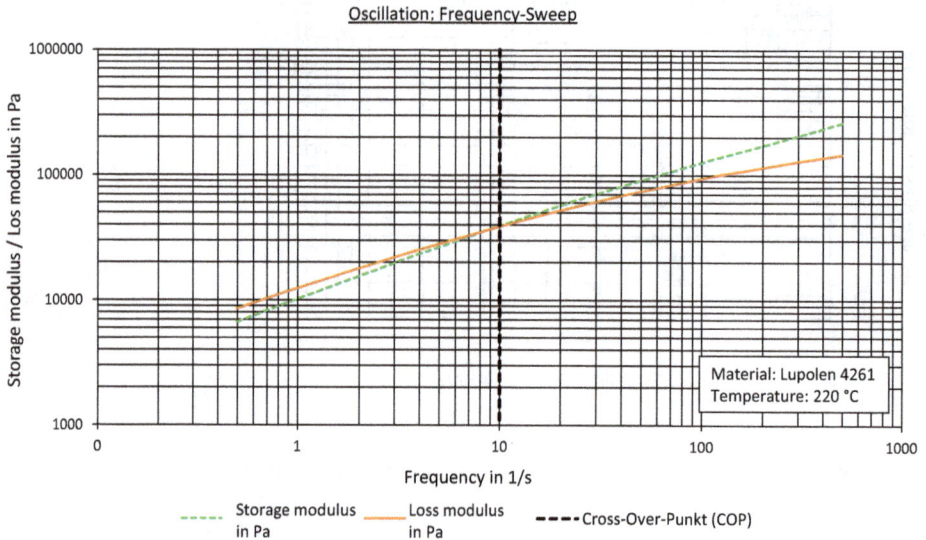

Figure 6.24 Storage and loss moduli measured with the frequency sweep; double-logarithmic plot for a Lupolen 4261 at $T = 220\,°C$

Figure 6.26 shows the measurement of the storage and the loss moduli as well as the resulting complex viscosity as a function of the angular frequency, measured using an oscillation rheometer (frequency sweep) for different temperatures. It is clear that an increasing temperature leads to a downward shift in the COP. Furthermore, the values in the viscous range for the storage and the loss moduli are further apart than in the elastic range. In general, the values are smaller at higher temperatures.

Oscillation: Frequency-Sweep

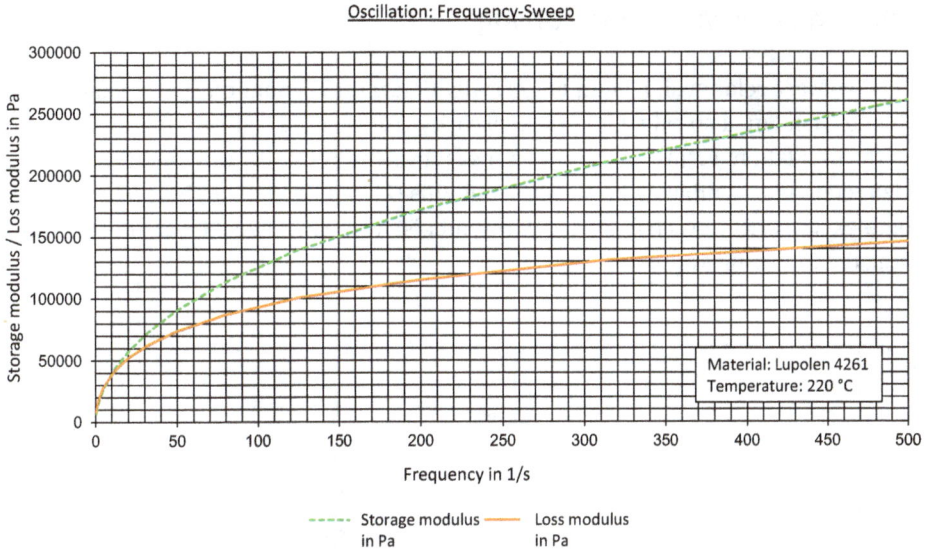

Figure 6.25 Storage and loss moduli measured with the frequency sweep; double-linear plot for a Lupolen 4261 at $T = 220\,°C$

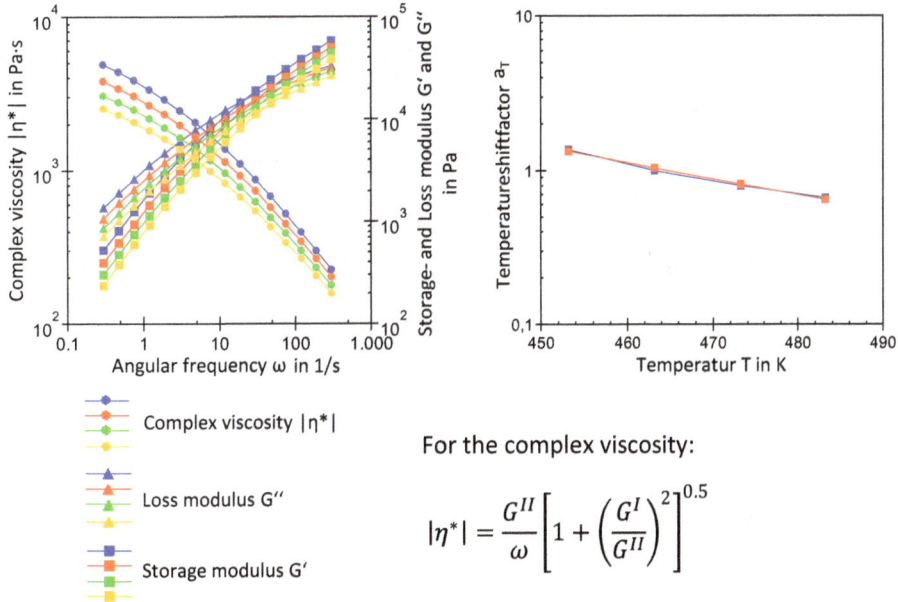

For the complex viscosity:

$$|\eta^*| = \frac{G^{II}}{\omega}\left[1 + \left(\frac{G^{I}}{G^{II}}\right)^2\right]^{0.5}$$

Figure 6.26 Influence of temperature on the storage and the loss moduli

6.5.6 The Cox-Merz Relationship

The Cox-Merz relationship is an empirically determined equation that makes it possible to calculate a so-called complex viscosity from the storage and the loss moduli via the oscillation. Figure 6.27 shows a measurement of the storage modulus G' (blue line/rectangular symbols) and the loss modulus G'' (red line/triangular symbols). The complex viscosity (green line/circular symbols) was calculated from these values using the following approach.

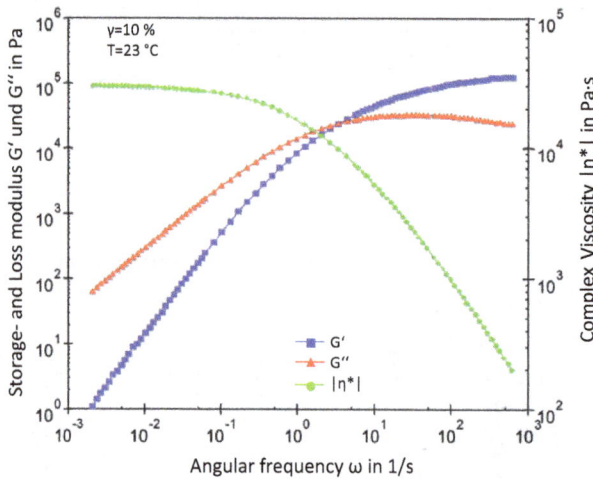

Figure 6.27 Storage and loss moduli and calculated complex viscosity using the Cox-Merz relationship for a plastic

The complex viscosity $|\eta^*|$ is calculated using the Cox-Merz rule. The relationship is as follows:

$$|\eta^*| \equiv \frac{|G^*|}{\omega} = \frac{1}{\omega}\left[G'^2 + G''^2\right]^{0.5} \tag{6.28}$$

or

$$|\eta^*| = \frac{G''}{\omega}\left[1 + \left(\frac{G'}{G''}\right)^2\right]^{0.5} \tag{6.29}$$

The complex viscosity corresponds to the viscosity $\eta = f(\dot{\gamma})$ if the angular frequency ω is set equal to the shear rate $\dot{\gamma}$:

$$|\eta^*| = \eta \text{ for } \omega = \dot{\gamma} \tag{6.30}$$

The application of the Cox-Merz rule is shown in Figure 6.28 for a polyethylene (PE-HD) melt.

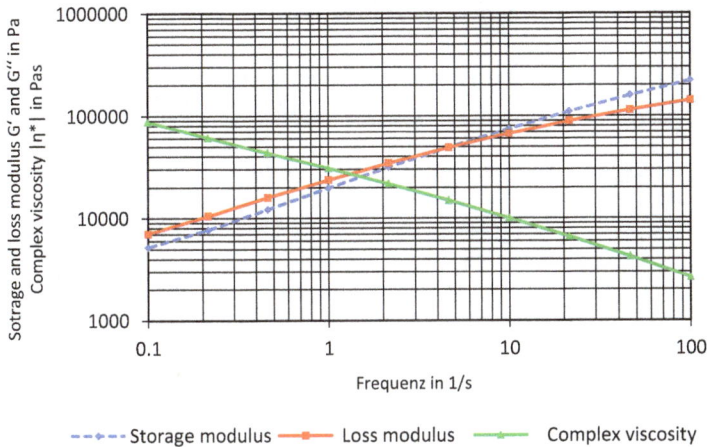

Figure 6.28 Storage and loss moduli as well as calculated complex viscosity using the Cox-Merz relationship for a Lupolen 4261 (PE-HD)

The validity of the empirical Cox-Merz rule can be confirmed experimentally for polymer melts and solutions in many cases. In principle, however, the validity must always be checked. Figure 6.29 shows the three measurements using rotational (red), oscillation (yellow), and high-pressure capillary (blue) rheometers for a polystyrene PS 158K. The rotational one gives good values for the viscosity for low shear rates. The comparison of the measured points of the high-pressure capillary rheometer (HPCR) for high shear rates shows that the Cox-Merz relationship also leads to a good agreement with the measured values in this case. A flow curve over a wide shear rate range is obtained by superimposing the rotation measurements with those of the oscillation rheometer and the HPCR. At low shear rates, the viscosity (zero viscosity) is determined by rotation. The measurement results of the oscillation rheometer are well suited for the transition range of the flow curve, and those of the HPCR are best suited for the high shear rate range.

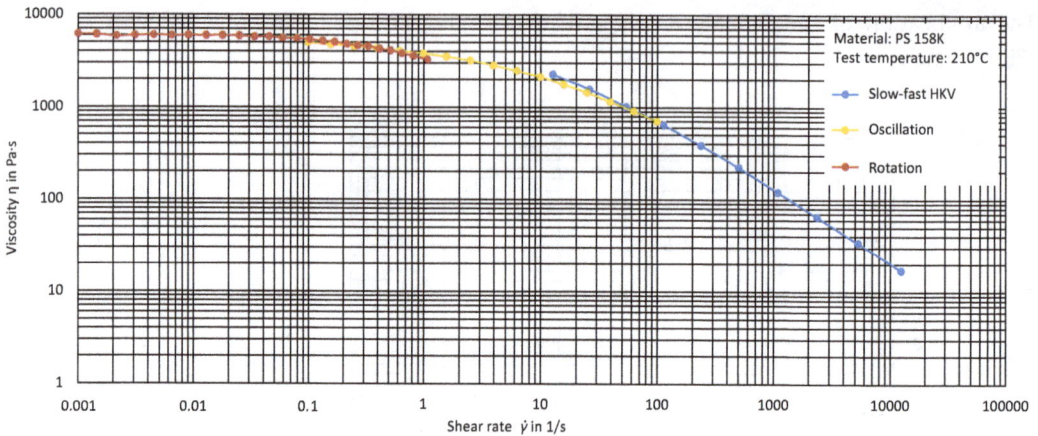

Figure 6.29 Measurement results of different rheometers (rotational, oscillation, and HPCR)

6.5.7 Relaxation Test Using a Rotational Rheometer

As already shown in Chapter 3, the viscoelastic properties of plastics are also a function of time. For a given constant deformation, the stress decreases as a function of time. This behavior is described by the combination of the Maxwell and the Kelvin–Voigt models. Rotational rheometry can be used to measure this time-dependent behavior of plastics. The so-called relaxation test is carried out for this purpose.

To determine the stress relaxation of a material, the sample is deformed a certain amount and the decrease in stress is recorded over a longer period of time. The stress relaxation rate is the slope of the curve at each point. Figure 6.30 shows the result of such a measurement for an extrusion of PE-HD Lupolen 4261 AG for different temperatures. It can be clearly seen that the relaxation modulus decreases with increasing time. This is a very long-chain product. For this reason, the mobility of the molecular chains is not as great, and relaxation does not take place as quickly. In principle, this behavior can be observed in all thermoplastics and varies from plastic to plastic (depending on the molecular structure, among other things) and is also a function of temperature. An increase in temperature shortens the relaxation times by increasing the mobility of the molecules or molecule segments. The relaxation time can play a role in the design of extrusion molds (ironing/compensation zone). Stress relief (relaxation) takes place in the ironing (compensation) zone. The so-called "swelling" is mainly affected in this zone and should be compensated by it [5, 6]. The relaxation time of the plastic melt is also taken into account for the die-swelling effect when designing the nozzle length [7].

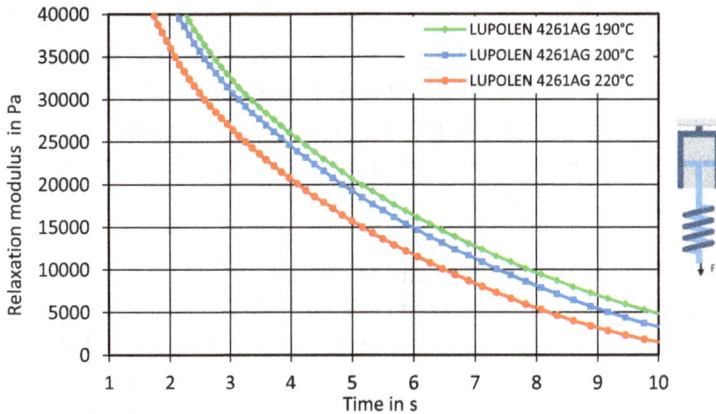

Figure 6.30 Relaxation test using rotational rheometry

In rheology, the relevance of elastic effects is assessed using the Deborah number or the Weissenberg number. The Deborah number is defined as follows:

$$De = \frac{\lambda}{t_p} \tag{6.31}$$

The variable λ is the relaxation time of the material and t_p is the process time, that is, the average residence time of a volume element in a flow channel section, for example. The Deborah number can be used to estimate easily the material behavior of the plastic melt in the processing area. Large values correspond to very long relaxation times—that is, the material behavior corresponds to that of an elastic solid. Measurements in the viscous flow range lead to Deborah numbers that are close to zero. If the relaxation time corresponds to the duration of the material stress, neither the elastic behavior nor the viscous flow dominates. In such a case, Deborah numbers of approximately 1 are obtained and the plastic behaves in a viscoelastic way.

The relationship between the Weissenberg number, the Deborah number, and the viscoelastic properties of the plastic can be illustrated in the Pipkin diagram [8].

Figure 6.30 shows that the elastic properties predominate in the case of the PE-HD Lupolen 4261 AG measured here, as the relaxation time is quite long.

6.5.8 The Large Amplitude Oscillatory Shear (LAOS) Theory

6.5.8.1 Historical Overview and Findings of the LAOS Experiments

The basic concept for carrying out LAOS experiments was introduced several years ago. As early as the 1960s to the 1970s, the first publications investigated nonlinear phenomena for a wide variety of viscoelastic materials under oscillating shear. The methods of Fourier transform analysis and stress wave analysis were proposed. At that time, technical problems, especially software and hardware limitations—such as computing power and torque resolution—hindered further progress. The first investigations were carried out with suspensions or polymer solutions.

MacDonald et al. [9] investigated the dependence of the complex viscosity for large-amplitude oscillating movements on four viscoelastic fluids using a Weissenberg rheogoniometer, including three polymer solutions and one polymer melt.

Tee and Dealy [10] carried out investigations on three polymer melts (PE-HD, PE-LD, and PS) to determine the nonlinear viscoelastic properties. To obtain a simplified characterization of the thermoplastics, they used closed-loop Lissajous curves of measured stress versus extension rates or stress versus extension. They concluded that the stress–extension rate loops were more pronounced than the stress–extension loops for these materials. At this point, they suggested that, based on experimental difficulties, only the first Fourier component could be accurately determined and the higher Fourier components had no direct relationship with traditional material functions. This was mainly due to the fact that experimental equipment, such as high-precision monochromatic sinusoidal excitations, high-performance analog-to-digital converter (ADC) cards, and improved torque transducers, had not been developed yet at that time. Nevertheless, even at this early stage of LAOS experiments, stress wave analysis and Fourier transform analysis were proposed as methods.

Dealy and Giacomin [11] investigated the behavior of various polymer melts under LAOS conditions using a special sliding plate rheometer, as shown in Figure 6.31 [12]. The special feature of this sliding plate rheometer is a special shear stress transducer that can be installed flush in the stationary plate, allowing the shear stress to be measured locally in a region of uniform deformation. This enabled them to conduct LAOS experiments on highly viscous polymer melts with a deformation amplitude of >. The measurement results were investigated using the Fourier transform analysis of the stress response, stress wave analysis, and constructive equation modeling.

Wilhelm et al. [13] developed a method for highly sensitive Fourier transform rheology by transferring the technique of NMR spectroscopy to oscillatory rheometry. To this end, they developed and applied highly sensitive detection methods. In particular, high-performance ADC cards with electrical and mechanical shielding and a special Fourier transform algorithm were used. This development made it possible to record high-resolution torque signals from standardized rheometers and to obtain Fourier transform spectra for complex fluids.

Ahn et al. [12] observed that the behavior of complex fluids in LAOS experiments can be divided into four primary categories. These four categories are Newtonian, linear viscoelastic, nonlinear viscoelastic, and elastic [14].

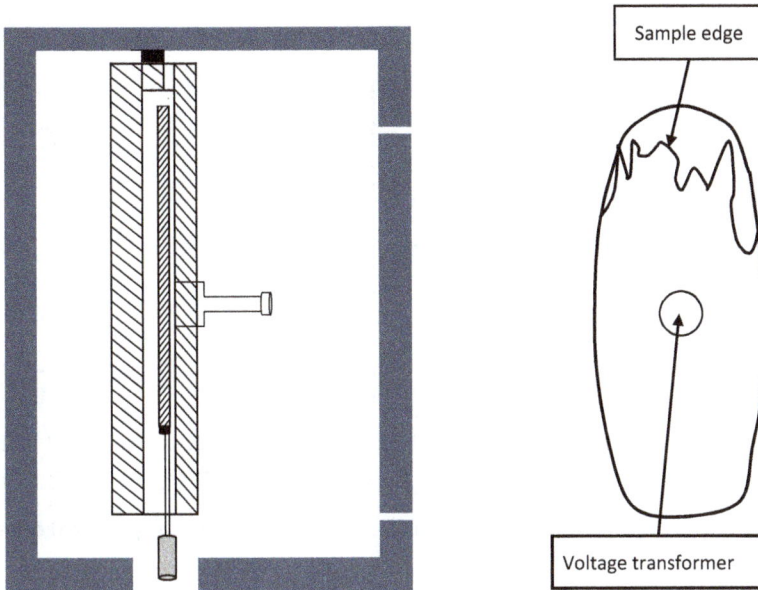

Figure 6.31 Cross section with essential elements of a sliding plate rheometer with elastic shear stress transducer

In 2011, another methodology was developed to quantify the nonlinear viscoelastic properties during LAOS experiments. Cho et al. [12] proposed a nonlinear stress decay method (or SD method, for short). In this method, the nonlinear stress behavior is decomposed into a superposition of elastic and viscous contributions. This SD method leads to a very useful step towards a physical interpretation of the nonlinear stress behavior.

Ewoldt et al. [15] succeeded in quantifying the nonlinear viscoelastic material functions on the basis of the SD method using the first set of Chebyshev polynomials [12].

6.5.8.2 Evaluation Methods of LAOS Behavior

In addition to the investigation of LAOS behavior with the aid of Fourier transform (FT) rheology, there are other ways of evaluating this behavior. The starting point is always the measured stress signal of the sample. The so-called Lissajous diagrams can be used to evaluate the LAOS behavior. An example of such a diagram is shown in Figure 6.32. Stress–extension diagrams are illustrated for a solid-like (elastic) sample and a fluid-like (viscous) sample. The LAOS measurements also give the third harmonic, which is a measure of the viscoelasticity in itself. This third harmonic is represented as a viscoelastic stress. It is divided into an elastic (τ') and a viscous (τ'') component. It can be seen that, in this case, the LAOS data exhibits an extension-specific colloidal structural transition that cannot be measured with conventional methods using a rheometer [16].

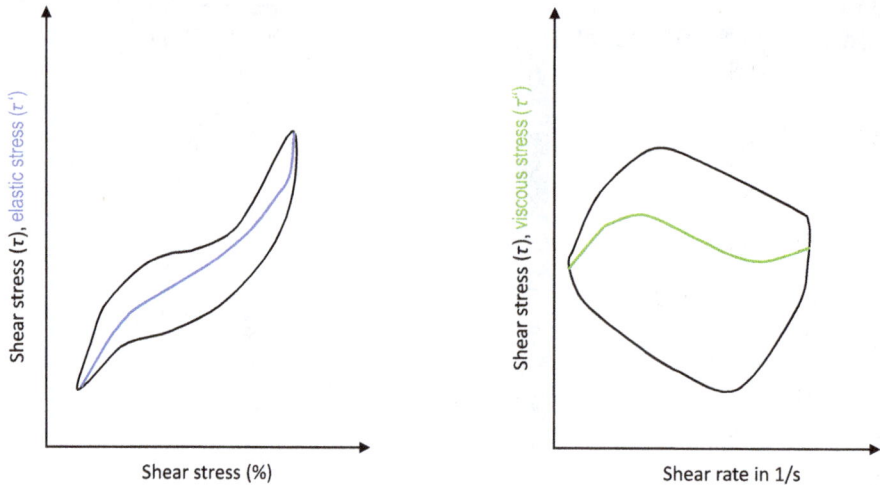

Figure 6.32 Stress–extension diagrams with a solid-like characteristic (left) and a fluid-like characteristic (right)

Another method of evaluating LAOS behavior is stress–extension wave analysis. An example of this is shown in Figure 6.33. Two solutions with different surfactants are plotted here. It can be seen that a sinusoidal extension is maintained, while the stress response has an offset to the extension and a distorted shape [16].

Wavelength-Analysis

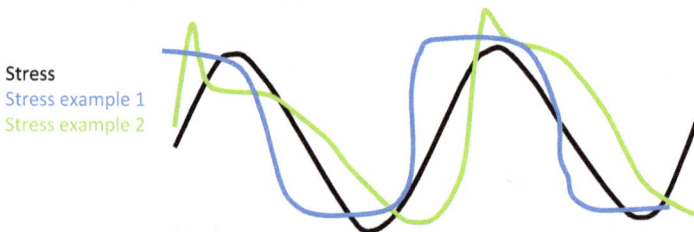

Stress
Stress example 1
Stress example 2

Figure 6.33 Example of the wave of a sinusoidal extension with examples of the resulting stress response

The difficulty of LAOS data interpretation lies in establishing a correlation between Lissajous plots, waves, and viscoelastic loading of the sample as well as the composition of the sample. There are already many studies that have defined structural differences between samples based on LAOS data, but a universal guide for presenting the results has yet to be established [16].

6.5.8.3 Classification of LAOS Behavior of Complex Fluids into Categories

It should be mentioned in advance that it is not exactly clear whether the classification into categories under LAOS conditions also applies to polymer melts or only to polymer solutions and suspensions.

Viscoelastic media have the ability to exhibit four qualitative behaviors under LAOS conditions. These four categories are:

- Viscous shear thinning and elastic shear softening (with $\eta_1' > \eta_2'$ and $G_1' > G_2'$)

- Viscous shear thinning and elastic shear stiffening (with $\eta_1' > \eta_2'$ and $G_2' > G_1'$)

- Viscous shear thickening and elastic shear softening (with $\eta_2' > \eta_1'$ and $G_1' > G_2'$)

- Viscous shear thickening and elastic shear stiffening (with $\eta_2' > \eta_1'$ and $G_2' > G_1'$) [17]

There is another practical proposal for describing the deformation behavior of viscoelastic media under LAOS conditions. These are:

- When G' and G'' both decrease

- When G' and G'' both increase

- When G' decreases and G'' increases before it decreases, as a weak overshoot of the deformation

- When G' and G'' both increase before they both decrease, as a strong overshoot of the deformation [17]

6.5.8.4 LAOS Experiments Using Fourier Transform Rheology

The term LAOS experiments—also known as experiments using oscillating deformation at large amplitudes—refers to the investigation of nonlinear viscoelastic material behavior with the aid of dynamic experiments. The curves of the stress signal to be identified and the specified deformation are compared. For this purpose, the sample is stimulated in a strictly sinusoidal manner and the deviations of the stress signal to be identified from the sinusoidal curve are analyzed.

One method of recognizing even minor deviations from the specified sine wave is to transform the signal to be identified from the time domain into the frequency domain. If a pure sinusoidal signal is transformed, a peak is observed at the frequency of the analyzed sinusoidal oscillation in the Fourier transform spectrum. However, since in LAOS experiments the deviations of the signal from the sinusoidal curve are analyzed, this signal can be regarded as the sum of several sinusoidal functions. Each individual sinusoidal function is represented as a peak in the FT spectrum. When examining the nonlinear viscoelastic behavior, in addition to the maximum at the main frequency, further peaks occur at multiples of the main oscillation frequencies. The relationship between the intensity I_n of an nth harmonic (in the magnitude spectrum, this corresponds to the magnitude of the Fourier variable at the corresponding frequency) and the intensity I_1 of the main frequency ω_0 represents the factor by which the frequency $\omega_0 \cdot n$ is included in the overall signal. This means that I_n/I_1 represents the same ratio as the ratio of the amplitude of the nth harmonic to the amplitude of the main oscillation.

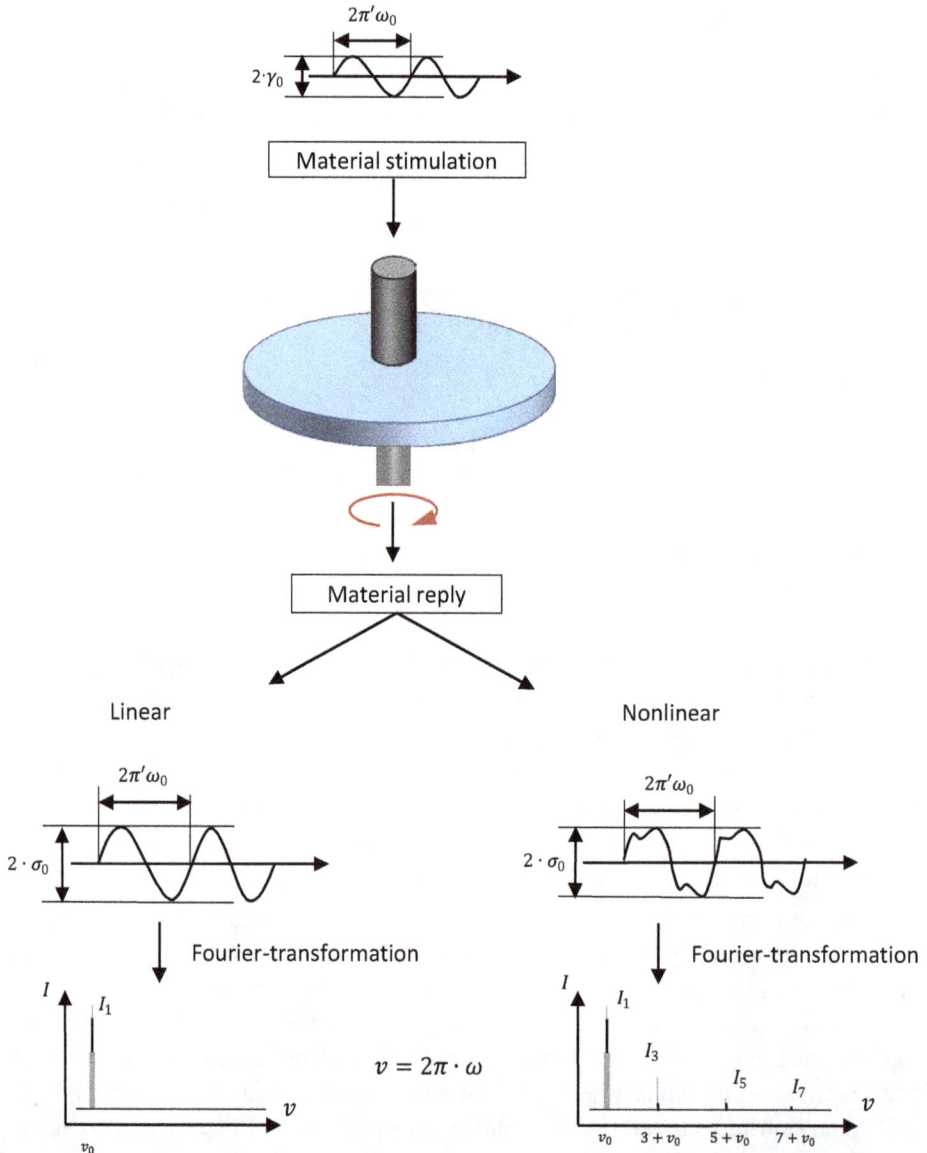

Figure 6.34 Schematic diagram of FT rheology

The harmonics of the shear stress at even multiples of the main frequency are not due to rheological properties of the materials but to phenomena such as wall slip in the rheometer. As a result, only the odd-numbered multiples of the main frequency, such as the ratio I_3/I_1, are used as a measure of the nonlinear viscoelastic behavior [18].

In Figure 6.34, the principle of Fourier transform rheology described above is shown schematically.

6.5.8.5 Possibilities and Limitations of LAOS Experiments on Polymer Melts Using FT Rheology

In the past, LAOS experiments have often been used to determine the nonlinear viscoelastic material behavior for the characterization of low-elasticity materials such as polymer solutions. When moving to materials with higher viscosity, as is the case with polymer melts, a number of experimental difficulties arise. On the one hand, the higher viscosity of the sample leads to a greater load on the measuring apparatus, which can lead to measurement inaccuracies, and on the other hand, the stability of the sample under the drastic conditions (large amplitudes) is of great importance. Research into highly viscous materials (polymer melts) has not yet progressed far enough to be able to extrapolate the behavior of polymer melts from measurements of polymer solutions [18].

The following sections summarize the extent to which LAOS experiments are at all suitable for investigating the nonlinear viscoelastic properties of polymer melts [18].

6.5.8.5.1 Material and Parameter Space

The material under investigation is anionically produced, monodisperse, narrowly distributed polystyrene. Table 6.3 lists the polystyrene types used, whereas the following studies relate to the polystyrene type PS145. Polystyrene is examined because this material has a narrow distribution. This has the advantage that complications caused by superimposed relaxation modes can be avoided as far as possible. At a reference temperature of $T_{ref} = 170°C$, the relaxation time is $\lambda_T = 1$ s, which makes both the viscous and the elastic range easily accessible by varying the frequency ω [18].

Since the frequency ω and the deformation amplitude $\hat{\gamma}_0$ at a given temperature are the most important parameters, these variables span the parameter space in which the LAOS experiments are investigated [18].

Table 6.3 Different Types of Polystyrene to Be Tested

	M_w [kg/mol]	M_n [kg/mol]	M_w/M_n [-]
PS644	644	623	1.03
PS350	350	330	1.06
PS330	330	300	1.1
PS230	230	223	1.03
PS213	213	206	1.03
PS145	145	141	1.03
PS120	120	116	1.03

6.5.8.5.2 Fracture Process as Parameter Space Limitation

The fracture process, also referred to in the literature as "edge fracture", is a phenomenon that can occur when performing oscillatory shear tests with large deformation amplitudes. If this fracture process occurs, it results in an intensive distortion of the measurement results, whereby this process represents a limit for the applicability of LAOS experiments.

In Figure 6.35, the frequency–deformation amplitude pairs for which no fracture was observed are marked with dots, while for the conditions in which the fracture process occurred, the locations are marked with an X. Thus, the area delimited by the dashed line describes experimental conditions that are not accessible for LAOS experiments due to the occurrence of melt fracture. The fracture process thus represents the parameter space boundary [18].

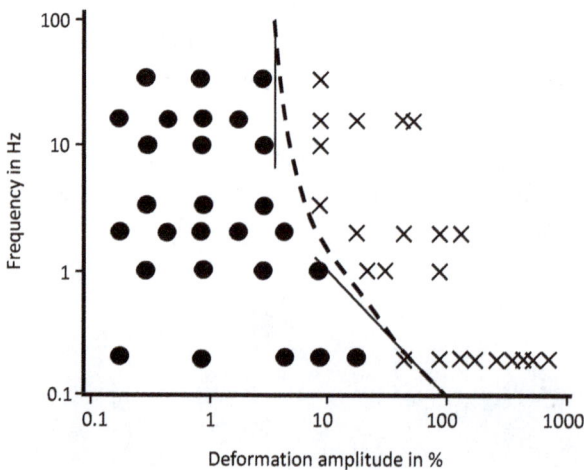

Figure 6.35 Parameter space of PS145 at 170 °C for a gap height of 1.5 mm

In order to form a theoretical basis for the calculation of the occurrence of the fracture process under LAOS conditions, the formula derived by Tanner and Keentok can be used [18]. It reads:

$$N_{2,c} = \frac{3.8 \cdot \Gamma}{h} \qquad (6.32)$$

$N_{2,c}$ represents the critical second normal stress difference, above which the fracture process occurs. The critical second normal stress difference can be calculated from the surface tension Γ of the material and the gap height h set on the rheometer. This equation was derived empirically for the case of constant shear rates. Furthermore, Equation 6.32 refers to a normal stress difference that, on the one hand, is difficult to access experimentally and, on the other hand, does not allow direct conclusions to be drawn about the test parameters ω and γ_0. However, if it is assumed that the second normal stress difference can be expressed by means of the first normal stress difference and that this can be calculated from the measured modulus value, Equation 6.33 [18] results:

$$\gamma_{0,c} = 4.36 \cdot \left(\frac{\Gamma}{h \cdot G'(\omega_c)} \right)^{\frac{1}{2}} \qquad (6.33)$$

With Equation 6.33, the critical deformation amplitude $\gamma_{0,}$ which occurs above the fracture process, can be calculated as a function of the critical angular frequency ω_c. In Figure 6.35, the calculated limit for PS145 is shown as a dashed line. In this experiment, the splitting height of $h = 1.5$ mm and a surface tension of $\Gamma = 25.9$ mJ/m^2 for polystyrene at a temperature of 170 °C were assumed. The values shown in Figure 6.35 represent the asymptotes for high and low frequencies respectively. The limit value for high frequencies $(G'(\omega) \rightarrow G_N^0>)$ results as [18]:

$$\gamma_c \approx 4.36 \cdot \left(\frac{\Gamma}{h \cdot G_N^0} \right)^{\frac{1}{2}} \propto \omega_c^0 \qquad (6.34)$$

The plateau module $(G_N^0 = 2 \cdot 10^5$ Pa$)$ was assumed to be the maximum value of $G'(\omega_c)$. This shows that at high frequencies the critical deformation amplitude $\gamma_{0,c}$ has a constant value and no longer depends on the frequency. The limit value for low frequencies $(G'(\omega) \rightarrow \psi_{10} \cdot \omega^2)$ results as [18]:

$$\gamma_c \approx 4.36 \cdot \left(\frac{\Gamma}{h \cdot \psi_{10} \cdot \omega_c^2} \right)^{\frac{1}{2}} \propto \omega_c^{-1} \qquad (6.35)$$

If one compares Equation 6.34 with Equation 6.35, the dependency between the critical deformation amplitude $\gamma_{0,c}$ and the critical frequency ω_c is described by a power law with an exponent of −1.

When comparing the theoretical considerations presented here (Equation 6.32 to Equation 6.35) with the experimental results from Figure 6.35, they show good agreement. However, further experimental investigations are necessary, especially on other materials, as the derivation of Equation 6.33 is based on an approximation [18].

6.5.8.5.3 Device-Related Sources of Error

Despite the parameter space limitation due to the fracture process, it should be possible to identify nonlinear effects, especially using the FT rheology method. In principle, this method is only limited by the signal-to-noise ratio. However, this can be improved by increasing the number of measurement cycles for the calculation of a spectrum. Nevertheless, the question arises as to how reliable the measured signal is and to what extent it reflects the physical behavior of the polymer melt [18].

6.5.8.5.4 Motor Accuracy

Before the LAOS behavior can be evaluated using the FT rheological measurement results, it must be ensured that the recorded nonlinear viscoelastic behavior is solely or predominantly due to the behavior of the polymer melt. If the excitation signal already shows strong deviations from the sinusoidal behavior, it is no longer possible to differentiate whether and to what extent the occurring harmonics originate from the rheometer itself or from the nonlinear viscoelastic behavior. As a result, the actual specified deformation was also recorded as a function of time during the time tests and then Fourier-transformed. The spectra obtained, both for the measurement signal and for the measured actual values of the deformation, were then used to determine the I_3/I_1 values. In Figure 6.36, the I_3/I_1 values determined are plotted together with the I_3/I_1 relationships of the stress signal.

It can be seen in this figure that the input signal deviates from pure sinusoidal behavior over almost the entire parameter space. Strong deviations are particularly noticeable at high frequencies and low deformation amplitudes. At low frequencies, the I_3/I_1 values also decrease up to a frequency $f = 1$ Hz. If frequencies $f < 1$ Hz are considered, a renewed increase in the harmonics generated by the rheometer can be seen. Based on this, the oscillations generated by the rheometer represent the smallest deviation from the sinusoidal shape at a frequency $f = 1$ Hz. Further FT rheological experiments are therefore carried out at this frequency [18].

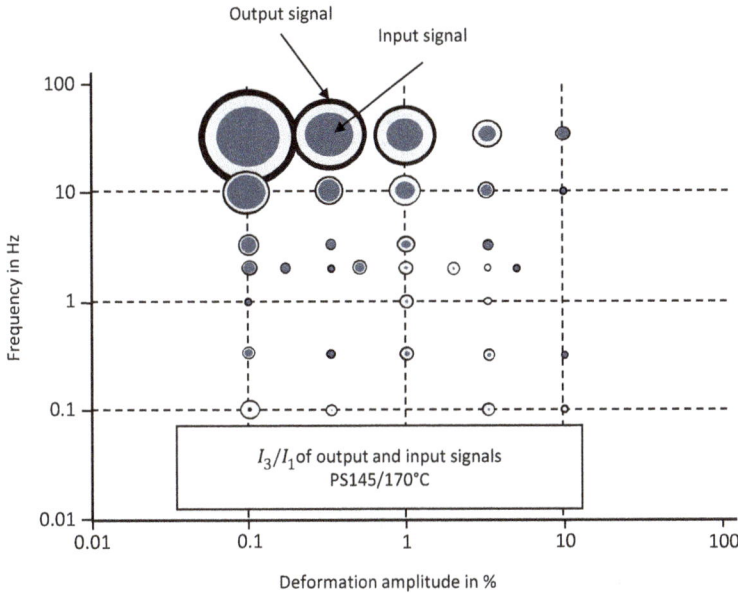

Figure 6.36 Comparison of the I_3/I_1 values between input and output signal of PS145 at 170 °C as a function of frequency and deformation amplitude

Resolution Limit of the Torque Sensor

In order to discuss an exact deformation dependence of I_3/I_1, further measurements were carried out at the specified frequency $f = 1$ Hz with a variation of the deformation amplitude. As shown in Figure 6.37, the I_3/I_1 values were plotted as a function of the deformation amplitude up to the onset of the fracture process [18].

It can be seen in this figure that before and after the deformation of approximately 6 %, the deviations of the rheometer from the sinusoidal behavior are relatively large compared to the measured values. This shows that for PS145 there is not, as expected, a monotonic increase in nonlinearity with increasing material stress, but rather a maximum in the nonlinear viscoelastic behavior at approximately 1 % of the deformation amplitude. Furthermore, the material behavior becomes more and more similar to linear viscoelastic behavior with increasing deformation amplitude.

This makes it possible to check the extent to which the measurement curves reflect the physical behavior of the material after other materials have been measured using the same method [18].

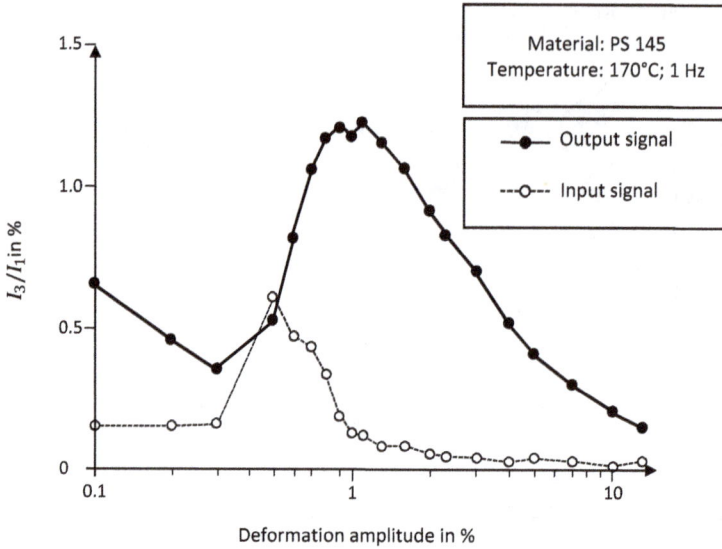

Figure 6.37 Comparison of the I_3/I_1 ratio of PS145 at 170 °C and 1 Hz as a function of the deformation amplitude

6.5.9 Coaxial Cylinder Systems

In a coaxial cylinder system, a cylindrical measuring solid is located in an outer cylinder. One of the two cylinders is set in a rotary motion, whereby a shear flow is formed in the gap between the cylinders. There are basically two different cases. A rotating inner cylinder with radius r_i is referred to as a Searle system, while a rotating outer cylinder with radius r_a is a Couette system (Figure 6.38). The shear rate of the flow in the gap is determined by the rotational speed n of the respective cylinder:

$$\dot{\gamma} = \pi \cdot n \cdot \frac{r_a + r_i}{r_a - r_i} \tag{6.36}$$

The shear stress is expressed in a measurable torque M:

$$\tau = \frac{M}{2 \cdot \pi \cdot r_i \cdot L} \tag{6.37}$$

Here, L stands for the length of the inner cylinder. The rotational speed or torque can be easily varied in a coaxial cylinder system. This allows the entire flow curve to be determined on a relatively small sample volume. If a corresponding shear rate is specified and kept constant by a control system and the resulting shear stress is measured, this is referred to as a controlled shear rate (CSR) system. In contrast to this are the controlled shear stress (CSS) systems, in which a torque (shear stress) is specified and the resulting velocity or shear rate is measured. CSS systems have a number of advantages. Firstly, they enable the direct determination of a yield point, and secondly, the control requirements for torque control are significantly lower than those for velocity control [1].

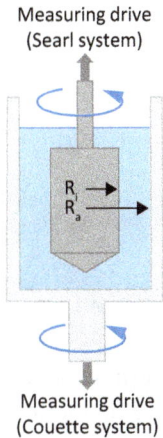

Figure 6.38
Coaxial cylinder system

Measuring Principle

The sample is deformed in a measuring gap between two coaxial cylinders, one of which rotates to generate a Couette flow. It is controlled either via the velocity (controlled shear rate, CSR) of the inner cylinder (Searle system) or the outer cylinder (Couette system) or via the torque (controlled shear stress, CSS). The variables measured are velocity and torque.

Advantages

- High shear rates possible
- Homogeneous shear rate distribution
- Less sensitive to sedimentation phenomena in suspensions

Disadvantages

- End effects with necessary correction
- Formation of Taylor vortices possible (Searle type)
- Precise calibration and measuring gap control required

6.6 Capillary Rheometers

As the name suggests, a capillary rheometer is used to press a fluid through a capillary with a piston. The capillary can have a circular, annular, or slit-shaped cross section. In the process, a mass flow $\dot{m}$, which is measured accordingly, flows through the capillary. In low-pressure and high-pressure capillary rheometers, circular capillaries with a diameter d and a length L are generally used. The aim of these rheometers is to be able to make a statement about the flowability of the plastic.

6.6.1 Low-Pressure Capillary Rheometer

6.6.1.1 Determination of the Melt Flow Index (MFI) and the Melt Volume Rate (MVR)

Low-pressure capillary rheometers (Figure 6.39) are suitable for determining the flowability of plastic melts and for the quality control of the polymer. A numerical value is measured, the so-called melt flow index (MFI) or melt volume rate (MVR). This value indicates how much melt volume (MVR) flows through a standardized nozzle within ten minutes under defined conditions (temperature and load/weight). It is important to note that the flow properties of different plastic melts can only be compared correctly if the measuring conditions are identical, as plastics react very sensitively to temperatures and shear rates. The device can also calculate and output a so-called apparent viscosity at an apparent shear rate. As these are apparent values, they cannot be used for rheological calculations. In addition, the apparent calculation only results in a pair of values, that is, a point on the viscosity curve, and therefore cannot reflect the shear-thinning viscous flow behavior of the polymers [19].

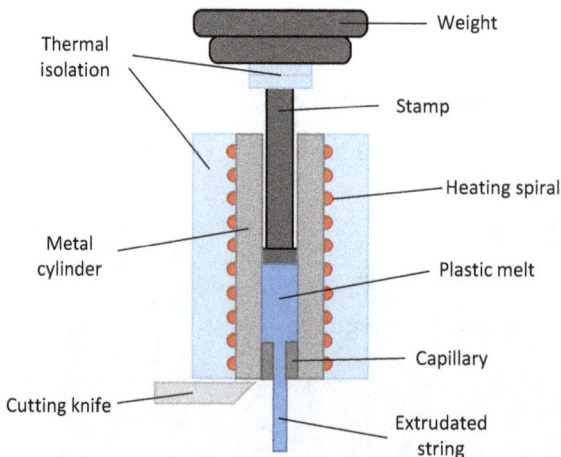

Figure 6.39 Low-pressure capillary rheometer principle (MVR/MFI)

The measurement method can be described as follows: Before measuring, it must be checked that the appliance has been cleaned appropriately. The cylinder and the plunger are preheated for 15 min. The cylinder is then filled with approximately 5 g of the granulate to be measured. The material is manually compressed under pressure using a compression rod. The aim is to allow the air bubbles in the sample to escape, as these would influence the measurement result. At the same time, care must be taken to ensure that hygroscopic plastics have been sufficiently dried beforehand, as water in the sample also distorts the measurement result. Furthermore, decomposition effects must be ruled out, that is, the filling and compaction process should not take too long and should be completed within 1 min. Plastics with fillers can also be measured using this method, where the fillers influence the measurement result. The plunger is then inserted unloaded and loaded with a specified mass after a dwell time of 4 min (complete melting to the specified temperature). The plunger sinks under the load and extrudes a melt strand through a nozzle. The measured value is determined using these strands, which are cut off within a specified time interval. Each strand should be free of bubbles and have a length of 10 mm to 20 mm according to the standard. At least three strands are weighed to an accuracy of 1 mg. If the differences in mass between the strands are greater than 15 %, the measurement should be rejected and repeated [19, 20].

The test temperatures can be found in the relevant standard. The same applies to the test load of the plunger, which is between 0.325 kg and 21.6 kg.

The melt flow index (MFI) value is given in g/10 min. This is the mass of the extruded strands that is produced within 10 min. The equation for calculating the MFI value is as follows:

$$MFI \, (T/m_0) = \frac{600 \cdot m}{\Delta t} \tag{6.38}$$

where
T is the test temperature [°C]
m_0 is the test load (mass) [kg]
m is the average value of the mass of the extruded melt strands per Δt [g]
Δt is the time interval for cutting off the extruded melt strands [s]

When specifying the MFI value, it must always be ensured that the measurement conditions are also specified. For example, the specification must read as follows: MFI(220 °C, 21.6 kg) = 24 g/10 min. This means that the measurement was carried out at a temperature of 220 °C and with a load of 21.6 kg.

For the melt volume rate value (MVR), the procedure is the same as for the MFI measurement. The difference is that the path of the die is measured. This is done with an accuracy of ± 0.1 mm and is used to determine the extruded melt volume. During the extrusion of the melt strand, the path is recorded during 10 min and the extruded volume is calculated from this. The equation for the calculation is as follows:

$$MVR\,(T/m_0) = \frac{600 \cdot A \cdot L}{\Delta t}$$

(6.39)

where

T is the test temperature [°C]
m_0 is the test load (mass) [kg]
A is the piston area (= 0.711 cm^2)
L is the plunger distance per Δt [mm]
Δt is the time interval for cutting off the extruded melt strands [s]

It must be ensured always that there are no bubbles; otherwise, measurement errors will occur.

The process is difficult to use for hygroscopic plastics that are sensitive to hydrolysis (PA, for example). The same applies to thermally sensitive plastics (PET, for example). Here, it is essential to ensure very precise drying and a short dwell time, as chain degradation can lead to a distortion of the measurement result. Furthermore, the water can act as a blowing agent, which reduces the viscosity.

If a comparison is made between the plastic granulate and the finished molded part, the section from the molded part must be crushed accordingly. If different measured values result from such a direct comparison, this may indicate a chain degradation, as the molecular weight is included in the zero viscosity with an exponent of approximately 3.4. The measurement of the MVR value can therefore also be used to detect degradation of the polymer.

6.6.1.2 Determination of the Apparent Shear Rate and Apparent Viscosity Using a Low-Pressure Capillary Rheometer

In principle, the values determined with the MFI/MVR device can also be used to calculate a shear rate and a shear stress and thus to determine the viscosity of the plastic melt. The following applies:

$$\tau = \frac{\frac{m_0 \cdot g}{A} \cdot D}{4 \cdot L}$$

(6.40)

The following applies too:

$$\dot{\gamma} = \frac{32 \cdot \dot{V}}{\pi \cdot D^3}$$

(6.41)

and the viscosity can then be calculated as:

$$\eta = \frac{\tau}{\dot{\gamma}} \tag{6.42}$$

where
m_0 is the test load (mass) [kg]
A is the piston area (= 0.711 cm^2)
g is the gravitational force (9.81 m/s^2)
D is the nozzle diameter [cm]
L is the nozzle length [cm]
$\dot{V}$ is the volume flow rate [cm^3/s]

However, if we compare these values with the results of the other rheometers, we will notice a difference. This is because the values for the shear rate and viscosity are not corrected. For this reason, we speak of apparent values. What this means is described in Section 6.6.2.3. The diagram in Figure 6.40 illustrates this relationship. The viscosity values determined with the low-pressure capillary rheometer (MVR) (red) are higher than the measured values of the high-pressure capillary rheometer (blue).

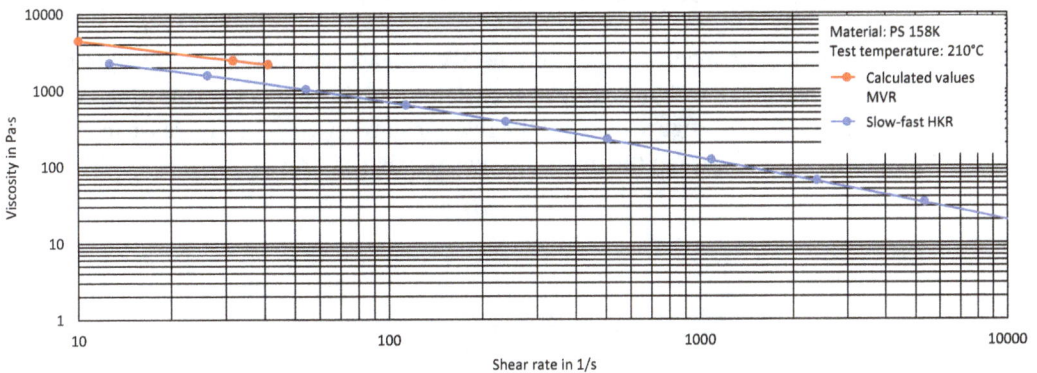

Figure 6.40 Comparison of the viscosities determined as a function of the shear rate with the MVR and the HPCR

6.6.1.3 Relationship between the MVR/MFI Value and the Molecular Weight

Studies of polyamide types using a rotational rheometer revealed a time- and temperature-dependent increase in viscosity at a constant low shear. The reason for this effect lies in the post-condensation of the polyamide. The time-dependent viscosity curve for PA11 is shown in Figure 6.41. The so-called zero viscosity—that is, the viscosity at a very low shear rate—was measured with a rotational rheometer at three different temperatures, as can be seen from the graph. At lower temperatures, the zero viscosity increases much more slowly than at high temperatures.

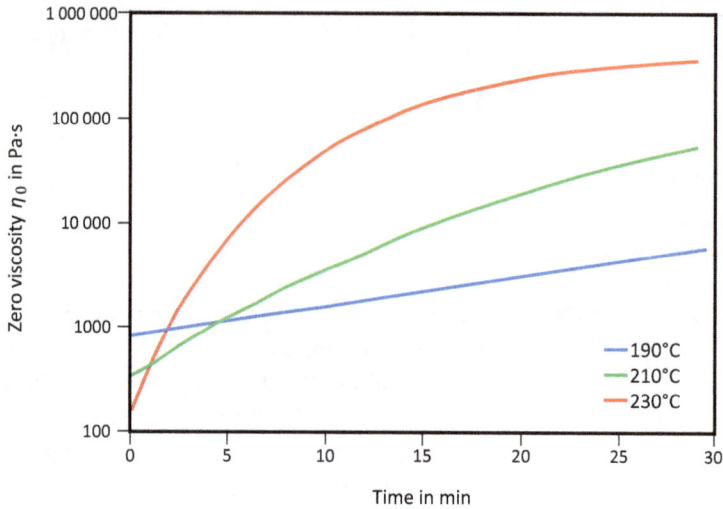

Figure 6.41 Time-dependent viscosity curve for PA11

In Figure 6.42, the mean values from the MVR measurements are assigned to the respective molecular weights, which were determined using gel permeation chromatography (GPC). This figure displays the relationship between the decrease in the MVR value and the increase in the molecular weight distribution.

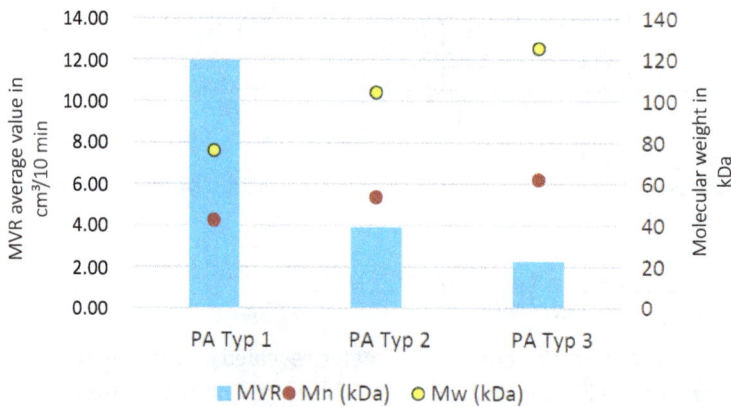

Figure 6.42 Relationship between MVR value and molecular weight for PA with different molecular weights

Figure 6.42 shows that the decrease in the MVR value is accompanied by an increase in the molecular weight averages. The number-average M_n increases significantly less than the weight-average M_w. The influence of the weight average is of greater importance for the viscosity. This experiment illustrates the relationship between viscosity increase and molecular weight. As the chain length (molecular weight) increases, the flow of the melt becomes more difficult.

6.6.2 High-Pressure Capillary Rheometer (HPCR)

With high-pressure capillary rheometers, the entire range of viscosities is measured; these values are of interest in practice when thin, long capillaries are used for low-viscosity media and correspondingly high pressures are used for high-viscosity media. In the discontinuous method, the required pressure is applied by gravity, external gas, or by means of a piston. In the continuous method, for example, a screw or pump generates the necessary volume flow rate. If the pressure is specified (given), the shear stress τ is applied (as a result). If a constant piston rate v_K is specified, the volume flow rate $\dot{V}$ is constant and the shear rate $\dot{\gamma}$ is applied.

High-pressure capillary rheometers are particularly suitable for determining the viscosity of plastic melts, as the granulate can be melted in the pre-chamber. Some of the heat generated by shearing leaves the capillary with the flowing material. For this reason, capillary rheometers can be used to measure viscosities at very high shear rates. The measuring range of commercially available high-pressure capillary rheometers for shear rates is $1\ s^{-1} \le \dot{\gamma} \le 10^5\ s^{-1}$.

A laminar layered flow (pressure flow) forms in the capillary and thus a parabolic pressure flow profile is created. The pressure is also recorded. If the viscosity is to be determined, the shear stress and the shear rate at the location must be known. In this case, the location with the maximum shear stress and the maximum shear rate—that is, the channel wall—is selected. Accordingly, the two variables to be determined are wall shear stress τ_W and wall shear rate $\dot{\gamma}_W$. The following applies for the viscosity:

$$\eta = \frac{\tau_W}{\dot{\gamma}_W} \tag{6.43}$$

The wall shear stress results from the applied or measured pressure difference:

$$\tau_W = \frac{\Delta p \cdot D}{4 \cdot L} \tag{6.44}$$

The wall shear rate can be calculated for a Newtonian fluid from the specified or measured volume flow rate:

$$\dot{\gamma}_W = \frac{32 \cdot \dot{V}}{\pi \cdot D^3} \tag{6.45}$$

As already described, the material previously melted in the heated cylinder is pressed through a capillary under the effect of the piston force F (see Figure 6.43). Either the mass flow is recorded by weighing or the volume flow rate is recorded by observing the lowering of the piston as a function of the pressing pressure.

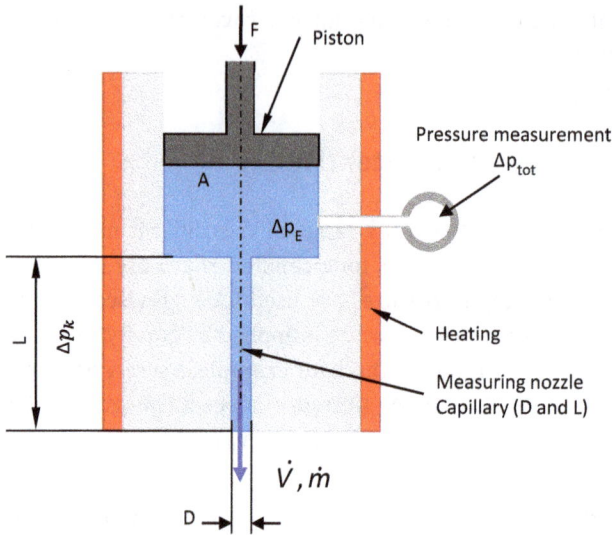

Figure 6.43 Measuring arrangement of a high-pressure capillary rheometer

The high-pressure capillary rheometer makes it possible to determine the flow curves (material laws) even at high shear rates. This measuring method is therefore well suited for determining the material data in the shear ranges relevant to processing. For this reason, Section 6.6.2.1 describes the procedure for determining the material data in more detail [21].

For a PMMA (polymethylmethacrylate), the measurement procedure with a high-pressure capillary rheometer and the subsequent evaluation procedure for determining the flow curve will now be shown in individual steps. A temperature $\vartheta = 225\,°C$ is assumed as the test temperature for the measurements.

6.6.2.1 Determining the Mass Flow–Pressure Function

First, the mass flow $\dot{m}$ at different molding pressures Δp_{tot} is determined, that is, measurement curves of the function $\dot{m} = f(\Delta p_{tot})$ are recorded. These measurements must be repeated with nozzles of different lengths and the same diameter; see Table 6.4.

Table 6.4 Mass Flow–Pressure Requirement Measured Values for a PMMA (Capillary Rheometer, $R = 0.5$ mm), Test Temperature $\vartheta = 225\,°C$

		L/D			
		2	4	8	16
Mass flow $\dot{m}$ [g/s]	Volume flow rate $\dot{V}$ [mm³/s]	Pressure losses Δp_{tot} [bar]			
5.4×10^{-3}	5	28	42	72	130
16.4×10^{-3}	15	47	67	108	188
27.1×10^{-3}	25	58	81	127	219
54.1×10^{-3}	50	78	106	160	270
81.2×10^{-3}	75	92	122	183	305
108.2×10^{-3}	100	105	137	203	334

In the example shown here, the nozzle lengths are $L = 2$ mm, 4 mm, 8 mm, and 16 mm. The nozzle radius is $r = 0.5$ mm. This results in the L/D values of 2, 4, 8, and 16. The variation of the nozzle lengths with the same diameter is required to correct the inlet pressure losses. This is described in Section 6.6.2.4.

6.6.2.2 Calculation of the Volume Flow Rate

If the mass flow is determined, this must be converted into a volume flow rate. The equation for this conversion is

$$\dot{m} = \rho \cdot \dot{V} \tag{6.46}$$

with
$$v = \tfrac{1}{\rho} \;\rightarrow\; \dot{V} = \dot{m} \cdot v\,(\vartheta)$$

The values for the temperature-dependent specific volume $v(\vartheta)$ can be obtained from the pvT diagram (Figure 6.44), or they can be calculated using the approach:

$$v\,(\vartheta) = v\,(\vartheta_0) \cdot [1 + \alpha \cdot (\vartheta - \vartheta_0)] \tag{6.47}$$

The values for the reference specific volume $v(\vartheta_0)$, the reference temperature ϑ_0, and the thermal extension coefficient α can be found in Table 6.5.

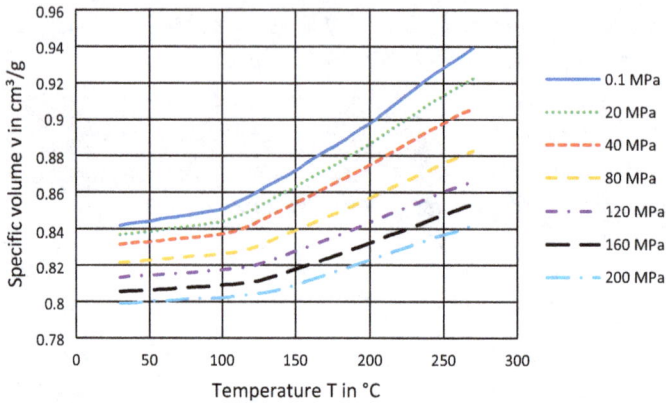

Figure 6.44 *pvT* diagram for PMMA (Röhm, type: Plexiglas® 7N)

Table 6.5 Constants for Various Plastics

Plastic type	ϑ_0 [°C]	$v(\vartheta_0)$ [cm³/g]	α [1/°C]
PE soft	115	1.249	6.9×10^{-4}
PE hard	131	1.262	6.9×10^{-4}
PP	186	1.318	6.1×10^{-4}
PVC	80	0.730	7.3×10^{-4}
PS	84	0.972	5.6×10^{-4}
PC	150	0.861	6.0×10^{-4}
PET	280	0.857	6.4×10^{-4}
PMMA	105	0.866	5.8×10^{-4}

6.6.2.3 Calculation of the Apparent Wall Shear Stress and the Apparent Wall Shear Rate

With these measured values (volume flow rate and pressure loss), the following equations can be used to calculate the *apparent wall shear stress* τ_{Wa}:

$$\tau_{\mathrm{Wa}} = \frac{\Delta p_{\mathrm{tot}}}{4 \cdot \frac{L}{D}} \tag{6.48}$$

and the *apparent wall shear rate* $\dot{\gamma}_{\mathrm{Wa}}$:

$$\dot{\gamma}_{\mathrm{Wa}} = \frac{32 \cdot \dot{V}}{\pi \cdot D^3} \tag{6.49}$$

Why are we talking about apparent values here?

Apparent Wall Shear Stress

The values for the wall shear stress are subject to an error, since Equation 6.48 includes the total pressure loss Δp_{tot}. However, the total pressure loss includes not only the pressure losses of the capillary, but also the inlet and outlet pressure losses (vortices, viscoelastic effects, extensional pressure losses) upstream and downstream of the capillary. Figure 6.45 shows qualitatively such a pressure curve.

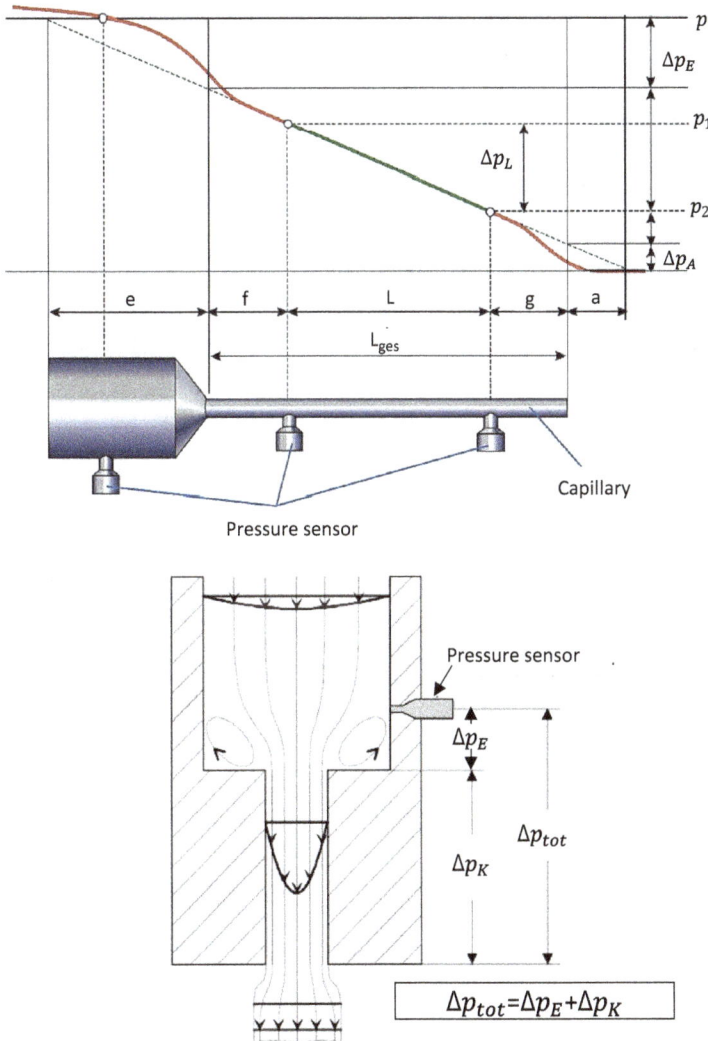

Figure 6.45 Pressure curve along a capillary

Apparent Wall Shear Rate

The wall shear rate is incorrect because Equation 6.49 for the calculation of $\dot{\gamma}$ only applies to Newtonian fluids. The shear-thinning viscous flow behavior of the plastic melts is therefore not initially taken into account.

For this reason, both values must be corrected in order to obtain the true flow curve for the medium. These corrections are called:

- Inlet pressure loss correction (Bagley correction)
    - Determining the true wall shear stress
- Weissenberg–Rabinowitsch correction
    - Determining the true wall shear rate

6.6.2.4 Determining the True Wall Shear Stress

First, the true wall shear stress is determined from the apparent wall shear stress. As already mentioned, the measured total pressure loss Δp_{tot} is made up of all the following pressure components:

- In the melt cylinder on the section between the pressure measurement hole and the inlet area of the capillary (nozzle)
- In the inlet and outlet area
- In the capillary

> **Excursus:**
> It would be better to measure the pressure directly in the capillary. However, for dimensional reasons, this is not possible with the high-pressure capillary rheometer (HPCR) using a round capillary. Capillary diameters of less than 2 mm are common with the HPCR. The pressure transducers suitable for this application usually have a diameter of around 8 mm at the end face. For this reason, the pressure measurement cannot be taken right up to the capillary without problems.

The rheometer flow required to determine the flow curve only forms in the capillary. As a result, only the pressure loss in the capillary can be used to determine the material data.

The following shows how pressure losses Δp_k, which occur in the capillary with the length L, can be determined graphically from the total pressure losses $\Delta p_{tot} = \Delta p_{in} + \Delta p_k + \Delta p_{out}$. The outlet pressure losses can generally be neglected, as they are very small. Furthermore, they cannot be determined with the high-pressure capillary rheometer. Therefore, it follows: $\Delta p_k = \Delta p_{tot} - \Delta p_{in}$.

The Bagley Correction

With the Bagley correction [22], the pressure loss Δp_k in the capillary is determined via the flow path L by plotting the total pressure loss Δp_{tot}, which is measured between the pressure measurement bore and the channel outlet, against the ratio of nozzle length to nozzle diameter L/D for a constant volume flow rate $\dot{V}$. The graph in Figure 6.46 shows such a representation. The result is a good approximation of straight lines that can be extended to $L/D = 0$. The pressure loss Δp_{in} that occurs upstream of the nozzle inlet can then be read off the ordinate. The value on the ordinate, that is, $L/D = 0$, corresponds to a measurement with a capillary of zero length. This must result in a pressure loss $\Delta p_k = 0$ bar, as no capillary is connected! Therefore, the pressure losses read can be assigned to the inlet pressure losses.

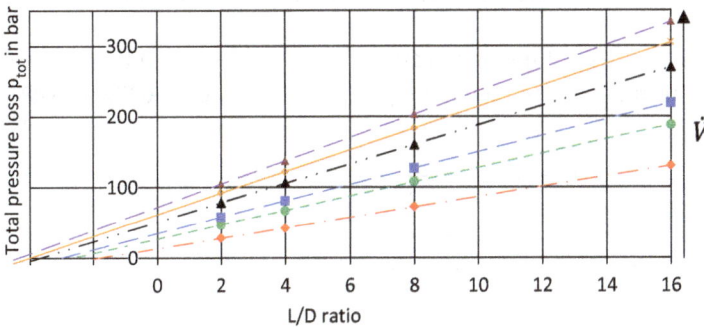

Figure 6.46 Total pressure loss as a function of the L/D ratio

In the following example, the Bagley correction is to be carried out for a volume flow rate $\dot{V} = 50\,mm^3/s$. First, the values from Table 6.4 are plotted in a representation $\Delta p_{tot} = f(L/D\text{ ratio})$. The diagram in Figure 6.46 now makes it possible to determine the inlet pressure losses for the different volume flow rates.

Figure 6.47 shows the Bagley correction. In the example considered here, the correction is carried out for a volume flow rate $\dot{V} = 50\,mm^3/s$. With the Bagley correction, for $\dot{V} = 50\,mm^3/s$, an inlet pressure loss of $\Delta p_{in50} = 50$ bar can be read off the ordinate.

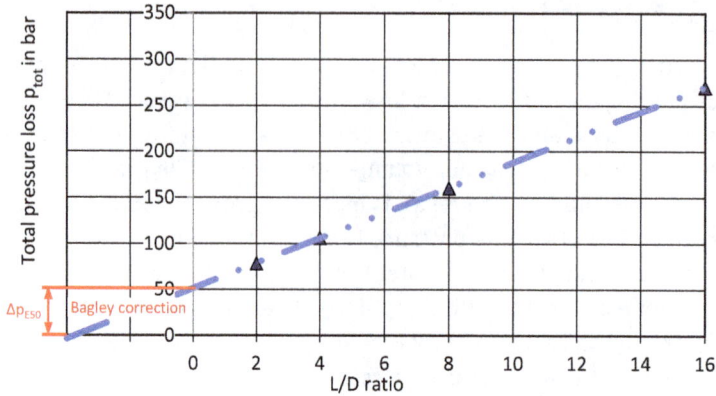

Figure 6.47 Bagley correction method for a volume flow rate $\dot{V} = 50\,\text{mm}^3/\text{s}$

As with the correction for the volume flow rate of $\dot{V} = 50\,\text{mm}^3/\text{s}$, the inlet pressure losses are read off the ordinate for the remaining volume flow rates; see Figure 6.48.

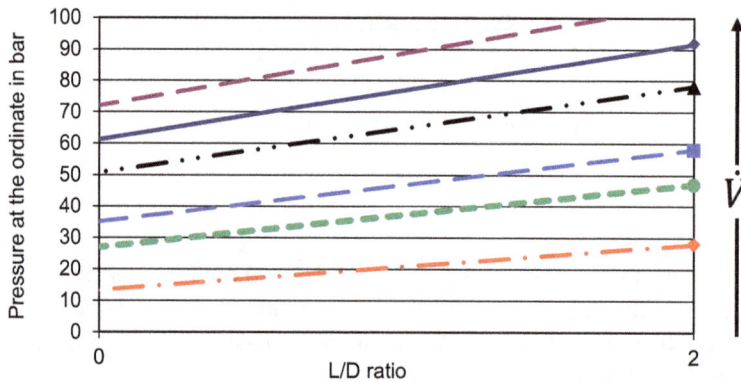

Figure 6.48 Determining the inlet pressure losses for the Bagley correction

In Table 6.6 the total pressure losses Δp_{tot} and the measured inlet pressure losses Δp_{in} are listed.

Table 6.6 Total Pressure Losses and Inlet Pressure Losses

Mass flow [g/s]	Volume flow rate [mm³/s]	L/D				
		2	4	8	16	
		Δp_{tot2} [bar]	Δp_{tot4} [bar]	Δp_{tot8} [bar]	Δp_{tot16} [bar]	Δp_{in} [bar]
5.4×10^{-3}	5	28	42	72	130	14
16.4×10^{-3}	15	47	67	108	188	27
27.1×10^{-3}	25	58	81	127	219	36
54.1×10^{-3}	50	78	106	160	270	50
81.2×10^{-3}	75	92	122	183	305	62
108.2×10^{-3}	100	105	137	203	334	72

The true wall shear stress can be calculated using Equation 6.50:

$$\tau_{Wt} = \frac{\Delta p_{tot} - \Delta p_{in}}{4 \cdot \frac{L}{D}} \tag{6.50}$$

This results in the example considered here (see Table 6.7):

Table 6.7 True Wall Shear Stress and Apparent Wall Shear Rate

Volume flow rate [mm³/s]	Δp_{in} [bar]	τ_{Wt2} [N/ mm²]	τ_{Wt4} [N/ mm²]	τ_{Wt8} [N/ mm²]	τ_{Wt16} [N/ mm²]	τ_{Wt} [N/ mm²]	τ_{Wa} [N/ mm²]	$\dot{\gamma}_{Wa}$ [s⁻¹]
5	14	0.175	0.175	0.181	0.181	0.18	0.26	51
15	27	0.250	0.250	0.25	0.252	0.25	0.41	153
25	36	0.275	0.281	0.284	0.286	0.28	0.49	255
50	50	0.350	0.350	0.344	0.344	0.34	0.64	509
75	62	0.375	0.375	0.378	0.380	0.38	0.74	761
100	72	0.413	0.406	0.409	0.409	0.41	0.83	1019

The true wall shear stress τ_{Wt} is calculated as the mean value from the respective wall shear stresses of the different L/D ratios, using the equation:

$$\tau_{Wt} = \frac{\tau_{Wt2} + \tau_{Wt4} + \tau_{Wt8} + \tau_{Wt16}}{4} \tag{6.51}$$

The apparent wall shear rate is calculated according to the equation

$$\dot{\gamma}_{Wa} = \frac{32 \cdot \dot{V}}{\pi \cdot D^3} \tag{6.52}$$

If the apparent and true wall shear stresses are plotted logarithmically as a function of the apparent wall shear rate, the result is as shown in Figure 6.49. It is clear that the curve for the apparent wall shear stress shifts downwards towards the true wall shear stress.

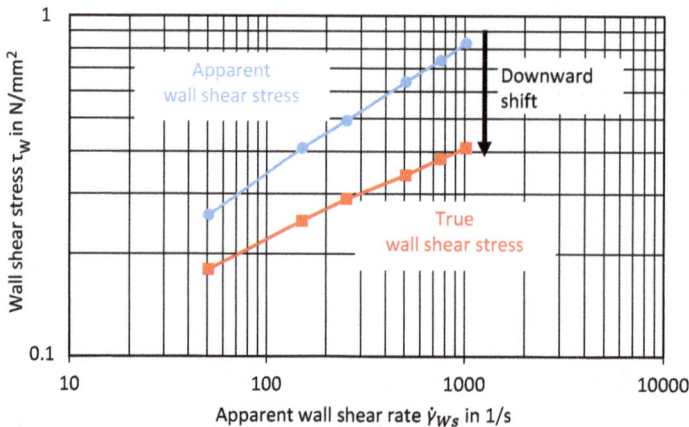

Figure 6.49 Apparent and true wall shear stress as a function of the apparent wall shear rate

Another way to determine the inlet pressure losses directly is to use a so-called zero nozzle (orifice). Figure 6.50 shows a high-pressure capillary rheometer with two coupled plungers driven by a common motor. On the left side is a round capillary with a defined length L and a diameter D. The right plunger presses the plastic mass inside through a capillary with the same diameter and an idealized length of 0 mm. This allows the inlet pressure losses on the right-hand side to be determined directly. However, this is based on the assumption that the capillary length on the right-hand side is zero.

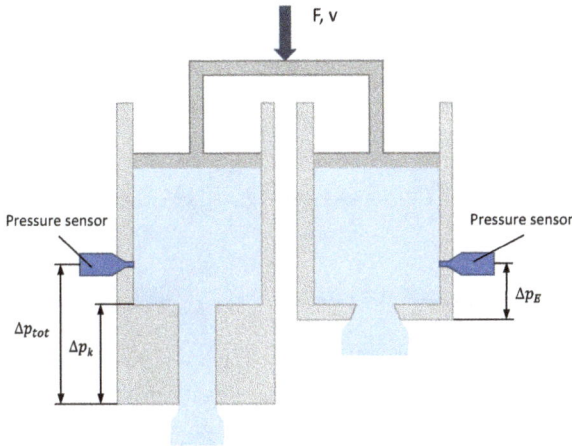

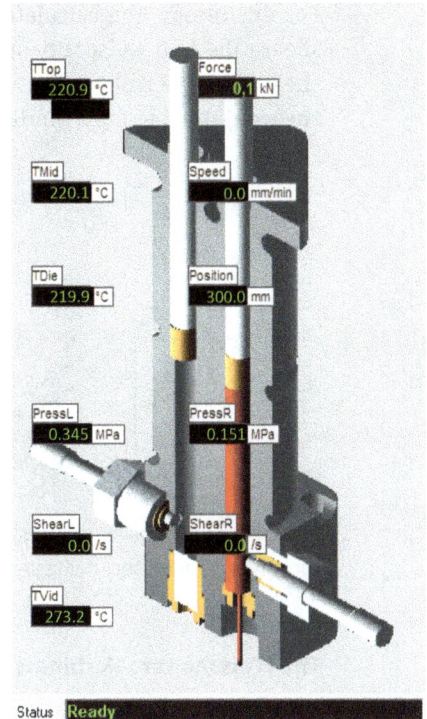

Figure 6.50 High-pressure capillary rheometer with two coupled plungers, the measuring capillary on the left and the zero nozzle (orifice) on the right

6.6.2.5 Determining the True Wall Shear Rate

In a first evaluation step, Newtonian flow behavior was assumed. For this reason, an "apparent wall shear rate" is obtained. In order to do justice to the shear-thinning viscosity of plastic melts, the apparent values must be converted into "true values", using correction methods.

One possibility for a correction is the Weissenberg–Rabinowitsch approach. This approach takes into account the shear-thinning viscous behavior of the melt. The velocity profile in the capillary has a maximum in the center of the capillary and drops to zero at the capillary wall (wall adhesion).

The gradient of the change in velocity (dv/dr) from the center of the flow channel to the wall is the so-called shear rate $\dot{\gamma} = dv/dr$. Polymeric materials show a much greater velocity gradient near the wall than Newtonian materials, especially at high volume flow rates. Therefore, at the wall, the shear rate of a shear-thinning viscous material is greater than that of a Newtonian material (Figure 6.51), that is:

$$\dot{\gamma}_{wall,Newtonian} < \dot{\gamma}_{wall,shear-thinning\ viscous}$$

For this reason, the calculated apparent wall shear rate is lower than the true wall shear rate. As the viscosity of shear-thinning viscous fluids decreases with increasing shear rate, the true viscosity is lower than the apparent viscosity. This behavior of shear-thinning viscous fluids will be described later for different polymers.

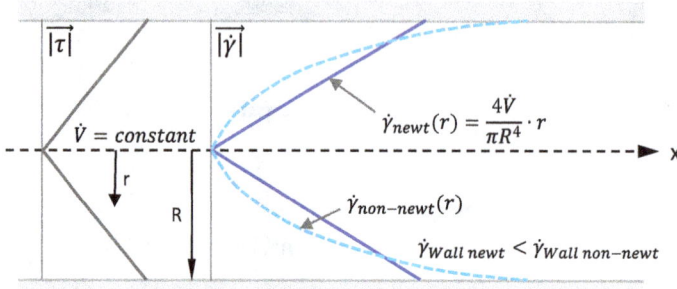

Figure 6.51 Shear rate and shear stress curve over the flow cross section for a Newtonian and a shear-thinning viscous fluid

The Weissenberg–Rabinowitsch Correction

As early as 1929, Weissenberg and Rabinowitsch were able to show that the true shear rate for any fluid with wall adhesion can be calculated from the apparent wall shear rate. This relationship is illustrated below [23].

The wall shear rate of a Newtonian fluid in a pipe flow is given by

$$\dot{\gamma}_{W,\text{Newtonian}}(R) = \frac{4 \cdot \dot{V}}{\pi \cdot R^3} \tag{6.53}$$

To derive the shear rate of a shear-thinning viscous fluid on the channel wall, the shear-thinning viscosity is taken into account using the power approach according to Ostwald and de Waele. The following applies:

$$\eta(\dot{\gamma}) = K \cdot \dot{\gamma}^{n-1} \tag{6.54}$$

For a pipe flow, this results in the shear rate at the channel wall when deriving the velocity curve for shear-thinning viscous media:

$$\dot{\gamma}_{W,\text{shear-thinning viscous}}(R) = \frac{\left(\frac{1}{n} + 3\right) \cdot \dot{V}}{\pi \cdot R^3} \tag{6.55}$$

It becomes clear that the true (shear-thinning viscosity) and the apparent (Newtonian) shear rates are proportional to each other. The following applies:

$$\dot{\gamma}_{true}(R) = c \cdot \dot{\gamma}_{apparent}(R) = \frac{3 + \frac{1}{n}}{4} \cdot \dot{\gamma}_{apparent}(R) \tag{6.56}$$

where
$\dot{\gamma}_{true}$ is the true shear rate
$\dot{\gamma}_{apparent}$ is the apparent shear rate

As a result, the second correction can now be made using the Weissenberg–Rabinowitsch approach and the equation for the true shear rate is:

$$\dot{\gamma}_{true} = \frac{3n + 1}{4n} \cdot \dot{\gamma}_{apparent} \tag{6.57}$$

n is the slope of the tangent of the true shear load (true wall shear stress) as a function of the apparent shear rate in the double-logarithmic plot at the point where the apparent shear rate is to be corrected.

$$n = \frac{\Delta \log (\tau_W)}{\Delta \log (\dot{\gamma})} \tag{6.58}$$

Newtonian flow is present for $n = 1$. This means $\dot{\gamma}_{true} = \dot{\gamma}_{apparent}$.

It should be noted that the value n is the slope at each point of the flow curve for a specific shear rate. Since the flow curve of a plastic is increasing and concave-down ("degressive"), a tangent must be created at each point of the flow curve in order to then determine the slope at this point, using a slope triangle. This procedure can be carried out using the auxiliary circle method, which is explained briefly below (Figure 6.52).

- A small circle (purple circle) is created at the point for which the slope is to be determined.

- The circle intersects the flow curve at two points. Two (red) circles of the same size are created through these two intersections.

- The two (red) circles intersect at two points. A line (red) is drawn through these intersections.

- A tangent (green) is now applied to the flow curve at a right angle.

- A slope triangle is then applied to the tangent and the slope n for this point is determined using Equation 6.58.

This procedure must be repeated for each point on the flow curve.

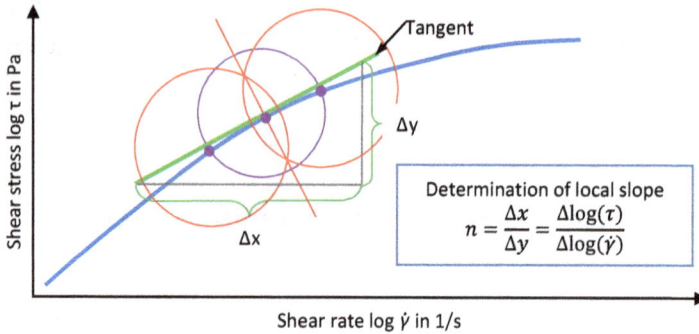

Figure 6.52 Determining the local slope n in the $\tau = f(\dot{\gamma})$ diagram using the auxiliary circle method

If this method is applied, the slope n of the tangent is obtained for the respective points of the apparent wall shear rates (volume flow rates). The true wall shear rate can then be calculated using the Weissenberg–Rabinowitsch equation.

Note: Only if the division (box size) in the double-logarithmic representation is the same size may the slope be read off and determined using a ruler. Otherwise, the values must be entered into Equation 6.58.

This results in the values given in Table 6.8.

Table 6.8 Determining the Slope of the Tangent and the True Wall Shear Rate

Volume flow rate [mm³/s]	τ_{Wt} [N/mm²]	$\dot{\gamma}_{Wa}$ [s⁻¹]	n [-]	$\dot{\gamma}_{Wt}$ [s⁻¹]
5	0.18	51	0.37	73
15	0.25	153	0.3	242
25	0.29	255	0.27	427
50	0.34	509	0.22	960
75	0.38	764	0.21	1483
100	0.41	1019	0.18	2180

Figure 6.53 shows the logarithmic plot of the true wall shear stress as a function of the apparent wall shear rate and the true wall shear rate. It is clear that the curve of the apparent wall shear rate, which applies to Newtonian materials, shifts to the right. The true wall shear rates for shear-thinning viscous materials are greater than the apparent wall shear rates of Newtonian media!

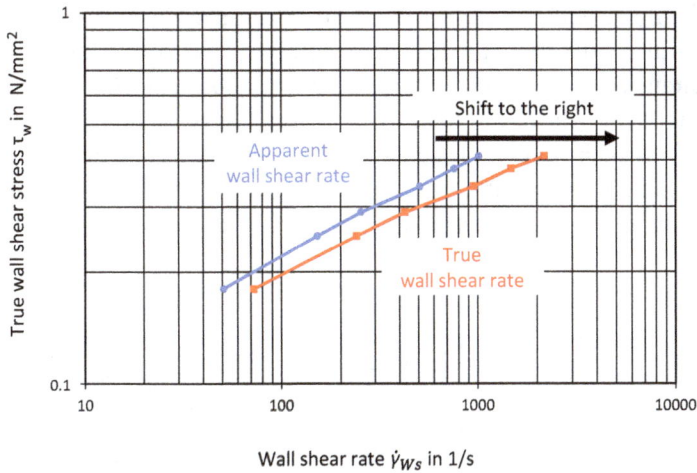

Figure 6.53 True wall shear stress as a function of the apparent and true wall shear rates

In summary, it can be stated that the curves in the plot shift horizontally downwards due to the Bagley correction and vertically to the right due to the Weissenberg–Rabinowitsch correction.

In this example, the flow curves were determined for the wall of the flow channel, as a simple evaluation is possible for this case. However, they apply in principle to all points of a flow. The true shear rate is then used to calculate the true viscosity.

The relationship between the shear stress and the shear rate then results in the second flow curve using Newton's law of friction, that is, the true viscosity as a function of the true shear rate. If the measured values are not plotted logarithmically but linearly, the representation shown in Figure 6.54 of the true viscosity as a function of the true wall shear rate is obtained.

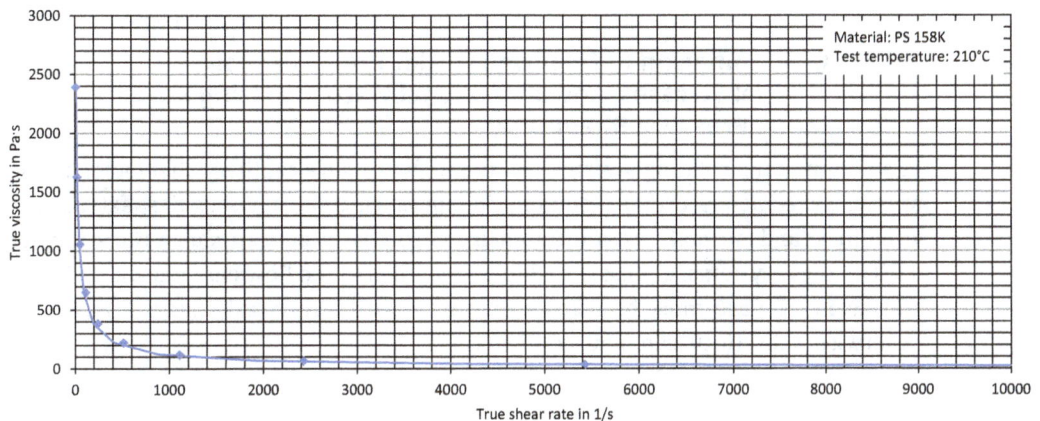

Figure 6.54 Linear plot of the true viscosity as a function of the true shear rate

It is clear that the viscosity depends greatly on the shear rate, especially at low shear rates. As the shear rate increases, the influence on the viscosity decreases. This actually contradicts the statement about the shear-thinning viscous flow behavior of plastics that speaks of Newtonian behavior at low shear rates and of shear-thinning viscous flow behavior at high shear rates. This apparent behavior results only from the logarithmic representation, as the graph in Figure 6.55 illustrates. The flow curve is plotted in this representation to simplify the mathematical description using the material laws (Chapter 8).

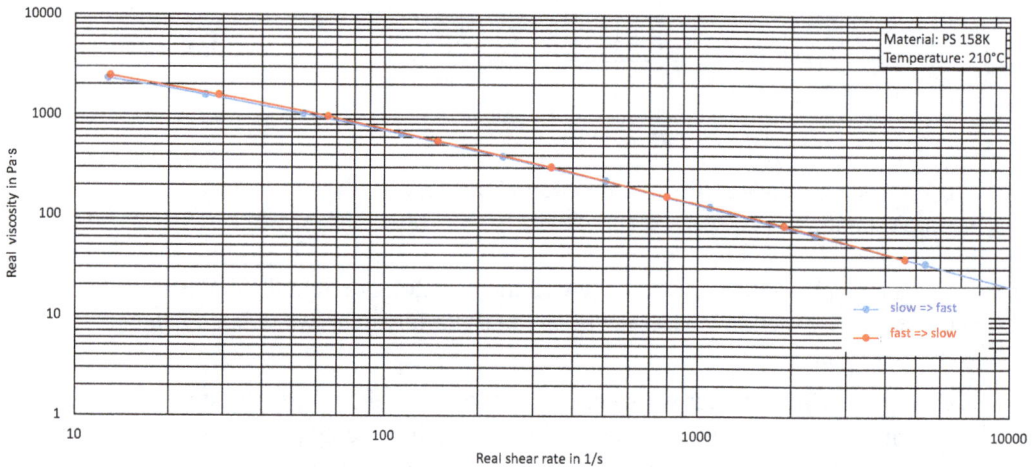

Figure 6.55 Logarithmic plot of the true viscosity as a function of the true shear rate

As part of the measurements, the effect of varying the shear rate on the viscosity values was also investigated. For this purpose, the test was switched from low shear rate to high values (slow–fast) and vice versa (fast–slow) in Figure 6.55. It is clear that there is no difference whether the measurement is started with low shear rates or high shear rates.

6.6.2.6 Determination of Inlet and Outlet Pressure Losses, Normal Stresses, and Extensional Viscosity Using an Inline Pressure Rheometer

In addition to the rotating systems and the high-pressure capillary rheometer, inline pressure rheometers are also suitable for measuring shear viscosities, extensional viscosities, and normal stress differences, particularly at high shear rates (Figure 6.56). In these systems, a pressurized flow is applied in the flow channel with width B and height H. Care must be taken to ensure that the ratio B/H satisfies $B/H \gg 10$. Otherwise, side effects can play a critical role.

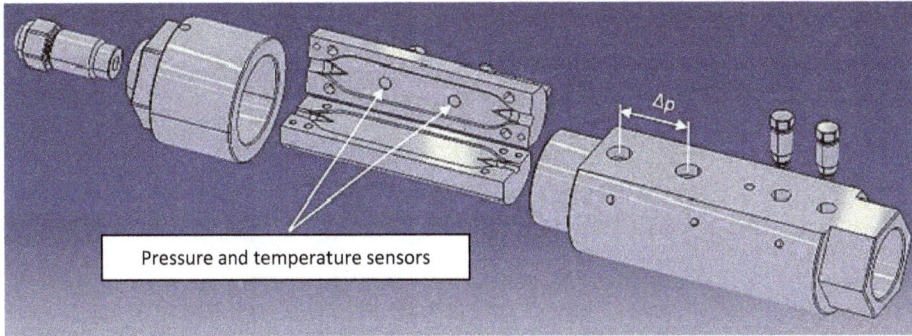

Figure 6.56 Inline rheometer nozzle with exchangeable slotted capillary with four pressure and temperature sensors

With these rheometers, it is possible to measure the pressure profile directly along a nozzle or slot using a pressure transducer (see Figure 6.57). Inserts with cross-sectional jumps can also be installed. In addition to the shear pressure losses, inlet and outlet pressure losses can also be determined for different geometries and plastics, which can be used to calculate the extensional viscosities and the first normal stress difference. An example of the evaluation on a cross-sectional jump and in a channel with a constant width is shown in Figure 6.57. The geometric data is as follows:

$$B_1 = 70 \text{ mm}; \ B_2 = (70; 40; 30; 20; 10; 5) \text{ mm}; \ H = 3 \text{ mm}; \ L = 59 \text{ mm}$$

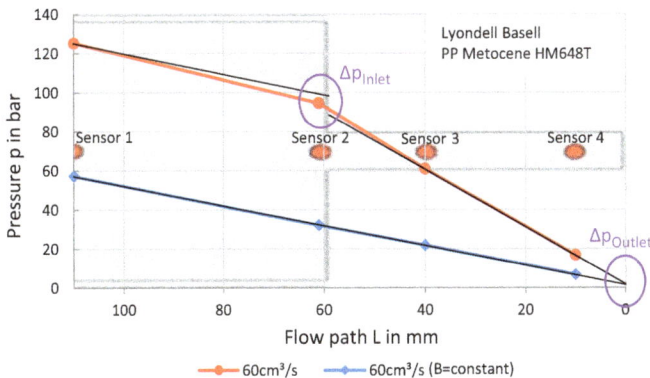

Figure 6.57 Possibilities for determining inlet and outlet pressure losses in a slotted capillary

The measurement results of the inlet pressure losses are shown in Figure 6.58.

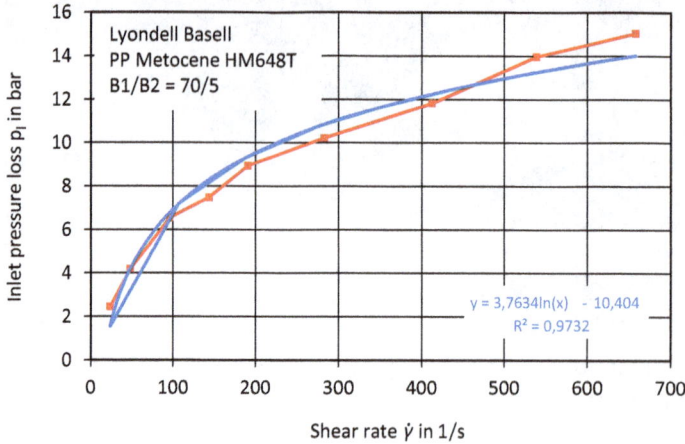

Figure 6.58 Inlet pressure losses as a function of the shear rate

Using the inlet pressure losses at the cross-sectional jump, the approach of Cogswell [24–26] can determine the extensional viscosity even at high shear rates (Figure 6.59). The calculation of the true wall shear rate at the channel wall is analogous to Equation 6.57 with Weissenberg–Rabinowitsch for a slit capillary:

$$\dot{\gamma}_{\text{wall, real}} = \frac{2n+1}{3n} \cdot \dot{\gamma}_{\text{wall, apparent}} \qquad\qquad (6.59)$$

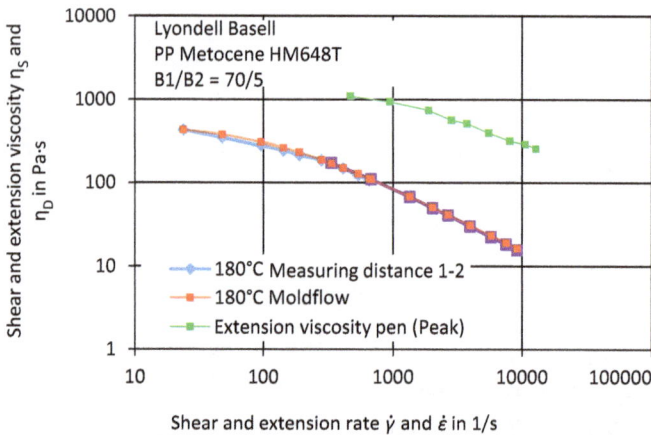

Figure 6.59 Extensional and shear viscosity determined with the inline rheometer nozzle and mold flow curve

It has also been shown that there is a direct correlation between the inlet pressure loss p_{in} and the wall shear stress τ_{W} in the developed flow [27]. The inlet pressure losses can therefore also be determined using this correlation. This relationship is plotted in Figure 6.60 for a polystyrene 495F.

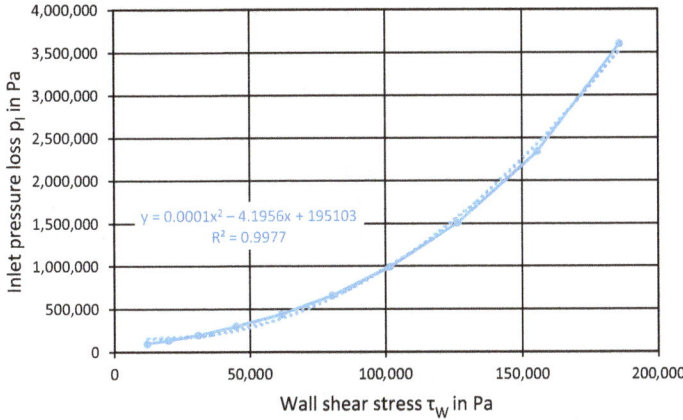

Figure 6.60 Inlet pressure loss as a function of wall shear stress for a polystyrene 495F

By extrapolating the pressure curve to the final length L (melt outlet), the so-called outlet pressure loss or the outlet pressure p_{out} can be determined. Figure 6.61 shows the pressure curve in a slit capillary. The pressure curve in the channel is recorded using four pressure sensors. It is extrapolated as in the case of the Bagley correction up to the melt outlet, in order to determine the outlet pressure losses at this point (flow path $L = 0$ mm).

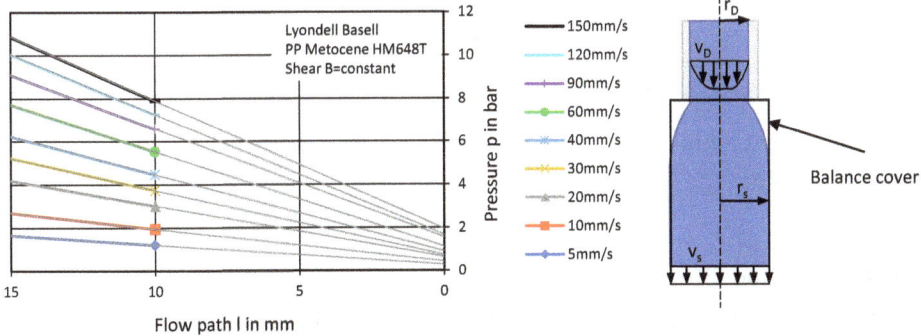

Figure 6.61 Extrapolated pressure drop at the nozzle outlet

If the outlet pressure p_W and the wall shear stress τ_W at the outlet are determined as a function of the shear rate (flow rate) at the channel wall, the first normal stress difference can be calculated (here for a slotted rheometer, with linear extrapolation of the outlet pressure) [28]:

$$N_{1,W} = p_W + \tau_W \frac{dp_W}{d\tau_W} \tag{6.60}$$

In Equation 6.60, p_W corresponds to the outlet pressure p_{out}, which can be determined graphically from Figure 6.61.

The great advantage of this measuring method is that, in contrast to rotation, large shear rate ranges can also be mapped.

The relationship between the outlet pressure, the wall shear stress, and the first normal stress difference as a function of the shear rate is shown in Figure 6.62. The graph also shows the first normal stress difference for low shear rates, measured using a rotational rheometer.

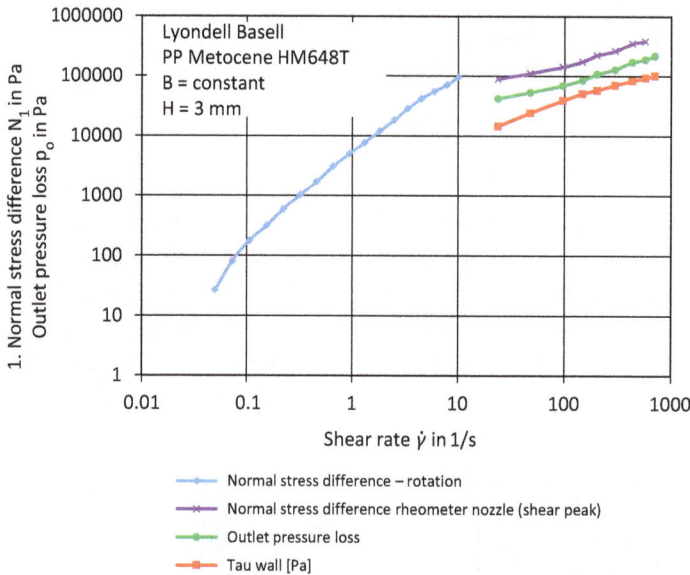

Figure 6.62 First normal stress difference and wall shear stress as a function of the shear rate

The Die-Swelling Effect

Another very interesting viscoelastic effect that can be observed in plastic melts is the extension of strands as they flow through nozzles. This effect, also known as die swelling (Figure 6.63), is more pronounced,

- The shorter the nozzle
- The higher the flow rate
- The smaller the nozzle diameter

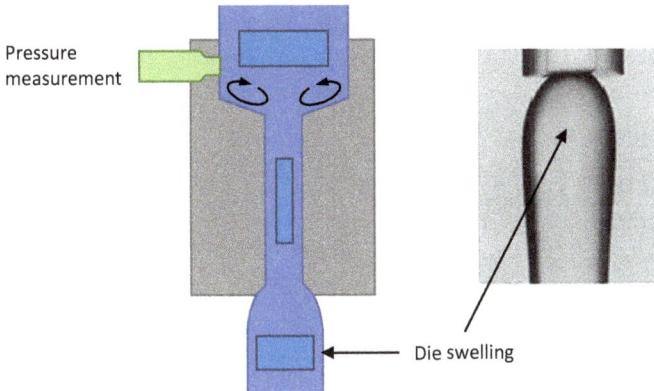

Figure 6.63 Strand widening (die-swelling effect)

As these images illustrate, a viscoelastic fluid expands its strand considerably after leaving a capillary, compared to the diameter of the capillary. At its thickest point, the diameter of the fluid jet is almost three times as large as the inner diameter of the flow channel. During the swelling process, this strand is shortened by a factor of nine compared to its length in the capillary. This is caused by considerable tensile stresses in the direction of flow. The design of extrusion molds, for example, is made considerably more difficult by such flow effects.

In order to prevent this effect and, if necessary, to make appropriate corrections to the mold (stockpiling), the swelling behavior must be determined. This is possible using capillary or oscillation viscometry.

One way of directly determining the strand extension is to use the high-pressure capillary rheometer (HPCR). For this purpose, the HPCR is equipped with a laser beam measurement or a camera to measure the strand extension (Figure 6.64).

$$\text{Strand extension} = \frac{D - d}{d} \cdot 100\% \tag{6.61}$$

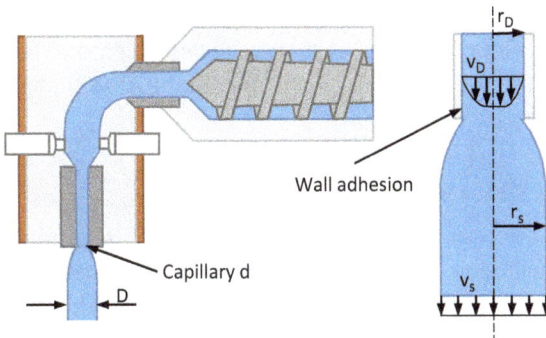

Figure 6.64 Strand extension (die swelling) in the high-pressure capillary rheometer

If a balance cover is drawn (see Figure 6.64 on the right) around the emerging jet, the maximum extension ratio of the jet can be determined by balancing and results in [28]:

$$\left(\frac{r_D}{r_s}\right)^2 = \frac{v_D{}^2}{v_s{}^2} - \frac{N_1 + 0.5N_2 - p_A}{\rho \cdot v_s{}^2} \tag{6.62}$$

where
v_D is the axial velocity over the nozzle radius r_D
v_s is the axial velocity of the strand with the radius r_s
p_{out} is the outlet pressure (see Figure 6.45)
ρ is the density of the fluid
N_1 is the first normal stress
N_2 is the second normal stress

The equation can be simplified for various assumptions (e.g., fully developed Poiseuille flow, no outlet pressure). In particular, it is possible to determine a critical nozzle radius $r_K = [12 \cdot (2 \cdot \Psi_1 + \Psi_2)/\rho]^{\frac{1}{2}}$, below which an extension (r_s) of the strand is a prerequisite [29].

The extension of the strand is described by the ratio of the nozzle diameter D_D to the strand diameter D_s after relaxation:

$$B_{ex} = \frac{D_s}{D_D} \tag{6.63}$$

For small Reynolds numbers, $Re < 2$, even Newtonian fluids exhibit a small strand extension; therefore, the measured values of B_{ex} must be corrected accordingly.

$$B = B_{ex} - 0.13 \tag{6.64}$$

On the basis of an integral model, it was shown that the corrected strand extension B can be quantitatively correlated with the first normal stress difference N_1, if the values B and N_1 are determined for the same wall shear stress τ_W:

$$N_1^2 = 8 \cdot \tau_W^2 \cdot \left[B^6 - 1\right] \tag{6.65}$$

Even if the equation is not valid for all applications, it shows the dependence of the strand extension on the first normal stress difference, $B \sim \sqrt[3]{N_1}$. In the case of polymer melts, B can even assume values of 2 and more.

Derivation of Normal Stresses via Other Variables

In addition to direct measurement, the normal stresses can be estimated using other measured variables. For example, the following relationship applies between the first normal stress coefficient at a shear rate D and the storage modulus $G'(\omega)$ derived from oscillatory measurements:

$$\Psi_1 (D) = \frac{2 \cdot G' (\omega)}{\omega^2} \tag{6.66}$$

for $D = \omega$

However, this relationship is only valid for very low or vanishing (approaching zero) shear rates or frequencies. The measurement of the storage modulus was discussed in Section 6.5.5.

One way of calculating the first normal stress difference can also be obtained by measuring the shear stress after a shear jump $\Delta\gamma_g$ (Lodge–Meissner relationship):

$$N_1 = \tau_{xy} \cdot \Delta\gamma_g \tag{6.67}$$

6.6.2.7 Determination of the Pressure-Dependent Viscosity Using an Inline Rheometer Nozzle

If the inline rheometer nozzle is used in combination with a flow spiral, the pressure-dependent viscosity increase can be calculated directly via the differential pressure of the pressure transducer with increasing back pressure as the flow path length in the flow spiral increases (Figure 6.65). The advantage of this measuring method is that the injection molding machine can be used to map the real range of the manufacturing process (temperatures, shear rates, etc.).

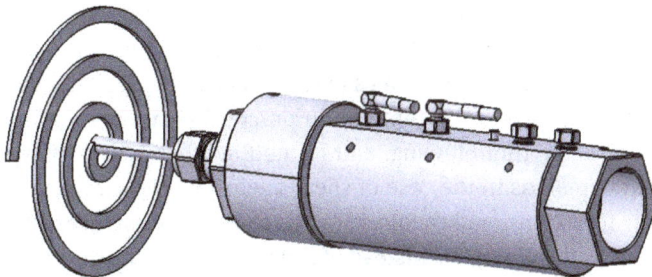

Figure 6.65 Inline rheometer nozzle with flow spiral for determining the pressure-dependent viscosity

Figure 6.66 shows the recorded pressure at sensor 1 and sensor 2 during the filling of the flow spiral. It can be seen that the pressure difference between the two sensors increases despite the constant injection rate. With otherwise constant boundary conditions (channel geometry, volume flow rate, temperature, etc.), this increase can only be explained by an increase in the pressure-dependent viscosity. The pressure-dependent viscosity was calculated directly and is shown in Figure 6.66 as well. It is clear that the viscosity value almost doubles. This point is discussed in more detail in Section 7.3.3.

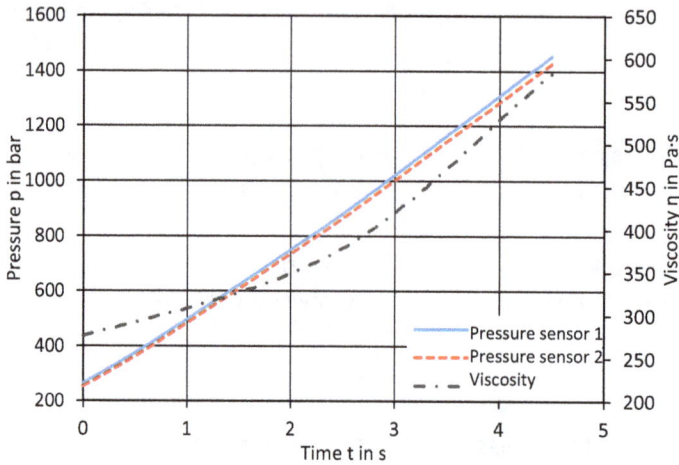

Figure 6.66 Pressure at sensors 1 and 2 in the inline rheometer nozzle and calculated viscosity

6.7 Extensional Rheology

6.7.1 Origin and Definition of Extensional Viscosity

The previous sections have mainly dealt with the material behavior under shear stress. In the following, other types of stress such as extension and compression will be discussed. Extensional stresses are not uncommon in practical processes. For example, they are found in spinning, film blowing, and in the flow through cross-sectional jumps (hot runner systems). As in the case of shear rheology, a few important variables are defined below for extensional rheology. In this case, the resistance is not expressed as shear stress but as tensile stress. Instead of shear or shear rate, the terms extension and extensional rate are used. Consequently, the viscosity value is not defined as the shear viscosity, but as the extensional viscosity. Figure 6.67 shows an example of a bar subjected to tensile stress. The application of the tensile force F leads to a change Δl in the length of the material [1].

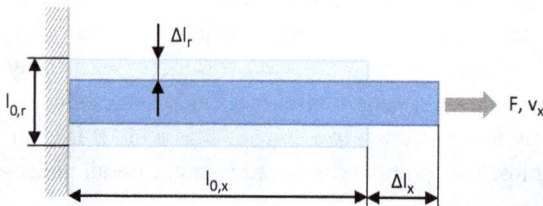

Figure 6.67 Principle representation of the extension of a bar

$$\varepsilon_x = \ln\left(\frac{l_{0,x} + \Delta l_x}{l_{0,x}}\right) \tag{6.68}$$

$$\varepsilon_r = \ln\left(\frac{l_{0,r} + \Delta l_r}{l_{0,r}}\right) \tag{6.69}$$

In relation to the total length, the so-called extension ε results and its derivation with respect to time provides the extensional rate.

$$\dot{\varepsilon} = \frac{dv_x}{dx} = \frac{1}{l_x} \cdot v_x \tag{6.70}$$

The tensile stress is defined as:

$$\sigma = \frac{F}{A} \tag{6.71}$$

where
F is the force
A is the sample cross section

The quotient of tensile stress and extension for an elastic solid provides the so-called modulus of elasticity E. This is greater than the shear modulus describing the shear. The reason for this is that the extension in one direction requires compression of the bar in the other two spatial directions due to the conservation of volume. This triple load is reflected in a modulus of elasticity that is three times higher than the shear modulus. The transverse contraction or Poisson's ratio μ describes the ratio of the compression in the two transverse directions to the extension in the tensile direction:

$$\mu = \frac{-\varepsilon_y}{\varepsilon_x} \tag{6.72}$$

For an ideal solid, $\mu = 0.5$. In practice, many real materials exhibit values of less than 0.5. Extension leads to a reduction in the total volume. An extreme example of this is cork, which shows almost no necking effects in the cross section when stretched.

Chapter 5 has already shown that, as in the case of shear, a viscous fluid exhibits a resistance to extensional deformation that is proportional to the extensional rate. The proportionality factor is accordingly called extensional viscosity:

$$\eta_D = \frac{\sigma}{\dot{E}} \tag{6.73}$$

The extensional viscosity is always greater than the shear viscosity. The relationship between the extensional viscosity and the shear viscosity is described by Trouton: $\eta_D = 3 \cdot \eta_0$ (Figure 6.68). As already mentioned in Chapter 5, this relationship only applies to non-Newtonian materials at very low deformation rates and therefore not in the range of higher shear rates, at which the processing of plastics usually takes place.

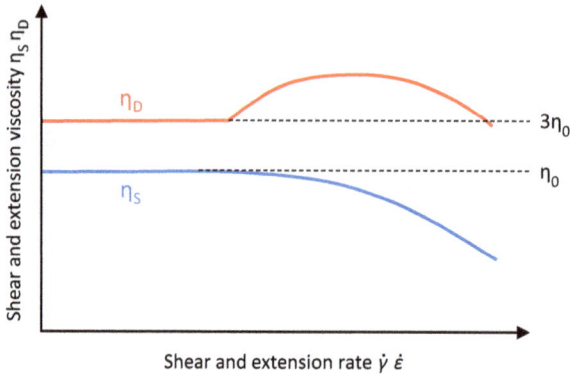

Figure 6.68 Qualitative progression of shear and extensional viscosity as a function of deformation rates

In the bar test shown in Figure 6.67, the material is stretched in the direction of one axis, which corresponds to uniaxial extension. Multiaxial types of extension will be discussed below.

6.7.2 Measurement of Extensional Viscosities

Chapter 5 has already dealt with some manufacturing processes in which extensional flows can also play a role. For this reason, some rheometers with which variables such as extensional viscosities can be determined will be presented below.

6.7.2.1 Measurements with Uniaxial Extension

Figure 6.69 shows some simple rheometers that can be used for uniaxial extension.

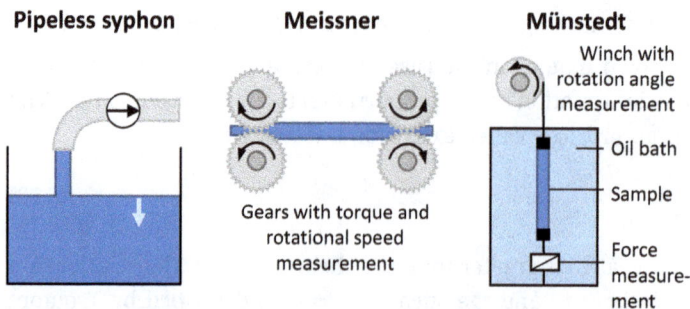

Figure 6.69 Extensional rheometer with uniaxial extension

The tubeless siphon (Fano flow) shown on the left can be used for low-viscosity media. Here, a fluid is drawn out of a container through a pipe with a radius R. By lowering the fluid level in the container or by raising the pipe, the distance between the pipe and the fluid surface is continuously increased. At a certain height h, the jet will break off. This height is a measure of the extensional viscosity. The following applies:

$$\eta_D = \frac{2 \cdot \pi \cdot R^2 \cdot h \cdot p_u}{\dot{V}} \qquad (6.74)$$

Since inertia, surface tension, and gravity effects are neglected, the derived extensional viscosity is only an apparent value.

6.7.2.2 Determining the Extensional Viscosity with the Rheotens Test

The pullout ability of plastic melts is of great importance for many manufacturing processes. For example, melt strengths play a major role in the spinning of fibers or in extrusion blow molding. For this reason, an extensional rheometer called Rheotens was developed in the 1970s. This device is able to determine the extensional viscosity as a function of the extensional rate by drawing the melt from a nozzle.

The method for determining the extensional viscosity of a plastic melt is relatively simple. The extruded melt strand is picked up by one or two rollers. The roller initially rotates at the same velocity as the extruded strand. This initial velocity is referred to as v_0. The velocity is then increased linearly. A load cell (see Figure 6.70, right) is used to measure the force on the rollers that is required to draw off the strand at the increasing velocity. The velocity of the two rollers can increase until the melt strand breaks off. The extensional rate can be calculated from the difference in velocity between the melt exit and the velocity at which the rollers move. The measured draw-off force can be used in conjunction with the cross section of the melt strand between the rollers to determine the draw-off stress. The ratio of the draw-off stress to the extensional rate ultimately provides the extensional viscosity [30, 31].

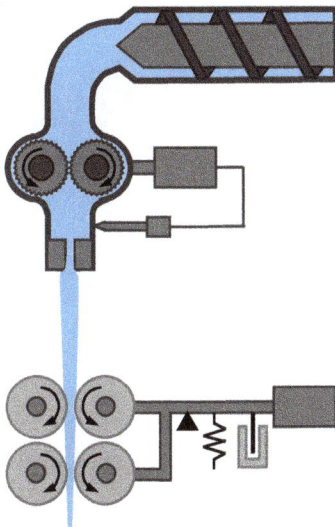

Figure 6.70
Measurement setup for Rheotens test

However, this process cannot be used for all types of plastic. One prerequisite is sufficient melt stiffness. The extensional viscosity of easy-flowing plastics, such as those often used in injection molding, usually cannot be determined using Rheotens. Other extensional rheometers must be used.

Figure 6.71 shows an example of the force–velocity diagram of a polypropylene (PP) melt. The draw-off force initially rises with increasing stretching and gradually changes to a horizontal curve at higher draw-off rates. The oscillations in the plateau area are known as draw resonances, which result from the fluctuation of the volume flow rate at a high draw rate.

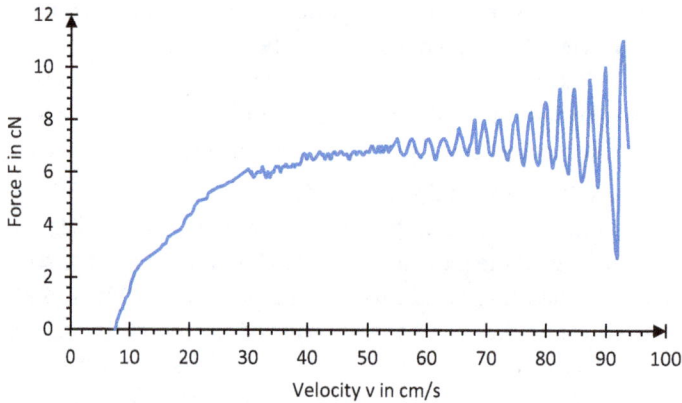

Figure 6.71 Draw-off force as a function of the draw-off rate

The extensional viscosity and extensional rate can be determined from the draw-off force and the draw-off rate, and can be calculated as shown in Figure 6.72.

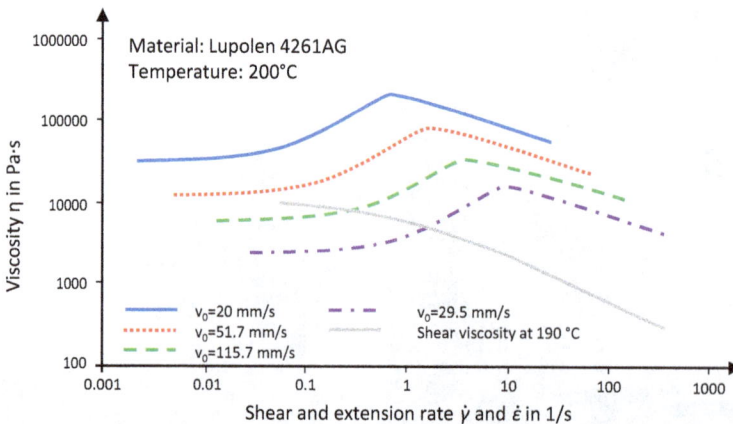

Figure 6.72 Extensional viscosity calculated from the Rheotens test as a function of the extensional rate for different extrusion rates v_0

The equations for calculating the extensional viscosity using Rheotens are listed below. The tensile stress from the draw-off force F is:

$$\sigma = \frac{F}{A} \tag{6.75}$$

Using the continuity equation, the line cross section in front of the take-off wheels is obtained:

$$A = \frac{A_0 \cdot v_0}{v} \tag{6.76}$$

where
v_0 is the initial velocity of the melt strand
v is the velocity of the wheels, and A_0 is the initial cross section

This results in the tensile stress:

$$\sigma = \frac{F \cdot v}{A_0 \cdot v_0} \tag{6.77}$$

The following applies to stretching:

$$\varepsilon = \frac{L_z - L_A}{L_A} \tag{6.78}$$

With $L = v \cdot t$, it follows:

$$\varepsilon = \frac{v_z \cdot t - v_A \cdot t}{v_A \cdot t} \rightarrow \dot{\varepsilon} = \frac{v_z - v_0}{v_0}$$

where
v_0 is the initial velocity of the melt strand
v_z is the velocity when the melt strand breaks off

The change in length results in:

$$l_w = v_0 \cdot \Delta t \tag{6.79}$$

Assuming constant acceleration a, the result is

$$l_z = v_0 \cdot \Delta t + \frac{1}{2} \cdot a \, (\Delta t)^2 \tag{6.80}$$

The change in length is now:

$$\Delta l = l_z - l_w = \frac{1}{2} \cdot a \, (\Delta t)^2 \tag{6.81}$$

where Δt can be determined using the following relationship:

$$\Delta t = \frac{v_z - v_A}{a} \tag{6.82}$$

The following applies to the calculation of extensional viscosity:

$$\eta_{ES} = \frac{\sigma_E}{\dot{\varepsilon}} \tag{6.83}$$

The extensional stress and extensional rate can also be determined directly using the following equations:

$$\sigma_E = \frac{v_L \cdot F}{v_0 \cdot \pi \cdot R_0{}^2} \tag{6.84}$$

$$\dot{\varepsilon} = \frac{v_L}{L} \cdot \ln \left(\frac{B R_0}{R_L} \right)^2 = \frac{v_L}{L} \cdot \left[\ln \left(\frac{v_L}{v_0} \right) + 2 \cdot \ln B \right] \tag{6.85}$$

where

σ_E is the extensional stress

$\dot{\varepsilon}$ is the stretching/extensional rate

L is the nozzle to roller pair distance

v_0 is the haul-off rate = extrusion rate

v_L is the draw-off rate

R_0 is the nozzle radius

R_L is the radius after nozzle outlet

B is the strand widening

6.7.3 Determination of Extensional Viscosity Using the Approach of F. N. Cogswell

According to F. N. Cogswell [24–26], an apparent extensional viscosity can be determined from the inlet pressure loss p_{in}. The extensional viscosity can be calculated from the quotient of the tensile stress and the extensional rate. The following applies:

$$\eta_{\text{extension}} = \frac{\sigma}{\dot{\varepsilon}} \tag{6.86}$$

According to Cogswell, the mean extensional rate is calculated from the wall shear rate in the nozzle and the angle of the streamlines in the nozzle inlet, relative to the central axis. This gives the extensional rate:

$$\dot{\varepsilon} = \frac{\dot{\gamma}_W}{2} \tan \Theta \tag{6.87}$$

Even if the real inlet angle in the flow channel were 90°, the melt flow would form its energetically most favorable flow shape and thus not follow this angle, but form its own inlet angle, which depends, among other things, on the volume flow rate and the viscosity of the melt. Figure 6.73 shows the melt streamlines for a nozzle inlet. It is clear that the streamlines (blue) do not follow the right-angled inlet geometry of the cross-sectional jump, but instead form a streamline angle of between 30° and 45° (see red lines) depending on the flow rate.

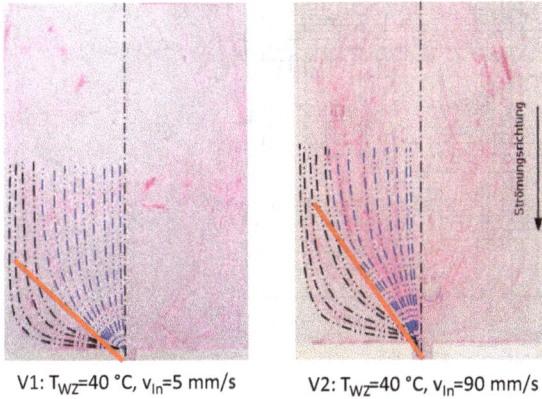

V1: T_{WZ}=40 °C, v_{In}=5 mm/s   V2: T_{WZ}=40 °C, v_{In}=90 mm/s

Figure 6.73 Comparison of the nozzle inlet flow lines

Even without knowing the angle, using the exponent n in the power law according to Ostwald and de Waele, the average extensional rate in the inlet is:

$$\dot{\varepsilon} = \frac{4 \cdot \dot{\gamma}_W \cdot \tau_W}{3 \cdot (n + 1) \cdot p_{in}} \tag{6.88}$$

Another approach for determining the extensional rate for a channel with a decreasing cross section was presented by H. Giesekus [2]:

$$\dot{\varepsilon} = -\frac{\dot{V}}{[A(x)]^2} \cdot \frac{dA(x)}{dx} \tag{6.89}$$

The mean tensile stress then increases:

$$\sigma = \frac{3}{8} (n + 1) \cdot p_{in} \tag{6.90}$$

Note that the extensional viscosity calculated with the above equations increases as the square of the inlet pressure drop.

Figure 6.74 and Figure 6.75 show the shear and extensional viscosity as a function of the shear rate measured with the high-pressure capillary rheometer for a polystyrene (PS) and a polyethylene (PE-HD). The extensional viscosity was determined using the approach of F. N. Cogswell.

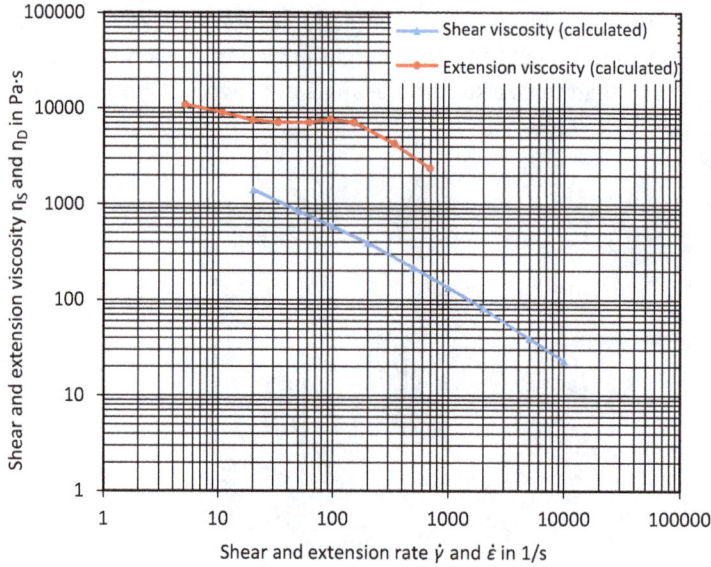

Figure 6.74 Shear and extensional viscosity as a function of shear and extensional rate for a PS 475F (piston diameter: 15 mm, nozzle diameter: 1 mm, nozzle length: 16 mm)

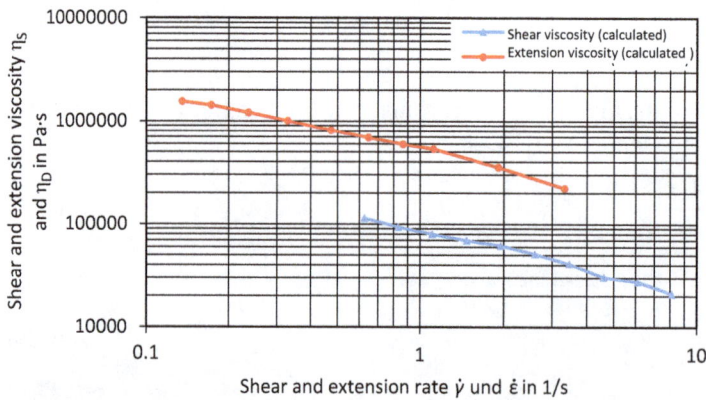

Figure 6.75 Shear and extensional viscosity as a function of shear and extensional rate for a PE-HD Lupolen 4261 AG (piston diameter: 15 mm, nozzle diameter: 1 mm, nozzle length: 16 mm)

While PS is an injection molding type with a relatively low viscosity, PE-HD is an extrusion blow molding type. The high melt stiffness required for the extrusion blow molding process results in a correspondingly high viscosity. At a shear rate of 10 s^{-1}, the viscosity of the PE-HD is still over 10,000 Pa·s. For comparison, the viscosity of PS at this shear rate is approximately 2,000 Pa·s.

At a shear rate of 10 s^{-1}, the extensional viscosity assumes a value of approximately 10,000 Pa·s for PS 475F and is therefore approximately 5 times greater than the shear viscosity. For Lupolen 4261, the value for the extensional viscosity is approximately 500,000 Pa·s at a shear rate of 1 s^{-1}, and the shear viscosity is approximately 80,000 Pa·s. The factor between these values is over 6.

The different shear rates on the x axis can be explained by the different processing methods. While the injection molding process tends to take place at high shear rates, the shear rate range for extrusion blow molding is between 1 s^{-1} and 10 s^{-1}.

6.8 Theory and Practice of Solution Viscometry

Solution viscometry can be used to measure the viscosity of a polymer in a solvent. The prerequisite for this is that the polymer can be completely dissolved in the solvent. Typical plastics for which this method is used are polyamides and polyethylene terephthalate (PET).

The test method allows a statement to be made about the average chain length of the macromolecules of the tested polymer. As already mentioned, the viscosity of the polymer increases with increasing chain length. This means that a statement about the quality (chain length) of the material can be made via an incoming goods inspection using solution viscometry. This is particularly important for PET, as there are different types, depending on the subsequent application. For example, the chains for bottling edible oil and still water tend to be short, making the material cheaper, while the chains for a bottle for carbonated water are longer in order to improve the burst pressure and barrier properties.

Furthermore, the method is also well suited for making statements about mechanical or thermal degradation of the polymer chains. If the viscosity value changes even in very small ranges, this indicates a degradation of the polymer chains. In the case of PET, for example, the permissible chain degradation during processing is specified. This limit value is 0.02 dl/g. Figure 6.76 shows the relationship between the dwell time and the measured solution viscosity (inherent viscosity). The value decreases with increasing residence time, that is, the chains are thermally degraded.

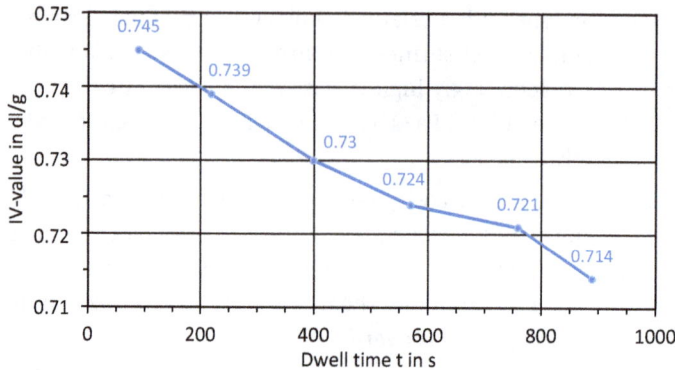

Figure 6.76 Inherent viscosity as a function of dwell time for a PET

The solution viscosity of polyamides is also often measured on the injection-molded part. The viscosity number of the granulate is compared with the viscosity number of the end product. A reduction in the viscosity number of 5 % is generally not considered critical. A reduction of more than 10 %, on the other hand, is rather critical for most applications.

In principle, this measuring method measures the time it takes for a defined quantity of fluid to flow through a capillary of known diameter and length. The driving pressure is the hydrostatic pressure in the fluid column. Overpressure can also be used to achieve higher shear rates. Regardless of the specific design, most U-shaped glass bodies often have spherical extensions, whose volume determines the amount of material to be measured. Measuring marks on the glass solid or precisely fixed sensors enable the temporal passage of the boundary layer of the product/air (meniscus) to be detected, allowing the flow time of such a limited volume of product to be measured with an uncertainty of less than 1/10 s.

The polymer is usually dissolved in a solvent at room temperature. As this process can take longer, the temperature is often increased to speed up the process. Care must be taken to control the temperature very precisely in order to avoid thermal influences on the measurement.

Hygroscopic plastics should be sufficiently dried beforehand, as the moisture can influence the measurement result. An example of this is shown for a PET in Figure 6.77.

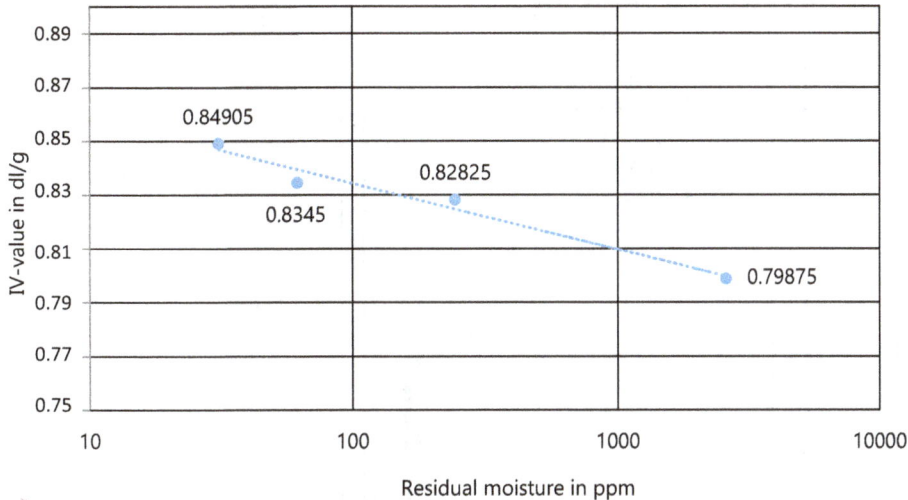

Figure 6.77 Influence of the residual moisture content on the solution viscosity for a PET type 1101

If a measurement of a filled polymer is to be carried out, the filler content must be determined beforehand (e.g., using thermogravimetric analysis (TGA)). After dissolving the sample, the fillers are removed from the solution. This can be done using filters, for example.

The capillary suitable for the measurement is selected taking into account the Hagenbach time correction [32]. The permissible minimum flow times depend on the capillary. If the minimum flow time is exceeded by more than 2%, the capillary used is not suitable for the measurement and the capillary with the next smallest diameter must be used [33].

Two basic viscometer types according to Ostwald and Ubbelohde are shown in Figure 6.78 [1, 19]. With both viscometers, the fluid to be tested is filled into the storage vessel (4) via the filling pipe (3). As the average pressure level of the Ostwald viscometer depends on the filling quantity, the prescribed measuring volumes must be strictly adhered to. A pipette is therefore used for filling. The sample is sucked into the pipe (2) for measurement. The time required for the meniscus to sink from the ring measuring mark M_1 to the ring measuring mark M_2 is measured. In Ubbelohde viscometers, the transition from the capillary (7) to the level vessel (6) is designed as a spherical cap into which an additional aeration pipe (1) opens. After filling the product to be measured into the vessel (4) via the pipe (3), the aeration pipe is closed.

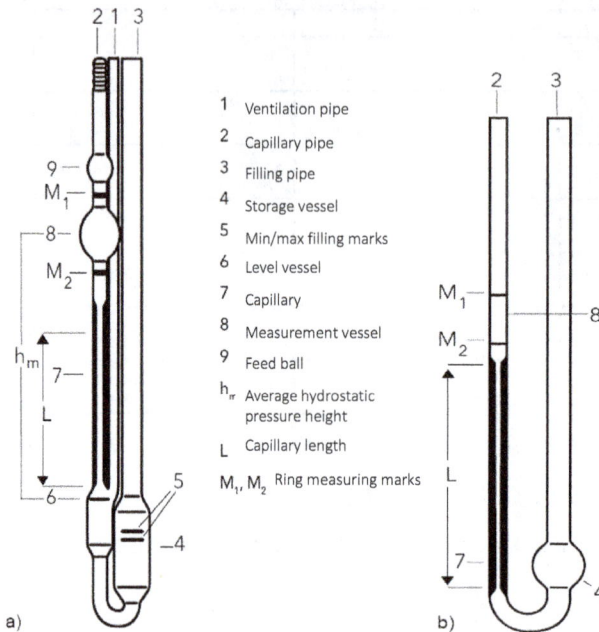

Figure 6.78 Solution viscometers according to Ubbelohde (a) and Ostwald (b) [34]

Depending on the operating mode, pushing or sucking, the level vessel (6), the capillary (7), the measuring ball (8), and at least half of the flow ball (9) are filled by overpressure on the pipe (3) or by suction via the pipe (2). After the venting pipe (1), the fluid column in the level vessel (6) breaks off. The so-called hanging level forms at the outlet of the capillary. For this reason, only a limited amount of sample, maximum and minimum fill marks (5), may be filled in. After venting the pipe (2), the measured material flowing out of the capillary runs off as a film on the inner wall of the level vessel (6). In this way, the hydrostatic pressure of the fluid column is independent of the amount of material filled in. The geometric design of the level vessel (6) also virtually eliminates the influence of surface tension on the measurement result.

The Ubbelohde viscometer also measures the time it takes for the fluid meniscus to sink from the ring measuring mark M_1 to the ring measuring mark M_2. With strongly colored, opaque fluid, the meniscus passage through the measuring marks cannot be detected visually due to pipe wetting. In this case, riser viscometers are used for manual operation (Figure 6.79).

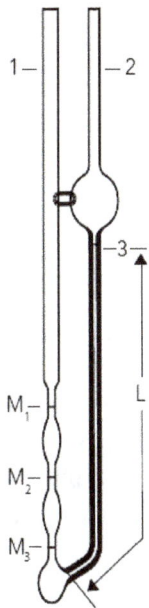

1	Ventilation pipe
2	Capillary pipe
3	Filling mark
L	Capillary length
M_1, M_2, M_3	Ring measuring marks

Figure 6.79
Structure of a Cannon-Fenske riser viscometer

The sample is filled into the spherical extension of the capillary pipe (2). The pipe (1) is closed during thermostabilization and opened at the start of the measurement. The flow time of the meniscus through the measuring marks M_1, M_2, and M_3 on the riser (1) is recorded as an image signal for the viscosity.

Dissolved polymers increase the viscosity of the solvent, even if they are only present in very low concentrations. This fact opens up a simple and frequently used method for determining molecular weights. The measurement of the relative viscosity increase in the capillary viscometer is experimentally uncomplicated. As it can be assumed to a first approximation that the densities of the pure solvent and the solution do not differ at low polymer concentrations, the relative viscosity η_{rel} is almost equal to the ratio of the transit times of solution and solvent in the viscometer:

$$\eta_{rel} = \frac{\eta_c}{\eta_0} \cong \frac{t_c}{t_0} \tag{6.91}$$

where
η_{rel} is the relative viscosity
η_c is the viscosity of the polymer solution with concentration c
η_0 is the viscosity of the pure solvent
t_c is the lead time of the solution
t_0 is the flow time of the solvent

The viscosity depends on the size and shape of the macromolecule, but also on the type of interaction with the respective solvent. On the one hand, the higher the molecular weight of a polymer sample (for the same weight), the more viscous the solution. On the other hand, the shape of the dissolved particles also has a decisive influence. The shape of the polymer beads (spheres, rods, etc.) as well as their arrangement and behavior in the solvent play a role here: In extreme cases, a distinction can be made between a non-flushed bead, in which the solvent is completely immobilized inside, and a freely flushed bead, in which only very weak interactions take place between the polymer and the solvent.

Solvation processes lead to an extension of the polymer clusters, so that the frictional properties of the solution change and thus also its viscosity. The hydrodynamic volume of the polymer particles is considered. In 1906, Albert Einstein derived a relationship between the viscosity of a solution and the volume fraction φ of the spheres for rigid unsolvated spheres distributed in a solvent. This is applied and explained here for polymer solutions. The volume fraction φ is the ratio of the hydrodynamic volume of the solute to the total volume.

$$\eta_c = \eta_0 \left(1 + \frac{5}{2}\varphi\right) \tag{6.92}$$

or

$$\eta_{sp} = \frac{\eta_c}{\eta_0} - 1 = \frac{5}{2}\varphi \tag{6.93}$$

where
η_{sp} is the specific viscosity

$$\eta_{sp} = \frac{\eta_c - \eta_0}{\eta_0} = \eta_{rel} - 1$$

To take into account the effects of the concentration, the specific viscosity η_{sp} of a solution is also described by a virial equation in which interactions between the solvated molecules are considered:

$$\eta_{sp} = \frac{5}{2}\varphi + B\varphi^2 + C\varphi^3 + \dots \tag{6.94}$$

where
B is the 2nd virial coefficient
C is the 3rd virial coefficient

Therefore, the increase in viscosity due to the solute depends on its hydrodynamic volume. To simplify matters, the reduced viscosity η_{red} is introduced to describe the contribution of the polymer to the viscosity.

$$\eta_{red} = \frac{1}{c} \cdot \frac{\eta_c - \eta_0}{\eta_0} = \frac{1}{c} \cdot \eta_{sp} \tag{6.95}$$

where
c is the concentration of the dissolved polymer

By graphically extrapolating the values measured for a concentration series of η_{red} to $c = 0$, a concentration-independent value is obtained:

$$\lim_{c \to 0} \eta_{red} = [\eta] \tag{6.96}$$

This limiting value $[\eta]$ is referred to as the Staudinger index, intrinsic viscosity, or limiting viscosity number. It has the unit ml/g and is therefore a specific volume. $[\eta]$ is the ratio of the hydrodynamic volume to the molecular weight.

If Einstein's model is to be applied to dissolved polymers, it must be taken into account that these are usually present as random balls. A characteristic property of balls is that their density decreases with increasing molecular weight. According to Kuhn's square root law, the following applies to an ideal ball in a dilute solution:

$$\rho_{eq} = K_\rho \cdot M^{0.5} \tag{6.97}$$

where
ρ_{eq} is the equivalent ball density (density of a sphere that causes an increase in viscosity equivalent to that of the ball under consideration)
K_ρ is a density-related constant
M is the molecular weight

The hydrodynamic volume of a solvated ball depends on the molecular weight in the same way. Therefore, the intrinsic viscosity $[\eta]$ is also a function of the molecular weight. The following applies to the case of the ideal ball:

$$[\eta] = \frac{\eta_{sp}}{c} = K \cdot M^{0.5} \tag{6.98}$$

where
K is an empirically determined constant
c is the concentration of the dissolved polymer

In a generalized form, the relationship of the Staudinger index to the molecular weight of the macromolecule is given by the Mark–Houwink equation (also known as the Kuhn–Mark–Houwink–Sakurada equation):

$$[\eta] = K \cdot M^\alpha \tag{6.99}$$

where
α is an exponent describing hydrodynamic interactions between solvent and macromolecule

The exponent α is influenced by the temperature, the type of solvent, and the structure of the solvated polymer. Thus,

- For compact spheres, $\alpha = 0$

- For a ball that has not been flushed, $\alpha = 0.5$

- For a partially flushed ball, $\alpha = 0.5$ to 1

- When the ball is completely flushed through, $\alpha = 1$

- For a chopstick, $\alpha = 2$

- Mostly, α lies between 0.65 and 0.75

The relationship between $[\eta]$ and M was discovered empirically by Hermann Staudinger during measurements on cellulose and cellulose derivatives. Werner Kuhn substantiated it with his description of the statistical tangle. Herman Mark formulated it in this form for the first time, and Roelof Houwink confirmed its validity by measuring various polymers in different solvents.

The molecular weight of a sample can be determined by viscosity measurements at different polymer concentrations using knowledge (tables) or after determining the proportionality factor K and the exponent α, which characterize the polymer–solvent system (Figure 6.80). The K and α values of many polymer–solvent systems have already been tabulated in detail in the literature. Otherwise, they can be accessed by measuring preparations of known molecular weights (with a narrow distribution): By plotting $\log[\eta]$ against $\log M$, a straight line with the slope α and the ordinate intercept $\log K$ is obtained [35–37].

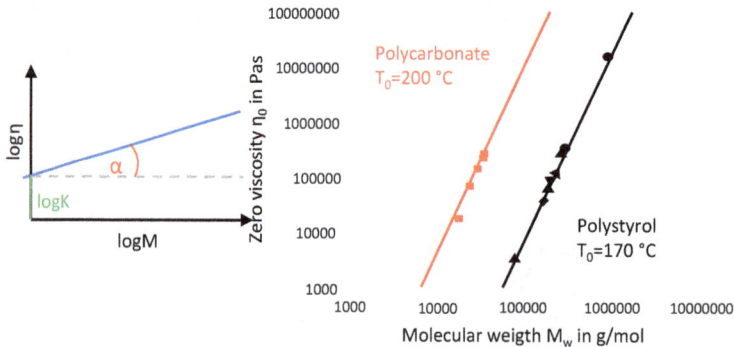

Figure 6.80 Influence of molecular weight on zero viscosity for different plastics

When calibrating using preparations of known molecular weights, the method employed to determine these molecular weights must be taken into account, that is, the type of mean value that results. To determine the viscosity average, preparations whose average molecular weight was determined using a method that yields the weight average (light scattering, sedimentation equilibrium, etc.) are often used. The value resulting from viscosity measurements should be classified accordingly.

Table 6.9 Symbols/Equations and Quantities Associated

Symbol/Equation	Quantity
η	Dynamic viscosity
$v = \dfrac{\eta}{\rho}$	Kinematic viscosity
$\eta_{rel} = \dfrac{\eta_c}{\eta_0}$	Relative viscosity, viscosity ratio
$\eta_{sp} = \dfrac{(\eta_c - \eta_0)}{\eta_0} = \eta_{rel} - 1$	Relative viscosity change, specific viscosity
$J_v = \dfrac{1}{c}\dfrac{(\eta_c - \eta_0)}{\eta_0} = \eta_{red}$	Staudinger function, viscosity number, reduced viscosity
$\eta_{inh} = \dfrac{\ln \frac{\eta_c}{\eta_0}}{c}$	Inherent viscosity
$J_g = \lim\limits_{c \to 0} \left[\dfrac{1}{c}\dfrac{\eta_c - \eta_0}{\eta_0}\right] = [\eta]$	Staudinger index, limiting viscosity number, intrinsic viscosity

6.8.1 Example Measurement of Solution Viscosity Using Polyethylene Terephthalate (PET)

As already mentioned in the previous section, measuring the structure-dependent viscosity profile with a high-pressure capillary rheometer or determining the MVR value using a low-pressure capillary rheometer with PET is very difficult to carry out due to the rapid absorption of moisture. Figure 6.81 illustrates the pronounced hygroscopic behavior of a PET.

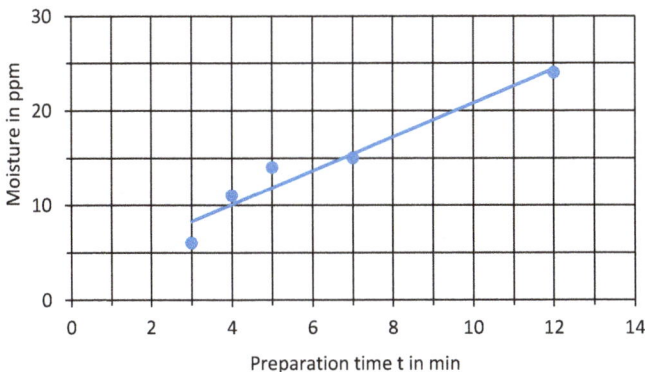

Figure 6.81 Water absorption of PET as a function of the preparation time

It is clear that the PET has already absorbed more than 20 ppm of water after a preparation time of approximately 10 min. Figure 6.82 shows how this moisture absorption affects the value of the MVR measurement.

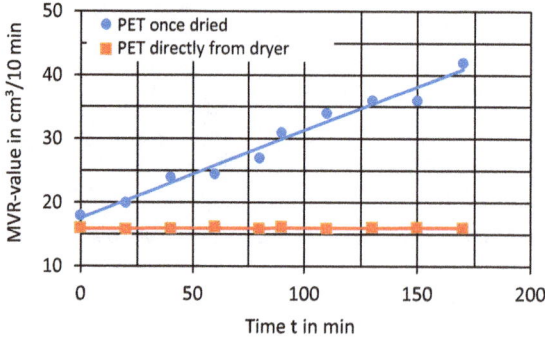

Figure 6.82 Effect of the moisture content of PET on the MVR measurement

The horizontal line shows the MVR measured values for different measurements at different times, which were carried out on the granulate that was taken directly from the air dryer for the MVR determination. The MVR value is constant, at 16 cm³/10 min. As the granulate is dry (5 ppm), the humidity has no effect on the measured value. In contrast, the rising line shows the influence of moisture on the MVR value. For the MVR determination, the granules were dried once to 5 ppm. The granulate was then removed from a container for each measurement. This allowed the granules to absorb moisture over time. As a result, the MVR value increases—that is, the viscosity decreases.

The decrease in viscosity with increasing moisture absorption can be explained by two effects (Figure 6.83). On the one hand, it is a meltwater (water vapor) mixture. The measured viscosity is therefore determined by the viscosities of the two media.

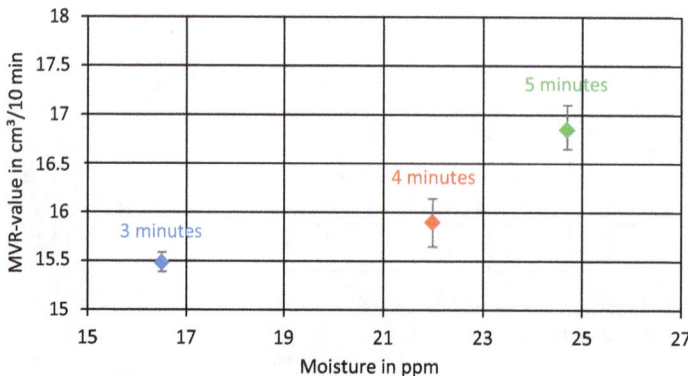

Figure 6.83 MVR value as a function of humidity and time

On the other hand, PET is also very sensitive to water at high temperatures (Figure 6.84). It tends to degrade as a result of a hydrolysis process—that is, the chain length is reduced.

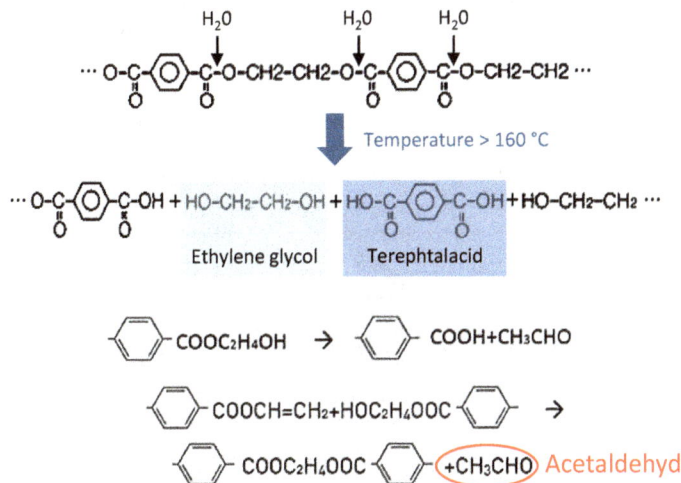

Figure 6.84 Example of the degradation of PET (hydrolysis and thermal)

An additional effect that must be taken into account during the measurement is that of the dwell time. Especially with thermally sensitive materials such as PET, the dwell time can have a very strong effect on the measurement result, as the polymer chains thermally degrade (i.e., are shortened). The influence of the dwell time on the MVR value is shown in Figure 6.85 for one PET type. It is clear that the curve is not linear, but exponential. This must be the case, as the molecular chain length is a variable included in the viscosity with an exponent of approximately 3.4.

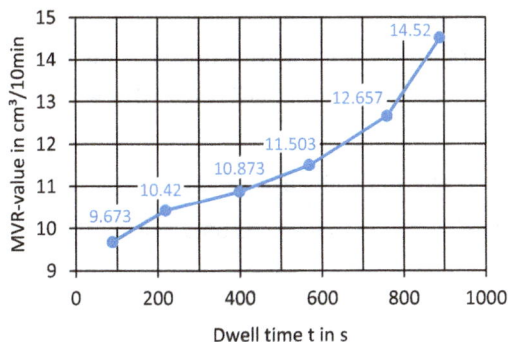

Figure 6.85
MVR value as a function of the dwell time

As the future properties of the part are also decisively defined by the molecular chain length of the PET, a suitable method that allows a statement to be made about this variable must be found. The diagram in Figure 6.86 shows the relationship between the molecular chain length and the intrinsic viscosity. It is clear that the so-called IV value and the chain length correlate directly. For this reason, this measurement method has become established as a quality control in the processing of PET.

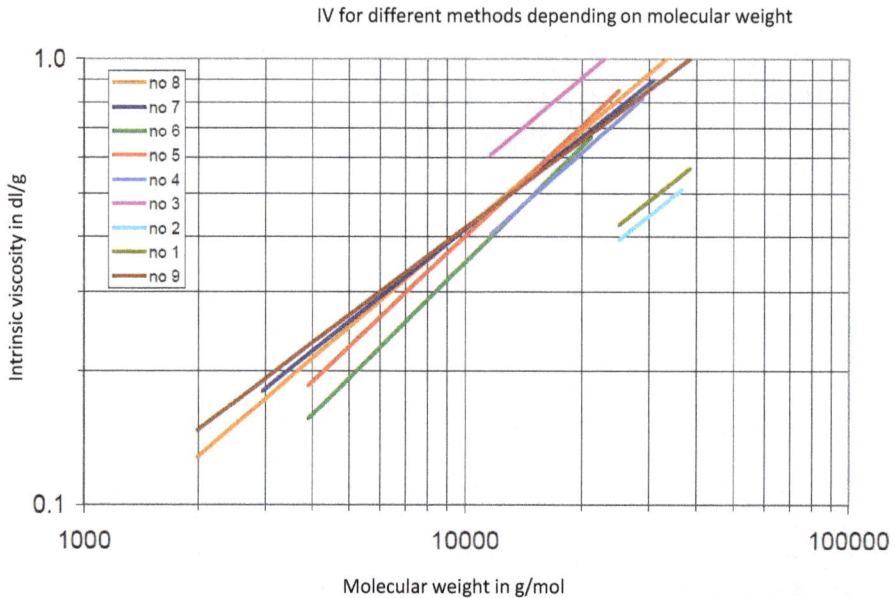

Figure 6.86 Correlation between molecular weight and solution viscosity (IV) for PET grades

The measurement method for determining the solution viscosity is standardized for different polymers in accordance with DIN EN ISO 1628 [38].

To determine the viscosity number for a polyethylene terephthalate (PET), the procedure is as follows (Figure 6.87), in accordance with the manufacturer's data sheets:

Prepare the 1 % solution:

- Weigh the dried PET (0.5 g) in dichloroacetic acid (50 ml)
- Dissolve the PET in the Erlenmeyer flask with magnetic rods on a warm magnetic plate (up to 10 h, 50 °C)

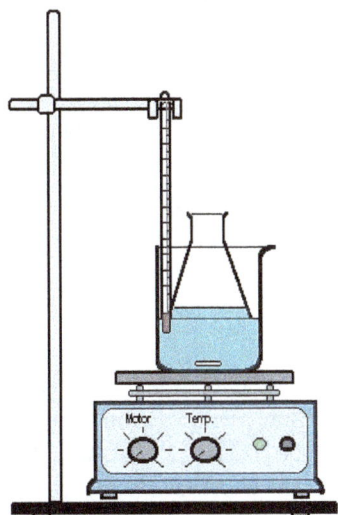

Figure 6.87
Experimental setup

In order to determine the intrinsic viscosity (IV), different solution concentrations are obtained with the prescribed solvent (in the case of PET, it is dichloroacetic acid), as shown in Figure 6.88.

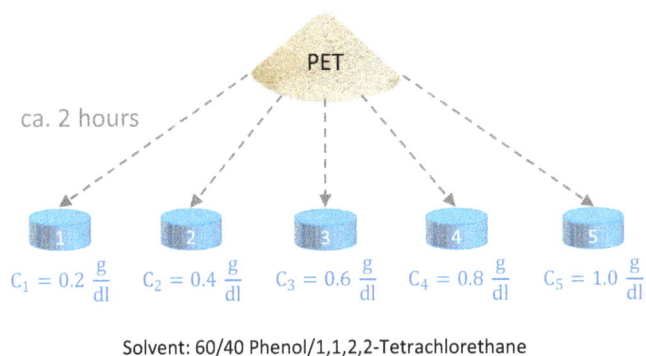

$C_1 = 0.2 \frac{g}{dl}$ $C_2 = 0.4 \frac{g}{dl}$ $C_3 = 0.6 \frac{g}{dl}$ $C_4 = 0.8 \frac{g}{dl}$ $C_5 = 1.0 \frac{g}{dl}$

ca. 2 hours

Solvent: 60/40 Phenol/1,1,2,2-Tetrachlorethane

Figure 6.88 Preparation of the different concentrations c according to ASTM D4603

In the next step, these concentrations are measured using an Ubbelohde viscometer. In other words, the time it takes for the concentrations to flow through the two way-marks A and B is recorded (Figure 6.89).

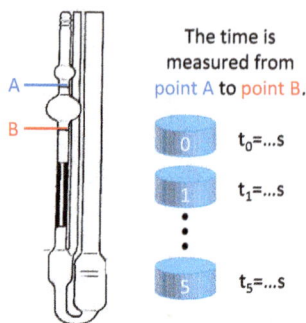

The time is measured from point A to point B.

$t_0=...s$

$t_1=...s$

$t_5=...s$

Figure 6.89
Timing procedure

To measure the solution viscosity using the Ubbelohde viscometer, different types of viscosities can be determined in a diluted polymer solution. Here, t is the time required for the solution to flow through the two ring marks M_1 (A) and M_2 (B). At t_0, it is the pure solvent that is measured. The individual concentrations that are measured are described by c.

Specific viscosity:

$$\eta_{sp} = \frac{t - t_0}{t_0} \tag{6.100}$$

Reduced viscosity:

$$\eta_{red} = \frac{\eta_{sp}}{c} \tag{6.101}$$

Relative viscosity:

$$\eta_{rel} = \frac{t}{t_0} \tag{6.102}$$

Inherent viscosity:

$$\eta_{inh} = \frac{\ln(\eta_{rel})}{c} \tag{6.103}$$

The measured values are then plotted in a diagram (Figure 6.90). The concentrations are plotted on the x axis and the calculated viscosities on the y axis. By extrapolating the values to the y axis, the inherent viscosity without solvent—that is, for the pure polymer—is obtained.

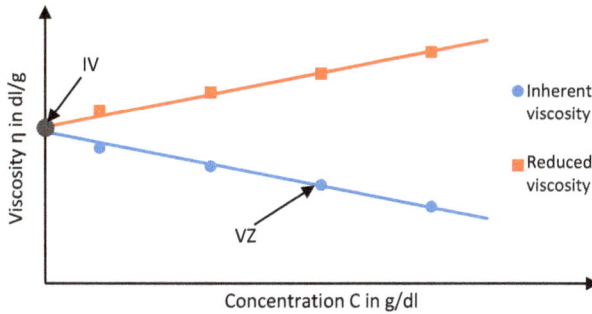

Figure 6.90 Huggins/Kraemer diagram

As the determination of the inherent viscosity is very time-consuming, raw material manufacturers for PET, for example, have opted for a single-point measurement. The concentration for the measurement to be carried out is specified and the so-called viscosity number is measured. The equation for calculating the *viscosity number* is as follows:

$$J_V = \left(\frac{\eta}{\eta_0} - 1 \right) \cdot \frac{1}{c} = \left(\frac{t}{t_0} - 1 \right) \cdot \frac{1}{c} \tag{6.104}$$

For the sample measurement on dried PET granulate, the following values were obtained in this example: $t = 75.42$ s, $t_0 = 42.37$ s, $c = 1$ g/dl. This results in a viscosity number of $J_V = 0.78$ dl/g.

As a measurement employing a solution viscometer is very time-consuming and the user has to handle hazardous chemicals, this measurement method is not very common in practice. In contrast, many plastics processors use MVR determination as a measurement method for incoming inspection. The correlation between the MVR value and the IV value has often been investigated in the past [33] and will be illustrated below using PET as an example. Figure 6.91 shows the correlation between the measured MVR value and the solution viscosity (IV). It is clear that the IV value increases exponentially as the MVR value decreases.

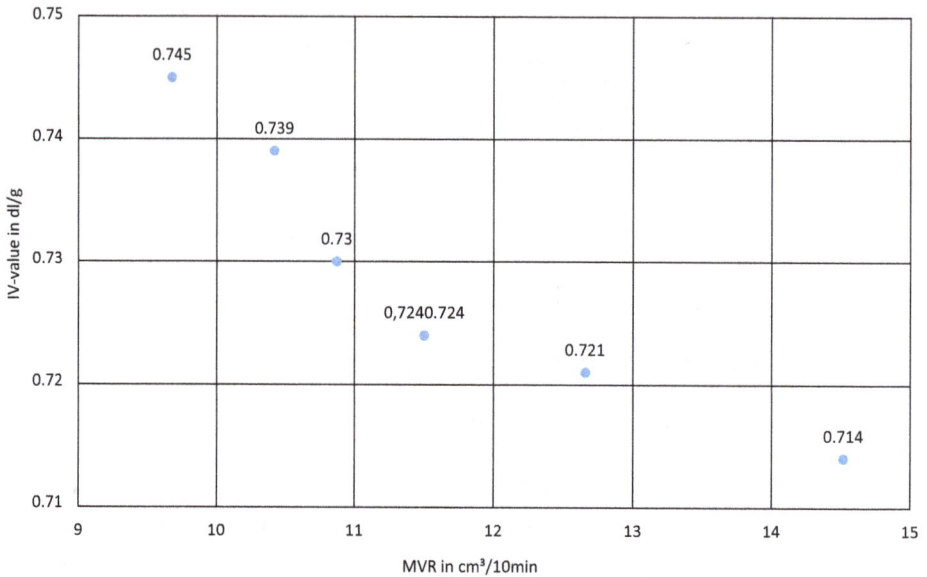

Figure 6.91 Correlation between MVR and IV value for PET 1101

J. F. Reilly et al. [39] established a correlation between the *MVR* value and the solution viscosity (*IV*). This is as follows:

$$IV_{corr} = A + B \cdot IV_{rheo} \tag{6.105}$$

$$IV_{rheo} = e^{\frac{2.30259}{5}\left(\log\left(\frac{1}{MVR}\right) - \frac{2953}{T(K)} + 1.99508\right)} \tag{6.106}$$

The constants are determined using at least three pairs of values (*IV* and *MVR*). These constants then apply to a test weight. Figure 6.92 shows how the constants *A* and *B* are determined. In this case, *A* = 0.3376 and *B* would have the value 3.0408. The diagram also contains the solution viscosity (IV) measured with the Ubbelohde device.

It is clear that although the MVR measurement provides an approximate value for the solution viscosity (IV), this is too inaccurate to carry out quality control using this measurement method, taking into account a maximum permissible chain degradation of 0.02 dl/g for PET.

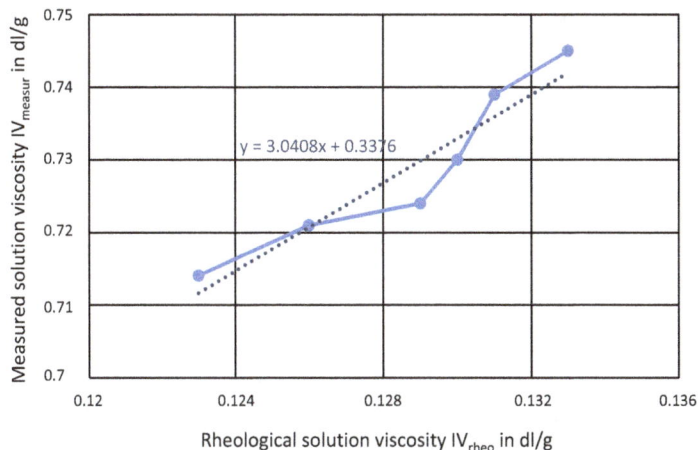

Figure 6.92 Determining the constants A and B

6.8.1.1 Information from Schott Instruments on the Measurement of Solution Viscosity

Sample Preparation/Performing the Measurement

Solution Viscosity (IV)

Weigh 200 mg (± 20 mg) of the unfilled polymer or such a quantity of the sample containing additives that it contains 200 mg of polymer into the 50 ml conical flask. Add 20 ml of solvent, add the stirring magnet, close the flask with the NS stopper and place it on the magnetic stirrer at 140 °C (± 10 °C). Dissolve the sample with constant stirring. As soon as the polymer is completely dissolved (dissolving time: 20 min to 40 min), immediately remove the flask from the magnetic stirrer and cool to RT in cold water. Separate the solution—if necessary—from glass fibers or other additives by centrifugation, filter through a glass filter crucible into the clean, dry capillary. Then clamp the capillary into the measuring stand, connect it to the AVS 360 and start the measuring process.

6.8.1.2 Determination of the *K* Value in Solution According to Fikentscher

The *K* value according to Fikentscher is determined, as in solution viscometry, by measuring the viscosity of polymer solutions. This value, also known as intrinsic viscosity, is often used in practice in the technical field to determine the molecular weight of polymers such as PVC. Under constant measurement conditions with regard to the solvent, solvent concentration, and temperature, the *K* value only depends on the average molecular weight of the polymers tested. Fikentscher derived the *K* value as a measure of the average molecular weight of a polymeric material from measurements of the relative viscosity of solutions of a polymer. The K value is calculated from this equation:

$$
\begin{aligned}
K &= 1000 \cdot k \\
&= 1000 \cdot \frac{1.5 \log(\eta_{\text{rel}}) - 1 + \sqrt{1 + \left(\frac{2}{c} + 2 + 1.5 \log(\eta_{\text{rel}})\right) \cdot 1.5 \log(\eta_{\text{rel}})}}{150 + 300c}
\end{aligned}
\tag{6.107}
$$

where
c is the concentration in g/100 ml
η_r is the relative viscosity of the solution

The k value calculated from the Fikentscher equation is multiplied by 1000 and given as the K value (= viscosity index).

To determine the k value, a sample portion is dissolved in a solvent and the viscosity of the solution is determined at a defined temperature, for example, in the Höppler falling ball viscometer. The viscosity of the pure solvent is then determined in the same viscometer. The measured viscosities and the concentration of the test sample are used in Equation 6.107 and the k value is calculated. The K value is obtained by multiplying the k value by 1000.

In rigid extrusion, suspension PVC with a very wide K value range of 57 to 68 is used. To improve the processing behavior, thermoplastic emulsion PVC with K values of 70 is used, especially for compact profiles. The favorable melting behavior improves the output rate as well as the surface quality and antistatic properties of the end product. The PVC types used in the production of window profiles have a K value of 65 to 68.

References

[1] Pahl, M.; Gleißle, W.; Laun, H.-M.: *Praktische Rheologie der Kunststoffe und Elastomere*. VDI-Verlag GmbH, Düsseldorf, 1995

[2] Giesekus, H.: *Phenomenological Rheology: An Introduction*. Springer, Berlin · Heidelberg, 1994

[3] Mezger, T. G.: *The Rheology Handbook*. Vincentz Network, Hanover, 2016

[4] Mezger, T. G.: *Applied Rheology: With Joe Flow on the Rheology Road*. Anton Paar GmbH, Graz, 2014

[5] Limper, A.: *Process Engineering in Thermoplastic Extrusion*. Hanser, Munich, 2012

[6] Michaeli, W.: *Extrusionswerkzeuge für Kunststoffe und Kautschuk: Bauarten, Gestaltung und Berechnung*. Hanser, Munich · Vienna, 1991

[7] Wolf, R.: Das unterschiedliche Schwellverhalten eines extrudierten Schmelzestranges bei Variation des Relaxationsspektrums. *Rheologica Acta*, 1983, Vol. 22, p. L

[8] Poole, R. J.: The Deborah and Weissenberg numbers. *Rheology Bulletin*, 2012, 53(2), pp. 32–39

[9] MacDonald, I. F.; Marsh, B. D.; Ashare, E.: Rheological behavior for large amplitude oscillatory motion. *Chemical Engineering Science*, 1969, Vol. 24, 10, pp. 1615–1625

[10] Tee, T.-T.; Dealy, J. M.: Nonlinear Viscoelasticity of Polymer Melts. *Transactions of The Society of Rheology*, 1975, Vol. 19, 4, pp. 595–615

[11] Giacomin, A. J.; Dealy, J. M.: Large-Amplitude Oscillatory Shear, in: Collyer, A. A. (Ed.): *Techniques in Rheological Measurement*. Springer, 1993, pp. 99–121

[12] Kyu, H.; Wilhelm, M.; Klein, C. O.; Cho, K. S.; Nam, J. G.; Ahn, K. H.; Lee, S. J.; Ewoldt, R. H.; McKinley, G. H.: A Review of Nonlinear Oscillatory Shear Tests: Analysis and Application of Large Amplitude Oscillatory Shear (LAOS). *Progress in Polymer Science*, 2011, Vol. 36, 12, pp. 1697–1753

[13] Wilhelm, M.; Reinheimer, P.; Ortseifer, M.: High sensitivity Fourier-transform rheology. *Rheologica Acta*, 1999, Vol. 38, pp. 349–356

[14] Pipkin, A. C.: *Lectures on Viscoelastic Theory*. Springer, New York, 1972

[15] Ewoldt, R. H.; Winter, P.; Maxey, J.; McKinley, G. H.: Large amplitude oscillatory shear of pseudoplastic and elastoviscoplastic materials. *Rheologica Acta*, 2010, Vol. 49, 2, pp. 191–212

[16] Murray, L. R.: Murray Rheology Consulting, January 11, 2018 [Online]. *https://www.murrayrheology.com/single-post/2018/01/11/LAOS* (accessed March 2, 2019)

[17] Metzger, T. G.: *Das Rheologie Handbuch: Für Anwender von Rotation- und Oszillations-Rheometern, Vol. 5*. Vincentz Network, Hanover, 2016, p. 177

[18] Kurt, M. M.: Methoden zur Charakterisierung des nichtlinear viskoelastischen Verhaltens von Polymerschmelzen, Albert-Ludwigs-Universität, Freiburg i. Br., 2007.

[19] Frick, A.; Stern, C.: *Praktische Kunststoffprüfung*. Hanser, Munich, 2010

[20] DIN EN ISO 1133-2: Determination of the melt mass flow rate (MFR) and melt volume flow rate (MVR) of thermoplastics

[21] VDI Guideline: VDI 2546 Capillary rheometry of plastic melts; representation of flow and viscosity curves and of inlet pressure losses. Düsseldorf, VDI Society Materials Engineering, 1977

[22] Bagley, E. B.: End Correction in the Capillary Flow of Polyethylene. *Journal of Applied Physics*, 1957, 28

[23] Giesekus, H.; Lange, G.: Die Bestimmung der wahren Fließkurven nichtnewtonscher Flüssigkeiten und plastischer Stoffe mit der Methode der repräsentativen Viskosität. *Rheologica Acta*, 1977, Vol. 16, 22

[24] Cogswell, F. N.: Converging flow of polymer melts in extrusion dies. *Polymer Engineering & Science*, 1972, Vol. 12, 1

[25] N. N.: Polymer Melt Rheology: A Guide for Industrial Practice. University Microfilms International, Ann Arbor, Michigan, 1992

[26] N. N.: Tensile deformations in molten polymers. *Rheologica Acta*, 1969, Vol. 8, 2

[27] N. N.: Characteristics for the processing of thermoplastics. VDMA Rheology. Hanser, Munich, 1986

[28] Böhme, G.: *Strömungsmechanik nicht-newtonscher Fluide*. Teubner Verlag, Stuttgart, 1981

[29] N. N.: *Fluid Mechanics of Non-Newtonian Fluids*. Vieweg + Teubner, Wiesbaden, 2000

[30] Göttfert, A.: Praktische Dehnrheologie von Polymerschmelzen. *Kunststoffe*, 11/98, p. l.

[31] Ryborz, H.: Influences of multiple processing of polyethylene on melt extension. *Plastverarbeiter*, 1979, 8, pp. 445–449

[32] Schurz, J.; Tomiska, M.: Die Hagenbach-Couette-Korrektur in der Kapillarviskosimetrie. *Monatshefte für Chemie*, 1966, 97, pp. 879–890

[33] N. N.: The Detection of the Intrinsic Viscosity (IV). Company publication of the Göttfert company

[34] N. N.: Visko-Fibel Theorie und Praxis der Viskosimetrie mit Glas-Kapillarrheometern. Company publication of SI Analytics

[35] Braun, D.; Becker, G.; Carlowitz, B.: *Die Kunststoffe: Chemie, Physik, Technologie – Kunststoff-Handbuch 1*. Hanser, Munich, 1990

[36] Dealy, J. M.; Larson, R. G.: *Structure and Rheology of Molten Polymers: From Structure to Flow Behavior and Back Again*. Hanser, Munich, 2006

[37] Kaiser, W.: *Plastics Chemistry for Engineers: From Synthesis to Application*. Hanser, Munich, 2015

[38] DIN EN ISO 1628-1:2012-10: Plastics – Determination of viscosity of polymers in dilute solution by capillary viscometer – Part 1: General principles (ISO 1628-1:2009 + Amd 1:2012)

[39] Reilly, J. F.; Limbach, A. P.: Correlation Melt Rheology of PET to Solution Intrinsic Viscosity. 115 Thousand Oaks Blvd, POB 709 Morgantown, Pa 19543, Kayeness Inc.

7 Viscometry—Influences on the Rheological Material Data

The material data for plastics determined in Chapter 6 depends on other influencing variables. These variables include

- Temperature
- Pressure
- Molecular weight or molecular weight distribution
- Fillers such as glass fibers
- Additives
- Time

This chapter will deal with the influence of these parameters on the material data (Figure 7.1).

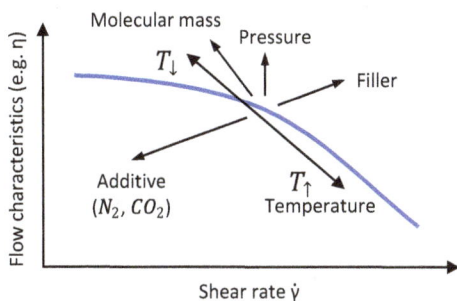

Figure 7.1 Influences on the rheological material data [1]

7.1 Influence of Dissipation

During the manufacturing process, the plastic melt is subjected to shear stress. As already described in Chapter 5, the shearing of the melt leads to a friction process between the individual fluid layers. This frictional process between the molecular chains is known as dissipation. The dissipation results in the heating of the melt.

From the energy equation (where ρ = density, c_p = specific heat capacity, λ = thermal conductivity, v_x = flow velocity in the x direction, τ_{xy} = shear stress in the x direction, and ϑ = temperature)

$$\rho c_p \cdot \left(\frac{\partial \vartheta}{\partial t} + v_x \frac{\partial \vartheta}{\partial x} \right) = \lambda \frac{\partial^2 \vartheta}{\partial y^2} - \tau_{xy} \cdot \frac{\partial v_x}{\partial y} \tag{7.1}$$

the following can be derived in a simplified way by neglecting heat conduction (adiabatic process), $\frac{\partial^2 \vartheta}{\partial y^2}$, and heat transport in the direction of flow, $\frac{\partial \vartheta}{\partial x}$:

$$\rho c_p \cdot \frac{\partial \vartheta}{\partial t} = -\tau_{xy} \cdot \frac{\partial v_x}{\partial y} \tag{7.2}$$

if $\tau_{xy} = \eta \cdot \dot{\gamma}$ and $\frac{\partial v_x}{\partial y} = \dot{\gamma}$, then:

$$\rho c_p \cdot \frac{\partial \vartheta}{\partial t} = -\eta \cdot \dot{\gamma}^2 \Rightarrow \Delta \overline{\vartheta}_{\text{diss}} = \frac{1}{\rho c_p} \eta \cdot \dot{\gamma}^2 \cdot \Delta t$$

This equation makes it clear that the shear rate is very strongly involved in the heat development of the melt. This means that if the shear rate is very high (surface layers), a lot of dissipated energy—and therefore heat—is generated. As viscosity is also a function of temperature, the viscosity will decrease due to the high temperature in the surface layers.

A rough estimate of the change in temperature as a function of pressure can be derived using the first law of thermodynamics.

$$Q_{12} + W_{12} = U_2 - U_1 + E_{a2} - E_{a1} \tag{7.3}$$

It is again assumed that this is an adiabatic process. This means that there is no heat exchange Q with the environment. The same applies to the change in external energy E_A. Thus, it follows (with $W_{12} = \Delta p \cdot V$ and $U_2 - U_1 = m \cdot c_p \cdot \Delta \vartheta$):

$$\Delta p \cdot V = m \cdot c_p \cdot \Delta \vartheta \tag{7.4}$$

The simplified relationship for the temperature increase due to an increase in pressure is then as follows ($m = \rho \cdot V$):

$$\Delta \vartheta \sim \frac{1}{\rho c_p} \cdot \Delta p \tag{7.5}$$

If the constant K is used for the material-dependent quotient $\frac{1}{\rho c_p}$, it is possible to obtain a rough estimate that can be employed to calculate the temperature increase of the plastic melt. For the constant K, a value of 0.05 can be assumed as a good approximation. This covers a wide range of plastic types. The simplified Equation 7.5 then reads:

$$\Delta\vartheta \sim 0.05 \cdot \Delta p \tag{7.6}$$

This means that a pressure increase of 100 bar leads to a temperature increase of approximately 5 °C.

Figure 7.2 shows the result of measurements carried out as part of an injection molding test. The measurements confirm the abovementioned temperature increase due to the increase in pressure.

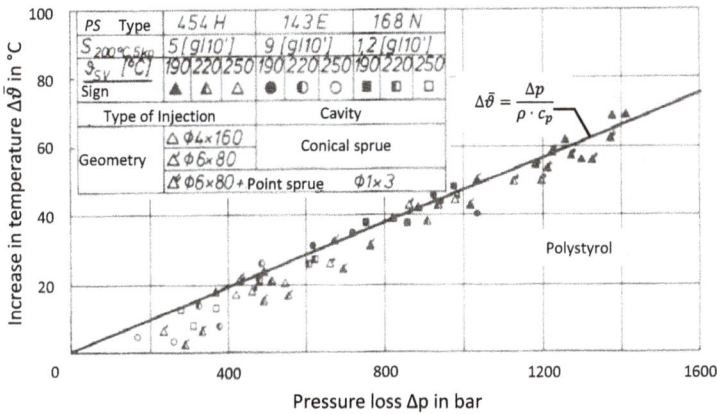

Figure 7.2 Temperature increase as a function of pressure loss for a polystyrene during injection molding [2]

In Table 7.1 some values for the pressure-dependent temperature constant K are listed. The values were calculated using the c_p and ρ values.

Table 7.1 Material Values *K* for Determining the Temperature Increase Due to Dissipated Energy

Plastic	K [°C/bar]
ABS	0.055
PA	0.038
PA with 50 % GF	0.026
PBT	0.044
PBT with 50 % GF	0.038
PE-LD	0.040
PE-HD	0.034
POM	0.025
PP	0.045
PP with 40 % talc	0.033
PS	0.055
PSU	0.047
PSU with 30 % GF	0.041
PESU	0.043
SAN	0.055
SAN with 35 % GF	0.044

7.2 Influence of Temperature on the Flow Curve

As the temperature increases, the toughness, that is, the viscosity of the plastic, decreases. The effects on the flow curves are shown in Figure 7.3. If we want to determine these flow curves for different temperatures using rheometry, a great amount of time is required. This measuring effort can be significantly reduced with the help of the temperature-invariant application of the flow curve.

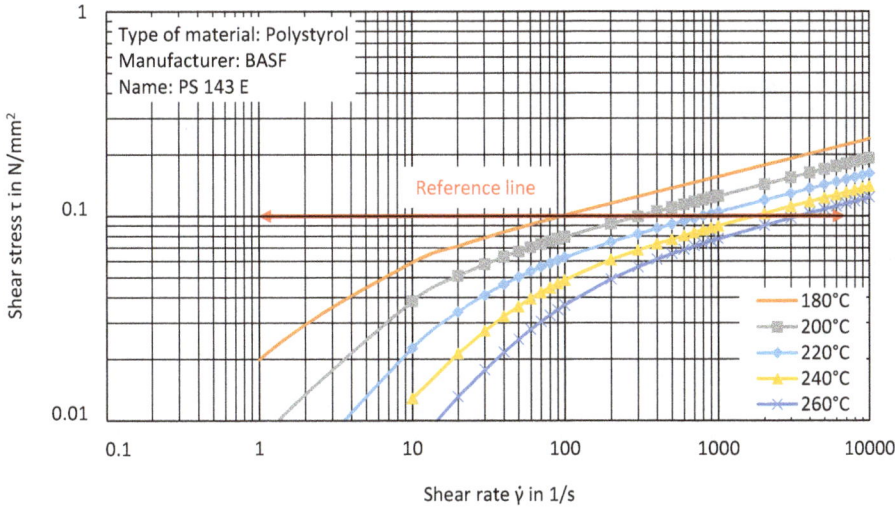

Figure 7.3 Shear stress curve as a function of the shear rate for different temperatures

The following section describes what is meant by a temperature-invariant application of the flow curve. For this purpose, the two graphs in Figure 7.3 and Figure 7.4 show the flow curves $\tau = f(\dot{\gamma})$ and $\eta = f(\dot{\gamma})$, respectively, for polystyrene.

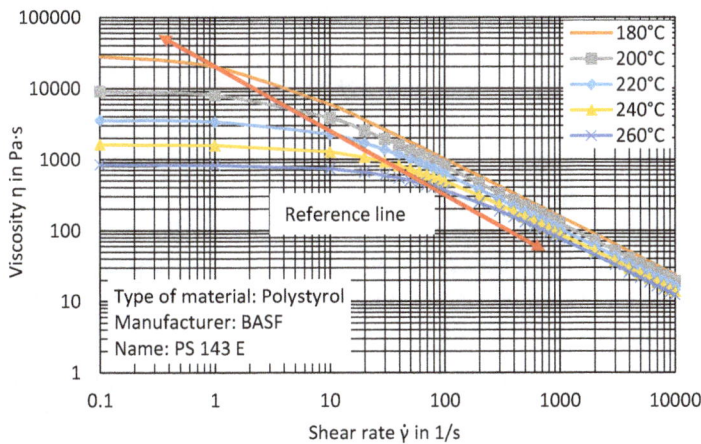

Figure 7.4 Viscosity curve as a function of shear rate for different temperatures

First, we look at the flow curves in the $\tau = f(\dot{\gamma})$ representation (Figure 7.3). If we move these flow curves along a horizontal line, we will see that the curves for the different temperatures are all absolutely congruent. This means that the shape of the curves does not change, and the curve is therefore not a function of the temperature.

Looking at the viscosity curve as a function of the shear rate (Figure 7.4) for different temperatures, it can also be seen that the shape of these curves does not change. The curves are again congruent for different temperatures. For this reason, these curves can also be shifted along a straight line. In this case, the shift line does not run horizontally, but as a diagonal in the logarithmic representation.

The following statement can therefore be made:

> Flow curves can be represented by a single flow curve for different temperatures. This flow curve is referred to as the master curve and refers to a reference temperature.

The temperature for which the master curve applies is freely selected. If possible, it should lie in the middle of the processing temperature range that is suitable for the plastic under investigation. For the polystyrene considered here, a temperature of 220 °C is selected for the master curve.

This means that the shear stress can be plotted as a function of the shear rate in a single curve. Starting from this master curve, which is valid for a specific reference temperature, the curves for the other temperatures can be determined by shifting this curve horizontally. For higher temperatures, the curve is shifted to the right. For lower temperatures, the curve is shifted to the left.

The viscosity curve as a function of the shear rate can also be represented as a master curve. Starting from this master curve, the curves for the other temperatures are shifted by moving the initial curve along the diagonal. For higher temperatures, the curve is shifted downwards and to the right along the straight line. For lower temperatures, the curve must be shifted upwards and to the left.

7.2.1 The Temperature Shift Factor

So far, it has been shown that the flow curves can be shifted along a straight line. What is still missing is the amount by which the curves for the different temperatures can be shifted along this straight line into a master curve.

The temperature dependence of the viscosity can be described with the help of the Boltzmann time–temperature superposition principle. For this purpose, a temperature shift factor a_T is defined.

> The factor for shifting the flow curves into a master curve is called the temperature shift factor a_T.

The temperature shift factor a_T is first calculated from the graphs. For the flow curve $\tau = f(\dot{\gamma})$, it is calculated as follows:

- First of all, a reference temperature $\vartheta_{reference}$ is defined. In our example, this is 220 °C.

- An arbitrary shear stress $\tau_{reference}$ is selected for this curve. For our example, this is 10^{-1} N/mm². The curve points for this shear stress are shown in Figure 7.3.

- The shear rates of the individual curves are determined in relation to this shear stress.

- The temperature shift factor a_T is calculated for each individual curve. As the curves can be shifted along a horizontal straight line, the relationship between the shear rate for the reference temperature and the shear rate for the desired temperature is as follows: $a_T = \frac{\dot{\gamma}_{reference}}{\dot{\gamma}_\vartheta}$

In this equation, $\dot{\gamma}_{reference}$ is the shear rate of the master curve for the selected reference point $\tau_{reference}$. The shear rate $\dot{\gamma}_\vartheta$ is the value of the curve that is to be shifted into the master curve. This is also read for the selected reference shear stress $\tau_{reference}$. Thus, the flow curve for the polystyrene 143 E can be used to determine the temperature shift factor for the respective temperatures (Table 7.2).

Table 7.2 Determining the Temperature Shift Factor a_T for Polystyrene 143 E from Figure 7.3 (Read Values) Reference Values: $\vartheta_{reference} = 220$ °C, $\tau_{reference} = 10^{-1} \frac{N}{mm^2}$

Temperature [°C]	180	200	220	240	260
Shear rate [s⁻¹]	100	400	800	2000	4000
a_T [-]	8	2	1	0,4	0,2

For the master curves themselves, $a_T = 1$ must apply.

The equation $a_T = \frac{\dot{\gamma}_{reference}}{\dot{\gamma}_\vartheta}$ states that:

- The curve shifts to the right for higher temperatures ($a_T < 1$).

- The curve shifts to the left for temperatures lower than the reference temperature ($a_T > 1$).

7.2.2 Temperature-Invariant Application of the Flow Curves (Master Curves)

Due to the congruence of the flow curves, it is sufficient to measure and map only one flow curve. This flow curve is also referred to as the master curve.

The master curve for the shear stress is plotted in a representation of $\tau = f(\dot{\gamma} \cdot a_T)$ (Figure 7.5). The product $\dot{\gamma} \cdot a_T$ is referred to as "normalized shear rate". When determining the master curve for the $\eta = f(\dot{\gamma})$ display, we proceed accordingly. The flow curve is plotted in the form $\frac{\eta}{a_T} = f(\dot{\gamma} \cdot a_T)$ (Figure 7.6). The quotient $\frac{\eta}{a_T}$ is referred to as "normalized viscosity". When displaying the shift of the master curve $\frac{\eta}{a_T} = f(\dot{\gamma} \cdot a_T)$, it should be noted that shifting the master curve along a straight line changes both the viscosity and the shear rate for the determined point.

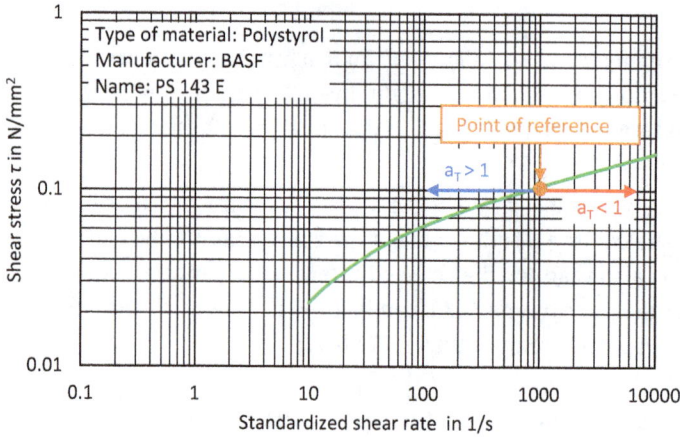

Figure 7.5 Master curve $\tau = f(\dot{\gamma} \cdot a_T)$ application, reference temperature $\vartheta_{\text{reference}} = 220\,°C$

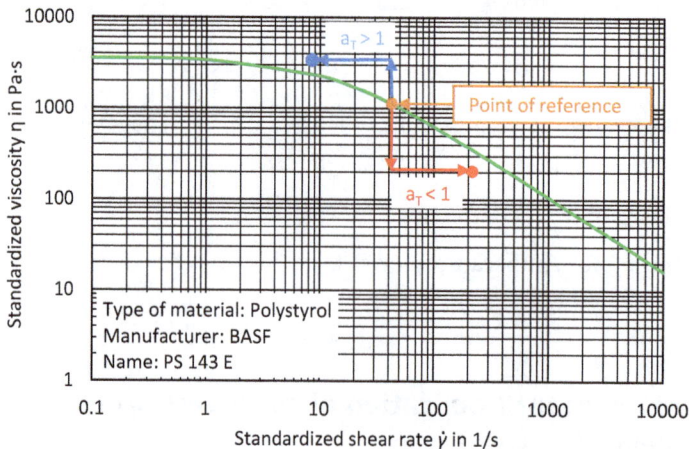

Figure 7.6 Master curve $\eta/a_T = f(\dot{\gamma} \cdot a_T)$ application, reference temperature $\vartheta_{\text{reference}} = 220\,°C$

If the master curve is given, the flow curves for other temperatures can be determined easily using the temperature shift factor a_T. The following applies:

$$\eta_{wanted} = a_T \cdot \eta_{reference} \tag{7.7}$$

and

$$\dot{\gamma}_{wanted} = \frac{1}{a_T} \cdot \dot{\gamma}_{reference}$$

The temperature shift factor a_T can be calculated using simple mathematical approaches (see Section 7.2.3).

Figure 7.7 shows the temperature shift factor as a function of the temperature or the reciprocal temperature. The plot also depends on the reciprocal temperature as this is included in the mathematical description of the temperature shift factor using an Arrhenius function.

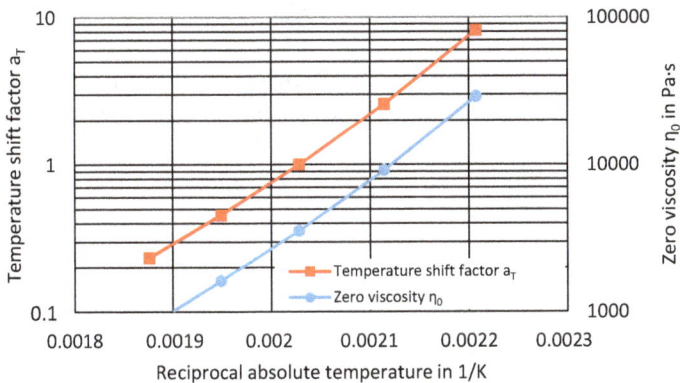

Figure 7.7 Temperature shift factor a_T and zero viscosity η_0 as a function of temperature for a polystyrene 143 E

The graph in Figure 7.7 shows the zero viscosity η_0 also plotted as a function of temperature. If we compare the two curves, we can see that their shape is identical. For this reason, the temperature dependence of the flow curve $\eta = f(\dot{\gamma})$ can also be plotted normalized to zero viscosity, that is, in the normalized form $\frac{\eta}{\eta_0} = f(\dot{\gamma} \cdot \eta_0)$. This procedure is illustrated in Figure 7.8.

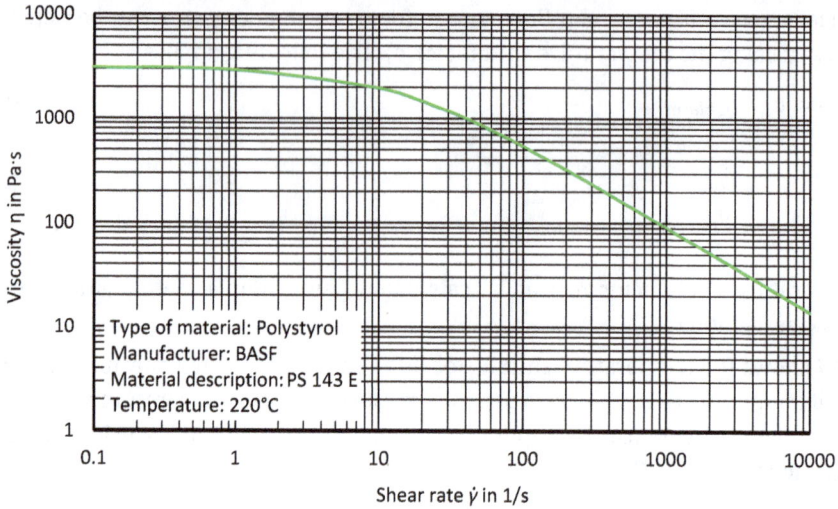

Figure 7.8 Viscosity as a function of shear rate for $T = 220\,°C$

For standardization, in the flow curve from Figure 7.8, several pairs of values are formed. The viscosity read from each pair of values is divided by the zero viscosity of the selected master curve, while the shear rate of each pair of values is multiplied by the same zero viscosity. These calculated pairs of values are shown in Figure 7.9 in normalized form.

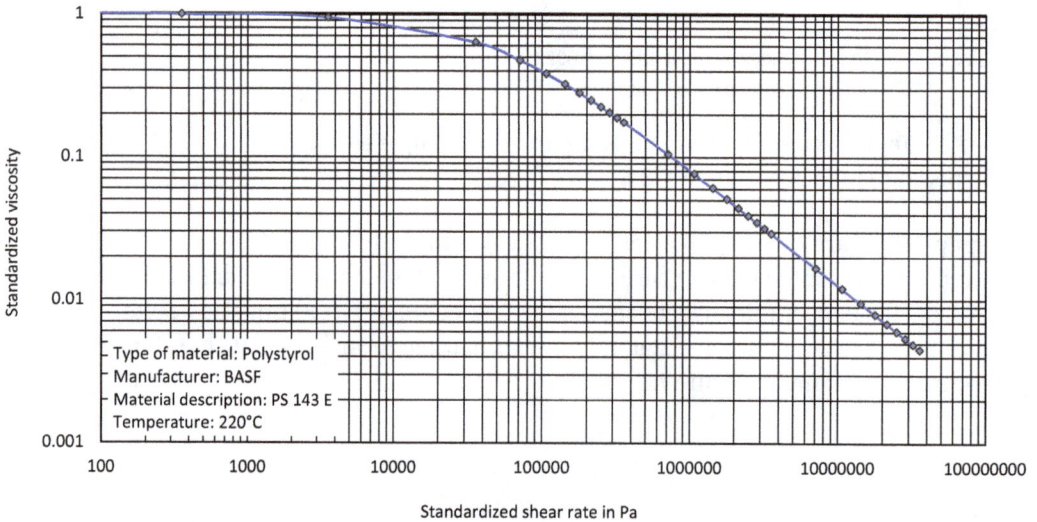

Figure 7.9 Normalized flow curve in the $\frac{\eta}{\eta_0} = f(\dot{\gamma} \cdot \eta_0)$ application for $T = 220\,°C$

Figure 7.9 thus shows a master curve normalized to zero viscosity at 220 °C. Each point of this master curve refers to the zero viscosity of the selected reference curve in Figure 7.8 for the selected temperature (here, for T = 220 °C, $\eta_0(T$ = 220 °C) = 3100 Pa·s). The normalized shear rate results from the product of the shear rate and the zero viscosity (i.e., $\dot\gamma \cdot \eta_0$). The normalized zero viscosity starts at the value 1 due to the normalization $\frac{\eta}{\eta_0}$ to 1.

7.2.2.1 Example of a Viscosity Determination for a Selected Shear Rate and Another Temperature

The following example uses the master curve in Figure 7.9 to determine the viscosity for a temperature of 200 °C and a specified shear rate of $100\,\mathrm{s}^{-1}$. The procedure is explained step by step:

- The specified master curve in Figure 7.9 refers to a temperature of T = 220 °C.

- This master curve can be used to determine any viscosity for any shear rate if the zero viscosities of the other temperatures are known. These measured zero viscosities are shown in Figure 7.7 for the material used here. *We are looking for* the viscosity for a temperature of 200 °C and a shear rate of $100\,\mathrm{s}^{-1}$.

- From the diagram (Figure 7.7) for zero viscosity as a function of temperature, the zero viscosity $\eta_0(T$ = 200 °C) = 9000 Pa·s can be read for T = 200 °C (0.00211 1/K). This value can also be taken from Figure 7.4.

- The normalized shear rate is now calculated for this temperature and a shear rate of $100\,\mathrm{s}^{-1}$:

$$\dot\gamma_{\mathrm{wanted}} = \dot\gamma \cdot \eta_0\,(T_{\mathrm{reference}}) = 100\ \mathrm{s}^{-1} \cdot 9000\ \mathrm{Pa}\cdot\mathrm{s} = 900,000\ \mathrm{Pa}$$

- This value is used to calculate the normalized shear rate of the master curve (Figure 7.9) on the x axis. Then, for this value, on the y axis a value is read for the viscosity normalized to the zero viscosity from the master curve. In our example, the value is 0.09.

- This value is then multiplied by the zero viscosity for the desired temperature, in this case, 200 °C. The zero viscosity for 200 °C is $\eta_0(T$ = 200 °C) = 9000 Pa·s.

- The following applies: $\eta_{\mathrm{wanted}} = \dfrac{\eta_{\mathrm{reference}}}{\eta_0\,(T_{\mathrm{reference}})} \cdot \eta_0\,(T_{\mathrm{wanted}})$

- Therefore: $\eta_{\mathrm{wanted}} = \dfrac{\eta_{\mathrm{reference}}}{\eta_0\,(T_{\mathrm{reference}})} \cdot \eta_0\,(T_{\mathrm{wanted}}) = 0.09 \cdot 9000\ \mathrm{Pa}\cdot\mathrm{s} = 810\ \mathrm{Pa}\cdot\mathrm{s}$

- If we check this value using Figure 7.4, we will find that it is correct.

This means that we only need a flow curve standardized to zero viscosity for a defined temperature and can then use the zero viscosities for other temperatures to determine all viscosities for this temperature and any shear rates. The temperature-dependent zero viscosities can be determined very easily, for example, using a rotational viscometer.

7.2.2.2 *Task*: Determination of the Viscosity for a Given Shear Rate Using a Master Curve

The flow curves for a polymer are given for the temperatures 200 °C, 220 °C, and 240 °C in the representation of shear stress as a function of shear rate (Figure 7.10). A master curve for this polymer at a temperature of 220 °C is also given (Figure 7.11). *We are looking for* the viscosity for a temperature of 240 °C and a shear rate of 400 s^{-1} using the master curve.

▪ First, determine the temperature shift factor a_T from the representation of the shear stress as a function of the shear rate.

▪ In the next step, carry out this exercise for a temperature of 200 °C and a shear rate of 100 s^{-1}. The value can be checked using Figure 7.10. Read off the shear stress for the required temperature and divide it by the specified shear rate. Pay attention to the units!

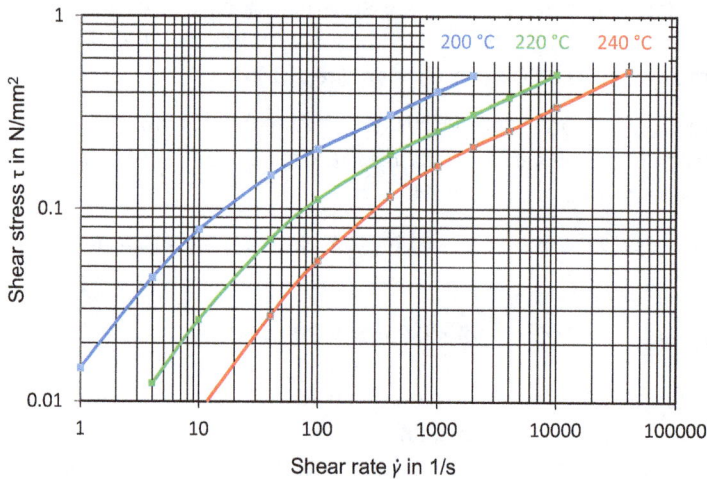

Figure 7.10 Wall shear stress as a function of shear rate

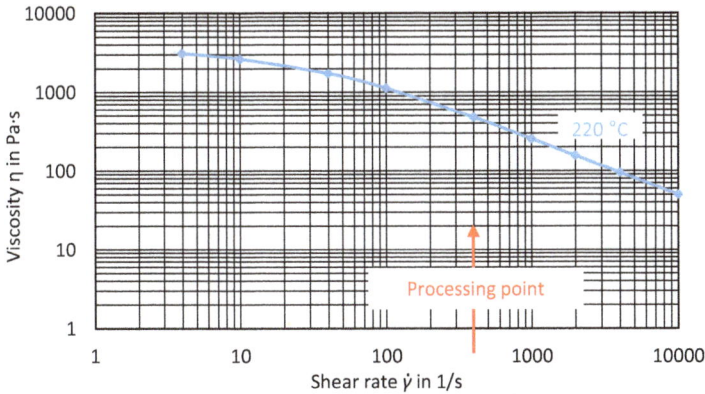

Figure 7.11 Viscosity as a function of shear rate

7.2.2.3 *Task*: Exercise on Temperature Shift Using Zero Viscosity

The three flow curves for different temperatures (200 °C, 220 °C, and 240 °C) are *given* (Figure 7.12).

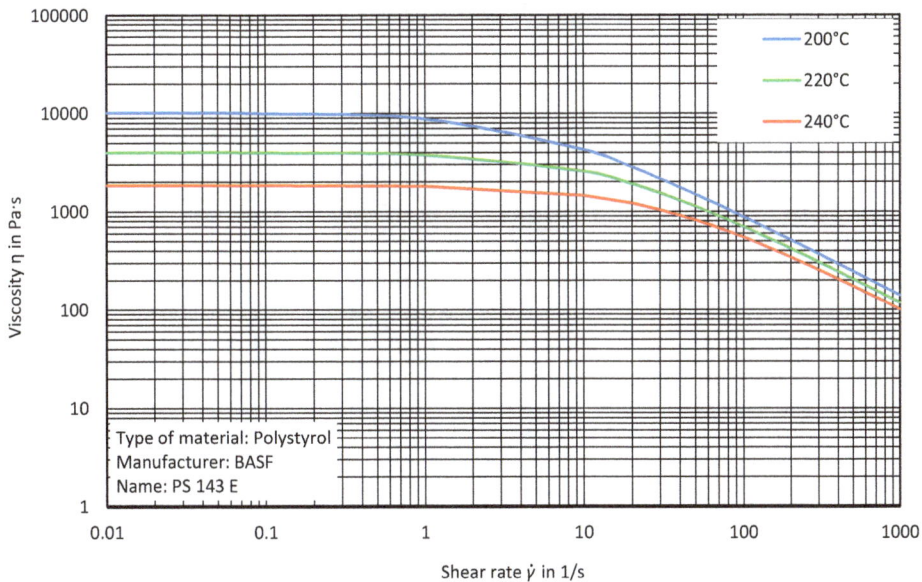

Figure 7.12 Viscosity curve as a function of the shear rate for different temperatures

- Read off the zero viscosities for all temperatures.

- Now read six pairs of values (viscosity and shear rate) from the flow curve for 220 °C. Remember that the complete flow curve should be mapped.

- The viscosities read off are then divided by the zero viscosity for 220 °C and the shear rates are multiplied by the zero viscosity at this temperature.

- Draw the normalized curve in the $\frac{\eta}{\eta_0}$ representation in a double-logarithmic plot.

- Now determine the viscosity for a temperature of 240 °C and a shear rate of 1000 s^{-1} from the master curve.

- The following applies:

$$\dot{\gamma}_{\text{wanted}} \cdot \eta_0 \left(T_{\text{wanted}}\right) = \dot{\gamma} \cdot \eta_0 \left(T_{\text{reference}}\right) \quad \text{and} \quad \eta_{\text{wanted}} = \frac{\eta_0 \left(T_{\text{wanted}}\right)}{\eta_0 \left(T_{\text{reference}}\right)} \cdot \eta_{\text{reference}}$$

7.2.3 Mathematical Description of the Temperature Shift Factor

The logarithmic plot of the temperature shift factor a_T as a function of temperature shows a linear function. For this reason, the temperature shift factor or the zero viscosity can be described as a function of temperature using relatively simple mathematical relationships. For example, an Arrhenius equation can be used for this purpose!

This relationship applies to the straight line of zero viscosity η_0:

$$\eta_0 \left(T\right) = \eta_0 \left(T_0\right) \cdot e^{\left[\frac{E_0}{R}\left(\frac{1}{T} - \frac{1}{T_0}\right)\right]} \tag{7.8}$$

The only material-specific parameter in this equation is the flow activation energy E_0 [kJ/mol]. It is material-specific, does not depend on the molecular weight, and is independent of the shear stress for thermorheologically simple polymers. E_0 is between 25 kJ/mol and 80 kJ/mol for polymer melts. It can be determined from the slope of the straight line in the Arrhenius diagram.

R is the universal gas constant, with $R = 8.314 \cdot 10^{-3} \frac{\text{kJ}}{\text{mol} \cdot \text{K}}$. All temperatures are absolute temperatures [K]. The flow activation energy is a measure of how strongly the viscosity reacts to temperature changes. T_0 is the reference temperature.

7.2.3.1 Arrhenius Function

The ratio of the zero viscosities describes the temperature shift factor a_T. The following applies: $a_T \equiv \frac{\eta_0(T)}{\eta_0(T_0)}$. This means that the temperature shift factor can also be described with an Arrhenius function. For semicrystalline plastics such as PE, PP, PA, PETP, and PEK, this equation reads:

$$a_T = e^{\left[\frac{E_0}{R}\left(\frac{1}{T} - \frac{1}{T_0}\right)\right]} \tag{7.9}$$

where
T is the current temperature in kelvin ($T \cong \vartheta + 273$)
T_0 is the selectable reference temperature in kelvin
E_0 is the flow activation energy
R is the universal gas constant

If Equation 7.9 is plotted logarithmically against the temperature, the result is the straight line already described. For amorphous plastics whose molten state begins above the glass temperature T_G, the linear relationship does not apply because the flow activation energy depends on the temperature.

It can be seen from Figure 7.13 that semicrystalline plastics spontaneously melt to a relatively low viscosity as soon as the crystallite melting temperature is exceeded, whereas the curve for amorphous plastics only behaves similarly to semicrystalline plastics well above the glass transition temperature.

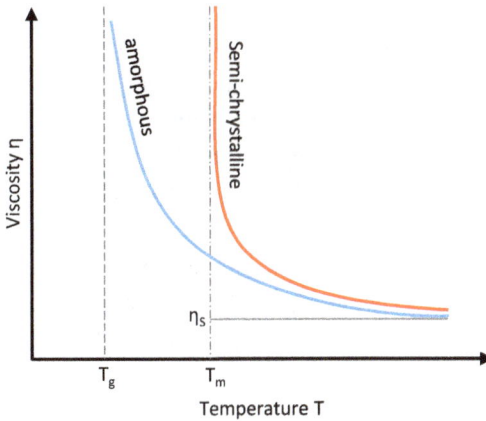

Figure 7.13
Temperature dependence of viscosity [1]

7.2.3.2 Equations of Williams, Landel, and Ferry (WLF Approach)

The equations of Williams, Landel, and Ferry, also known as the WLF approach, apply to amorphous and semicrystalline plastics. The WLF approach is used for this group of plastics as follows:

$$\log{(a_T)} = \log{\frac{\eta\,(T)}{\eta\,(T_0)}} = \frac{-8.86 \cdot (T - T_s)}{101.6\,\text{K} + (T - T_s)} \tag{7.10}$$

In this approach, T_s describes the so-called standard temperature. This temperature depends on the material and is approximately 50 °C above the softening temperature (glass temperature T_G; see Table 7.3) at $p = 1$ bar.

Table 7.3 Glass Temperatures of Different Plastics

Plastic	ϑ_G [°C]
PE	−40
PP	−40
PS	97
PMMA	110
PPO	−75

The WLF relationship is based on the assumption that the segment mobility of polymer chains in the vicinity of the glass temperature is primarily determined by the free volume, which increases approximately linearly with the distance to the glass temperature. At high temperatures far above the glass temperature T_G ($T > T_G + 190\,°C$), the curve of the WLF relationship approaches that of an Arrhenius relationship.

If one compares the values for the temperature shift factor a_T determined using the Arrhenius approach with those of the equation of Williams, Landel, and Ferry for a semicrystalline plastic, the values agree quite well, as Figure 7.14 shows for a polypropylene (PP). It is also clear that the Arrhenius equation depicts a straight line in this representation, while the WLF function has a curvature.

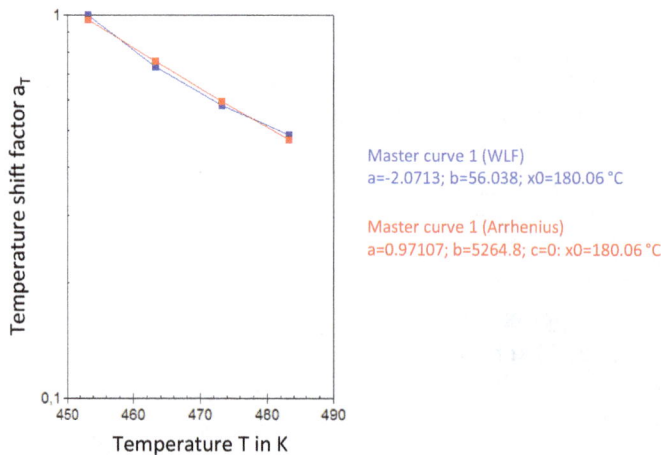

Master curve 1 (WLF)
a=−2.0713; b=56.038; x0=180.06 °C

Master curve 1 (Arrhenius)
a=0.97107; b=5264.8; c=0: x0=180.06 °C

Figure 7.14 Temperature shift factor a_T for a semicrystalline plastic (PP)

Figure 7.15 shows that the influence of temperature on viscosity is much more pronounced for amorphous plastics (e.g., PMMA) (the curve is steeper) than for semicrystalline plastics (e.g., PE).

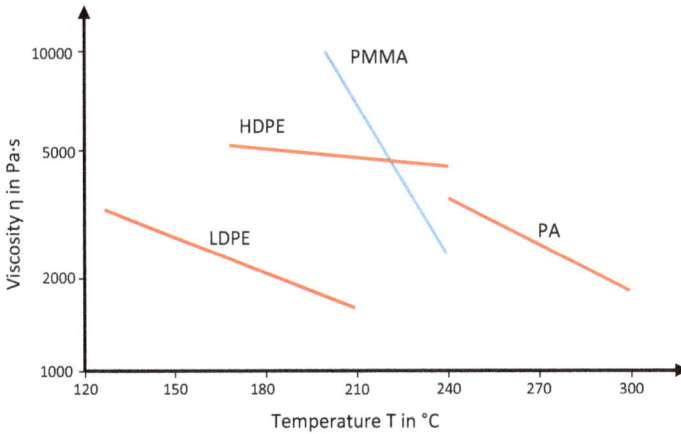

Figure 7.15 Temperature dependence of viscosity for amorphous and semicrystalline plastics

As a result, the temperature shift factor a_T in amorphous plastics also changes more markedly as a function of temperature than in semicrystalline plastics, as Figure 7.16 shows. It is also clear that this relationship for amorphous plastics is not linear, but displays a curvature. In contrast, the relationship of the temperature shift factor for semicrystalline plastics with the description according to an Arrhenius function is linear.

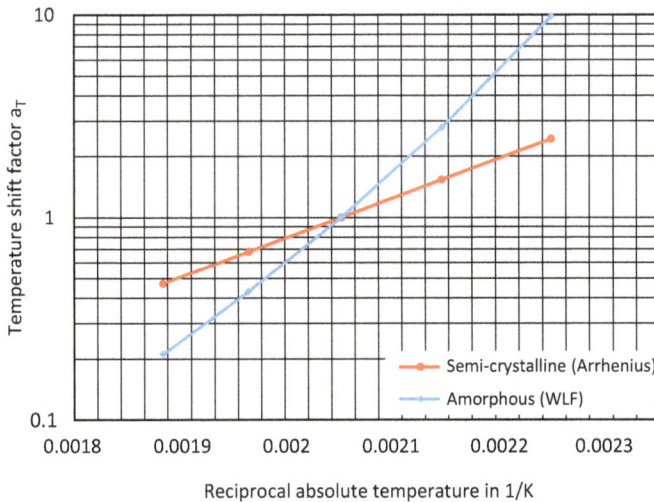

Figure 7.16 Temperature shift factor for an amorphous (PMMA) and a semicrystalline (PE) plastic as a function of the reciprocal absolute temperature

It is also possible to determine the temperature shift factor between two temperatures. The following applies to the shift from temperature 1 to temperature 2:

$$\eta\,(T_2) = a_{T2} \cdot \eta\,(T_s) \quad \text{and} \quad \eta\,(T_1) = a_{T1} \cdot \eta\,(T_s) \quad \rightarrow$$
$$\eta\,(T_2) = \frac{a_{T2}}{a_{T1}} \cdot \eta\,(T_1) = a_{T2,1} \cdot \eta\,(T_1) \tag{7.11}$$

$$\log\,(a_{T2,1}) = \log\,(a_{T2}) - \log\,(a_{T1}) \quad \rightarrow$$
$$\log\,(a_{T2,1}) = \frac{-8.86(T_2 - T_s)}{101.6\,\text{K} + (T_2 - T_s)} + \frac{8.86(T_1 - T_s)}{101.6\,\text{K} + (T_1 - T_s)}$$

If the reference temperature of the master curve is used for T_1, the WLF approach is as follows:

$$\log\,(a_T) = \frac{8.86\,(T_{\text{reference}} - T_s)}{101.6\,\text{K} + (T_{\text{reference}} - T_s)} - \frac{8.86\,(T - T_s)}{101.6\,\text{K} + (T - T_s)} \tag{7.12}$$

where
$T_{\text{reference}}$ is the reference temperature of the master curve
T_s is the standard temperature
T is the current temperature

With this approach, the flow curve for the required temperature can be determined for a given master curve (reference temperature).

Because of $\log\,(a_T) = \log \frac{\eta(T)}{\eta(T_0)} = \log\,(\eta\,(T)) - \log\,(\eta\,(T_s))$, $\log\,(a_T)$ represents the distance by which the viscosity curve $\eta(T_s)$ is shifted along a straight line in a double-logarithmic plot.

7.2.4 *Task*: Calculation of the Temperature Shift Factor

In the following, the temperature shift factor for a PP (Novolen 1100 H) will be calculated for the temperatures 190 °C, 235 °C, and 280 °C.

- *Calculate* the temperature shift factor for these temperatures first using the Arrhenius approach and then using the WLF approach.

- Then, enter the calculated values in a diagram (such as the one in Figure 7.17).

The values for calculating the temperature shift factors are indicated in Table 7.4.

Table 7.4 Material Data for a PP (Type Novolen 1100 H)

Arrhenius	$E_0 = 41.77$ kJ/mol	$R = 8.31 \cdot 10^{-3}$ kJ/(mol K)	$\vartheta_{\text{reference}} = 212.5$ °C
WLF	$\vartheta_S = +10$ °C	$\vartheta_G = -40$ °C	$\vartheta_{\text{reference}} = 212.5$ °C

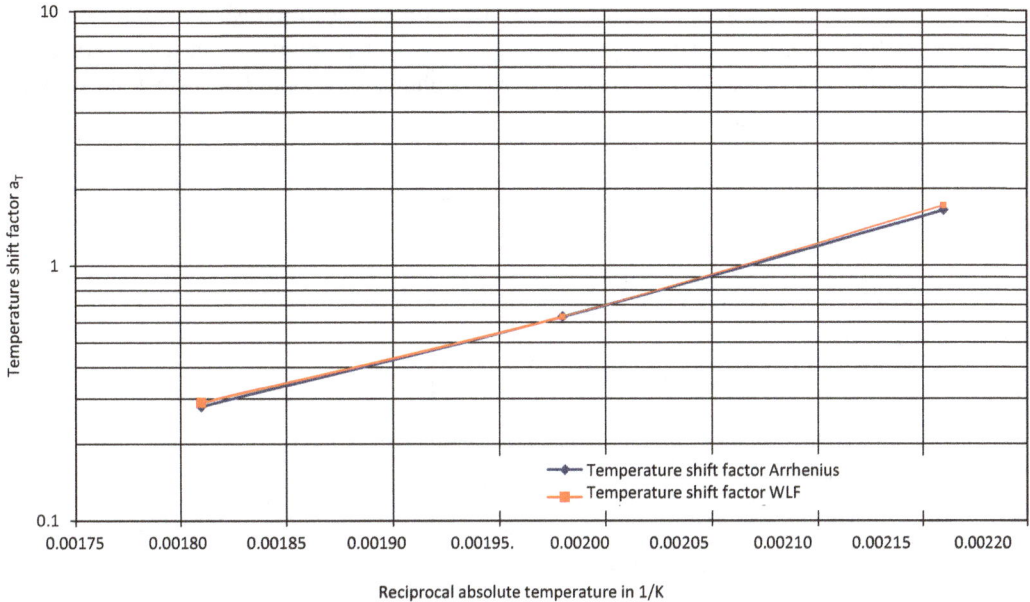

Figure 7.17 Calculated temperature shift factor a_T as a function of temperature

Note: The reciprocal absolute temperature is plotted on the *x* axis in Figure 7.17, as this value is used in the Arrhenius approach.

7.3 Thermorheological Variables

The flow curves cannot always be easily shifted into one another and thus made to coincide. In the following, such deviations from the thermorheologically simple relationships discussed so far will be addressed. In the case of standard plastics, three main effects lead to thermorheologically complex behavior:

- Presence of crystalline structure in the melt
- Change in the morphological structure due to the effect of heat
- Influence of additives or reinforcing fibers

7.3.1 Changes in the Morphological Structure Due to Heat

The first and the second effects in plastics (see list above) include the case of PVC, whose viscosity decreases with increasing dwell time at elevated temperatures (Figure 7.18). The same applies to PET, which is used for the production of preforms, for example. PVC melt also contains a certain number of unmelted crystallites because PVC is processed in a temperature range close to the solidification temperature for reasons of thermal stability (PVC is an amorphous plastic but has a small proportion of crystalline structure).

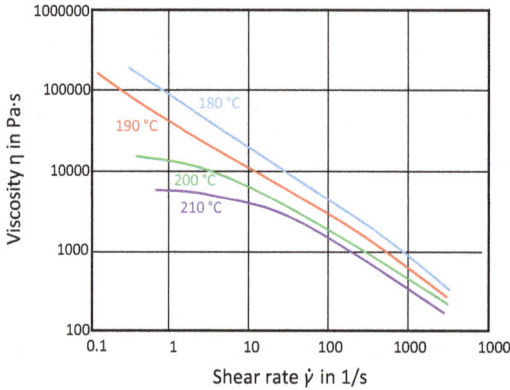

Figure 7.18
Temperature dependence of the viscosity function of a PVC melt

Since crystallites hinder the flow processes, the viscosity behavior is similar to that of a filled melt with an apparent yield point. Starting from a yield point at low temperatures, the flow behavior changes with increasing temperature to that of a homogeneous melt with quasi-Newtonian behavior, that is, pronounced zero viscosity.

In the case of thermally sensitive plastics such as PET, chain degradation can occur: the polymer chain is thermally degraded. As short chains are more flexible than long chains, the viscosity decreases with increasing residence time, especially at high temperatures. The flowability of the plastic melt increases and the necessary filling pressure decreases. From the injection molder's point of view, the filling process has improved (the necessary filling pressure has decreased). However, the mechanical properties of the end product, for example, are adversely affected. Molecular chain degradation results in undesirable cleavage products and reduces the burst pressure properties of the end product (bottle). For this reason, when processing thermally sensitive plastics, it is always important to pay attention to the dwell time in the screw and in the hot runner systems.

Figure 7.19 shows the measurement of zero viscosity as a function of time using rotation for a PET in contrast to a PP. It is clear that the viscosity of the PET decreases with increasing measurement time, that is, the chain length decreases due to the thermal load. In the case of PP, there is almost no change in viscosity as a function of time.

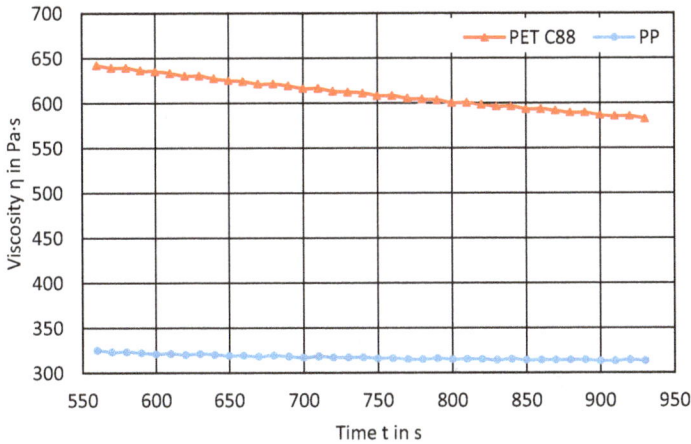

Figure 7.19 Zero viscosity as a function of dwell time for a PET and a PP

7.3.2 Fillers

The addition of fillers is not uncommon in plastics processing. For example, the mechanical properties (such as the modulus of elasticity) of plastics can be significantly improved by adding glass fibers. The same applies to thermal stability. Furthermore, the shrinkage behavior of plastics can be influenced by the addition of fillers.

These fillers have a major influence on the flow behavior. Glass fibers, in particular, must be mentioned here. Figure 7.20 and Figure 7.21 show the viscosity curves as a function of the shear rate for a polyethersulfone (PES). The viscosity curve in Figure 7.20 corresponds to the material without glass fibers. The viscosity curve in Figure 7.21 shows the same polymer type with 20 % glass fiber content.

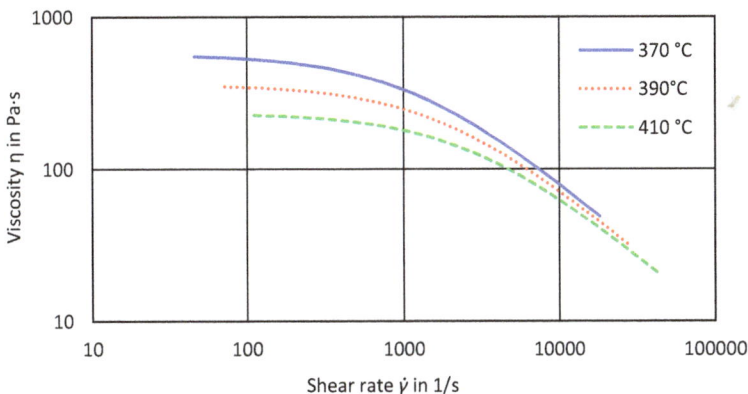

Figure 7.20 Viscosity curve as a function of shear rate for a PES without glass fibers

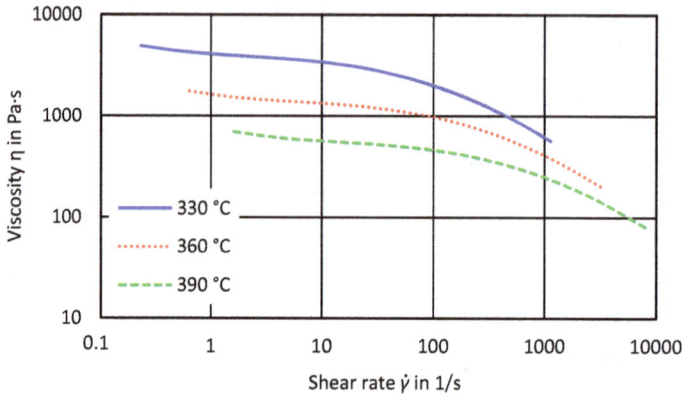

Figure 7.21 Viscosity curve as a function of shear rate for a PES with 20 % glass fiber content

The behavior of the flow curves of the filled melt differs from that of the unfilled melt both at low shear rates and at high shear rates. The viscosity of the unfilled PES is almost constant at low shear rates. This range is referred to as the Newtonian range and the value on the y axis as zero viscosity. In contrast, the viscosity of the filled plastic melt increases at low shear rates.

Figure 7.22 and Figure 7.23 show the viscosity curves for a polyamide (PA) without fillers and for the same PA with 35 % glass fiber content, respectively. It is clear that the flow curves differ both at low shear rates and at high shear rates, that is, in the shear-thinning viscosity range.

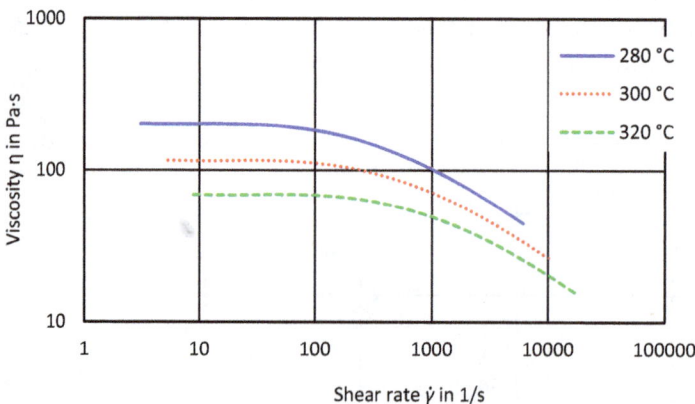

Figure 7.22 Viscosity curve as a function of shear rate for a PA 6.6 without glass fibers

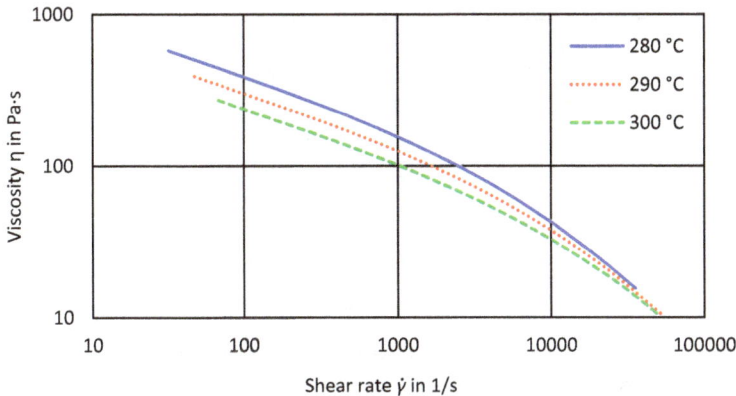

Figure 7.23 Viscosity curve as a function of shear rate for a PA 6.6 with 35 % glass fiber content

With increasing glass fiber content, the special behavior of the filled melt becomes more pronounced in the zero-viscosity range. Initially, as in the case of the Bingham material, a yield point must be overcome. Once the flow process has been initiated, the glass fibers are increasingly aligned in the direction of flow as the shear rate (shear stress) increases and the flow resistance (viscosity) decreases.

In the range of high shear rates, the filled plastics exhibit a pronounced shear-thinning viscous flow behavior. This means that the viscosity drops sharply in the high shear rate range depending on the filler content.

The mixture of the plastic with the filler (glass fiber, for example) is a solid–fluid mixture. This mixture is referred to as a suspension, as solid particles are dispersed in a fluid. The flowability of suspensions is influenced by the flow function of the carrier or matrix fluid as well as the properties and concentration of the suspended particles.

The flow processes in a particle–fluid system are illustrated in Figure 7.24. A pure fluid without additional particles with the viscosity η_M is sheared between two parallel plates with a distance S. Since the upper plate is moving and the lower plate is stationary, this is a pure drag flow with a linear velocity profile. The shear rate in the fluid between the plates results from the velocity gradient in Figure 7.24 (left). To simulate a suspension, z stiff plates of thickness h were distributed in the fluid. Their volume fraction is X. As the solids do not experience any change in velocity in this drag flow and only flow along with it, the gap width for the pure fluid is reduced by the value $z \cdot h$. This increases the actual, effective shear rate between the plates of the model suspension, which is shown in the right-hand diagram in Figure 7.24. In the case of shear-thinning viscous media, such as plastic melts, the increasing shear rate results in a decrease in viscosity [3].

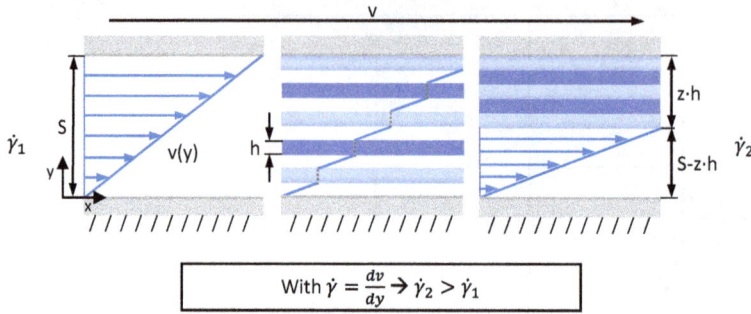

$$\text{With } \dot{\gamma} = \frac{dv}{dy} \rightarrow \dot{\gamma}_2 > \dot{\gamma}_1$$

Figure 7.24 Influence of rigid particles on the velocity profile between two parallel plates (the upper plate is moved with velocity *v*)

A simple, proven method according to Geisbüsch [2, 4] was developed to quickly estimate the viscosity of filled plastic melts from the viscosity of the unfilled melt, even without measurements. It is based on a simple model (see Figure 7.25). According to this, the shear rate in a filled melt is effectively a factor *k* higher than in the unfilled melt, as the filler particles are only passively entrained by the flow but are not sheared themselves [1] (Figure 7.26).

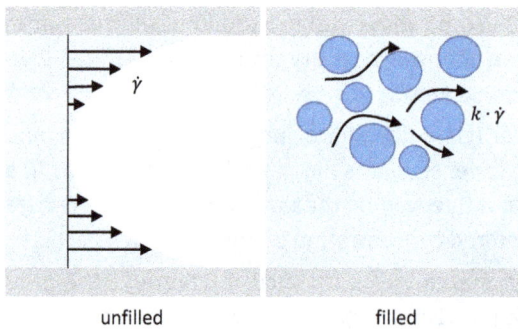

unfilled   filled

Figure 7.25
Shear rate increase in filled thermoplastics

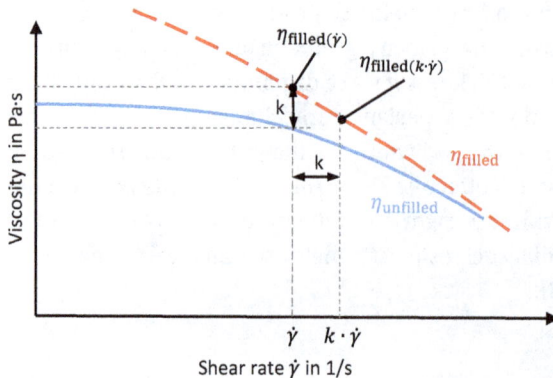

Figure 7.26 Calculation of the viscosity of a filled melt from the unfilled melt [1]

This gives the viscosity of the filled plastic melt $\eta_{\text{filled}} = f(\dot{\gamma})$, which is calculated from the geometry of the channel and the volume flow (see Figure 7.26):

$$\eta_{\text{filled}}(\dot{\gamma}) = \frac{\tau_0}{\dot{\gamma}} + k \cdot \eta_{\text{unfilled}}(k \cdot \dot{\gamma}) \tag{7.13}$$

Or, for large $\dot{\gamma}$:

$$\eta_{\text{filled}}(\dot{\gamma}) \approx k \cdot \eta_{\text{unfilled}}(k \cdot \dot{\gamma}) \tag{7.14}$$

Here, τ_0 is the yield point, which has a big impact at low shear rates but a negligible influence at real shear rates occurring in machines. In a semilogarithmic plot, the factor k is an almost linear function of the filler volume fraction, as shown in Figure 7.27 [2].

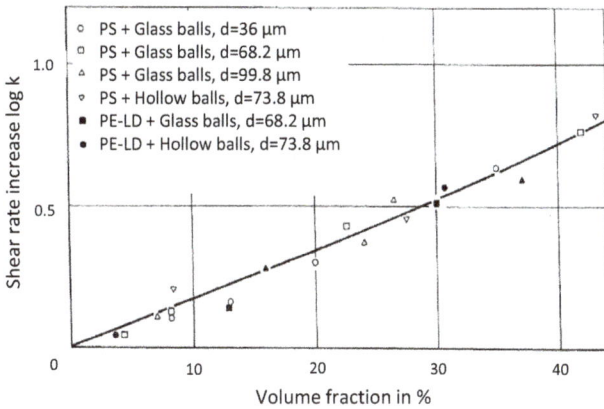

Figure 7.27 Shear rate increase as a function of filler content [1]

For small volume fractions (less than 10 % by volume), the Einstein–Gold relationship is used for relatively precise calculations, with the filler volume fraction c:

$$\frac{\eta_{\text{filled}}}{\eta_{\text{unfilled}}} = 1 + c^2 \tag{7.15}$$

7.3.3 Influence of Pressure

If a fluid is under pressure, it is compressed. The distance between the atoms or molecules decreases and the density increases. The reduced free volume between the molecular chains leads to increased interaction, which increases the viscosity. When a plastic melt flows, a pressure gradient forms in the direction of flow. The pressure is maximum at the start of the flow path and minimum at the flow front. As a result, the viscosity of the melt also decreases in the direction of flow.

If the shear stress is plotted as a function of the shear rate for different pressures, it can be seen that the curves are not congruent (Figure 7.28). The reason for this is that the pressure shift factor is also a function of the shear rate.

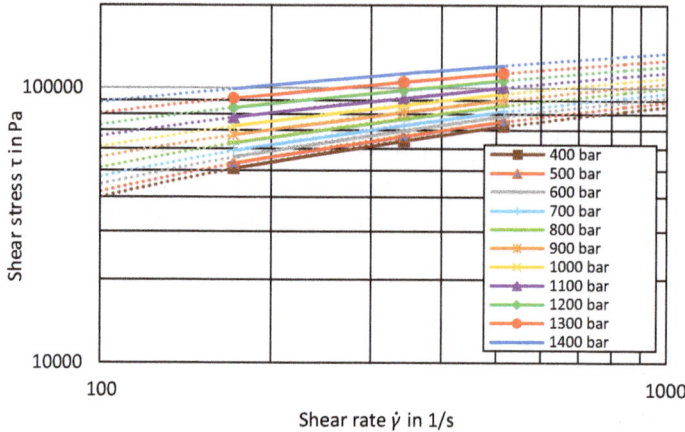

Figure 7.28　Shear stress as a function of shear rate for different pressures

As the curves are not congruent, a pressure shift factor a_p must be determined for each pressure and also for each shear rate. The equation for this can be as follows:

$$a_p\left(\Delta p\right) = \frac{\dot{\gamma}\left(\Delta p_{\text{reference}} = 0\right)}{\dot{\gamma}\left(\Delta p\right)} \tag{7.16}$$

where
$\dot{\gamma}\left(\Delta p_{\text{reference}} = 0\right)$ is the $\dot{\gamma}$ value of the reference curve for ambient pressure for a freely selected shear stress τ
$\dot{\gamma}\left(\Delta p\right)$ is the $\dot{\gamma}$ value of the curve for additional pressure

Mathematically, in many cases the pressure dependence can be described approximately using the Arrhenius relationship:

$$a_p\left(\Delta p\right) = e^{\left(\alpha_p \cdot \Delta p\right)} \tag{7.17}$$

where
α_p is the pressure coefficient
Δp is the pressure difference measured against atmospheric pressure ($\Delta p = p - p_0$)
The reference pressure is usually $p_0 = 1\,\text{bar}$

How high the influence of pressure on the viscosity can be is shown by measurements for a PMMA 6H. Between 0 bar and 500 bar, the viscosity increases by a factor of around 3. The pressure coefficient for this type of material is $\alpha_p = 2.34 \cdot 10^{-3}\,\frac{1}{\text{bar}}$. Furthermore, a relationship analogous to the Williams–Landel–Ferry equation can be used to determine the pressure shift factor. This is

$$a_p = e^{\left[\frac{c_1 \cdot (p - p_0)}{c_2 + (p - p_0)}\right]}$$ (7.18)

where
c_1 and c_2 are material constants

The effect of an increase in pressure can be compensated by an increase in temperature. For Plexiglas® 6N, the temperature must be raised by about 18 °C to compensate for the pressure effect of 500 bar. The following are approximate values for the influence of pressure on viscosity:

- The pressure coefficient for PE-LD is 3 to 4 times lower than that for PS

- For PE-HD, it is lower by a factor of 2

This means that the viscosity of PE-HD at a pressure of 200 bar increases by a negligible 3 % to 5 % compared to that for the pressure $p = 0$ bar.

As a rule, especially with amorphous plastics, the influence of pressure is not negligible. The decisive factor here is the molecular structure of the polymer.

Figure 7.29 shows the pressure shift factor a_p for a polystyrene (PS 158 K) between 500 bar and 1000 bar measured with an inline rheometer nozzle and subsequent flow spiral as a function of the shear rate. It becomes clear that the pressure shift factor not only depends on the pressure itself but also on the shear rate. As the shear rate increases, the dependence of the viscosity on the pressure decreases. This means that at very high shear rates, the influence of pressure on viscosity may be negligible.

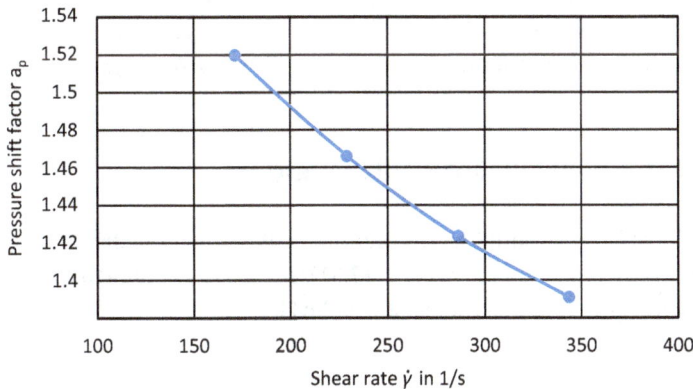

Figure 7.29 Pressure shift factor a_p between 500 bar and 1000 bar as a function of shear rate

Figure 7.30 shows the pressure-dependent viscosity as a function of the shear rate for PS 158 K. If the viscosities are compared at a shear rate of $400\,s^{-1}$, it can be seen that an increase in pressure from 500 bar to 1100 bar leads to an increase in viscosity of 50 %, that is, from 200 Pa · s to 300 Pa · s.

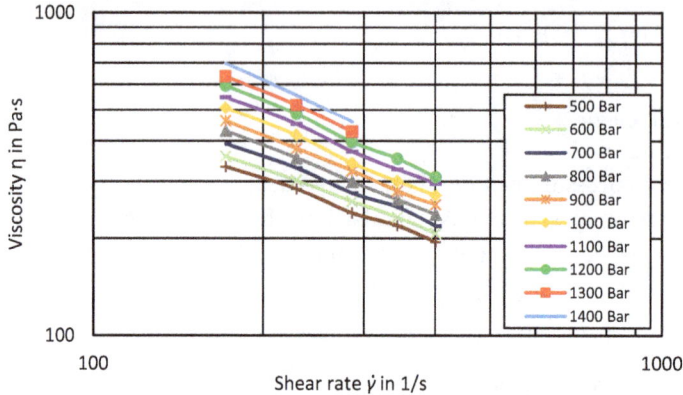

Figure 7.30 Viscosity as a function of shear rate for different pressures

When calculating filling processes in injection molds with simulation programs (Cadmould, Moldex, Moldflow, Sigmasoft, etc.), it is possible to use the pressure coefficient D_3 in the Cross–WLF approach to take into account the influence of pressure on viscosity. However, it quickly becomes apparent that this parameter is not included in most material data sheets of the simulation programs, as it is not trivial to determine. As a rule, the pressure dependence of viscosity is currently neglected in most calculation programs in order to keep the calculation effort low.

Another way to consider the influence of pressure on viscosity is to describe the standard temperature as a function of pressure in the WLF approach. The standard temperature results from the glass temperature ($T_S = T_G + 50\,\text{K}$), which in turn is pressure-dependent, as can be quickly seen from the pvT diagram. With increasing pressure, the glass temperature shifts to higher values (see Chapter 6, Figure 6.44).

7.3.4 Influence of the Average Molecular Weight

Basically, polymers have different chain lengths and molecular weight distributions; branching within the chains can also occur (Figure 7.31). All these phenomena have an effect on the rheological properties of the plastic melts [1, 5–11].

Average molecular mass

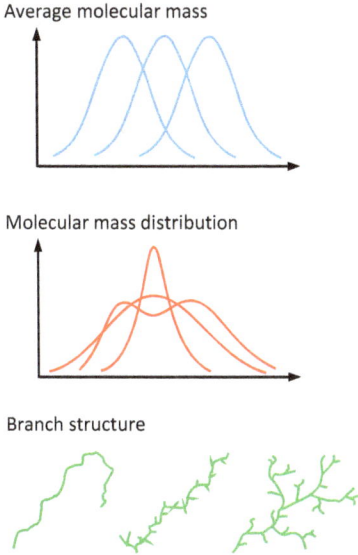

Molecular mass distribution

Branch structure

Figure 7.31
Molecular sizes and structures of polymers [1]

The properties of plastics and plastic melts are essentially defined by the average molecular weight $\overline{M}$ and the molecular weight distribution. With increasing chain length, the strength properties such as impact strength at low temperatures and dimensional stability under heat, among others, are improved. At the same time, the flow behavior of the plastic deteriorates, as shown in Figure 7.32.

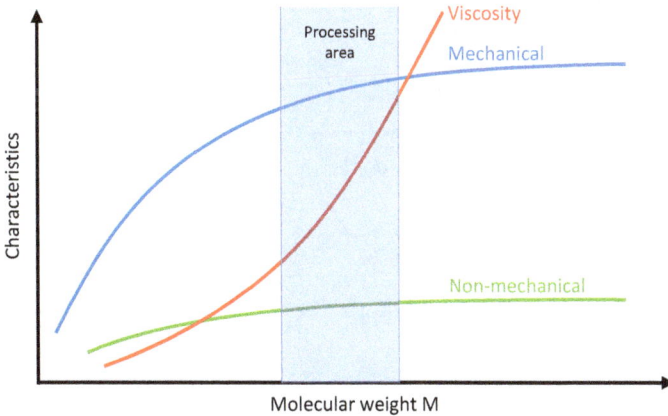

Figure 7.32 Influence of molecular weight on the physical properties of plastics [11]

The average molecular weight $\overline{M}$ for most plastics corresponds to values between 10^4 g/mol and 10^6 g/mol [1, 5]. Although this very large range is the exception rather than the rule (e.g., for PE), it illustrates the wide ranges that can occur in reality. The distribution of the molecular weight also varies greatly (Figure 7.33). A narrow distribution leads to a narrower thermal softening pattern. With a broad molecular weight distribution, the low molecular weight components have a positive effect on processing, as they act like a lubricant to promote the flow of the plastic melt.

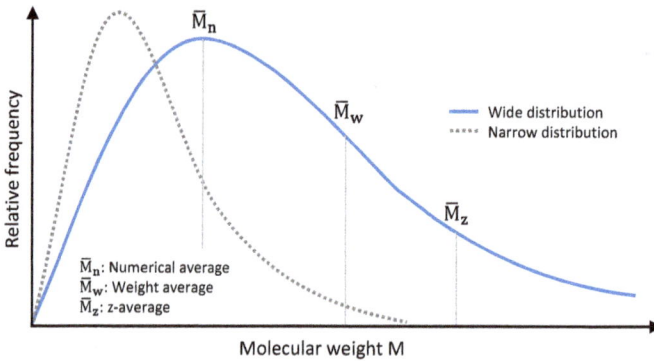

Figure 7.33 Molecular weight distributions for plastics [12]

As the molecular weight increases, the molecular chains become longer and the molecular clusters larger. As a result, the flow becomes more difficult as the molecular weight increases, since there is a greater resistance to movement (alignment). This also increases the viscosity. Figure 7.34 shows three types of polystyrene with different molecular weights.

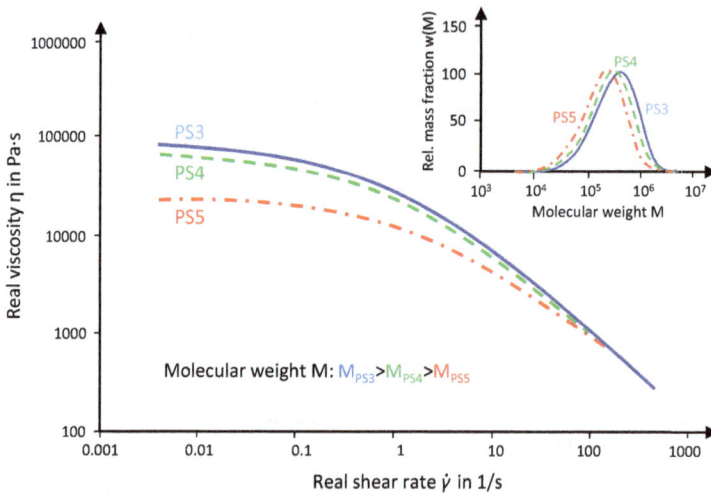

Figure 7.34 Influence of the average molecular weight of various polystyrenes on the viscosity curve at $T = 190\,°C$

It becomes clear that the viscosity, especially the zero viscosity, increases with increasing molecular weight. However, the viscosity curve is the same for all molecular weights. This means that only the zero viscosity (parameter A in the Carreau approach; see Chapter 8, Figure 8.4) and the transition from the Newtonian to the shear-thinning viscosity range (parameter B in the Carreau approach; see Chapter 8, Figure 8.4) change, while the slopes of the flow curves in the shear-thinning viscous flow range (parameter C in the Carreau approach; see Chapter 8, Figure 8.4) are identical. This means that, here too, the curves are congruent. For this reason, the flow behavior of a plastic melt can also be plotted as invariant with respect to the molecular weight. As with the temperature and the pressure, a molecular weight shift factor is introduced. This is

$$a_M = \frac{\dot{\gamma}_{reference}}{\dot{\gamma}_M} \tag{7.19}$$

where
$\dot{\gamma}_{reference}$ is the $\dot{\gamma}$ value of the master curve (reference curve)
$\dot{\gamma}_M$ is the $\dot{\gamma}$ value of the curve to be moved to the master curve for the same η value as for $\dot{\gamma}_{reference}$

The following relationship between zero viscosity and molecular weight is often given in the literature:

$$\eta_0 = k_0 \cdot M^{3.4} \tag{7.20}$$

where
k_0 is a material constant (related to η_0)
M is the molecular weight (here, viscosity-averaged molecular weight)

Figure 7.35 shows the relationship between the zero viscosity and the molecular weight for different plastics. With short chains, the melt can move freely, that is, the molecular weight of the plastic is subcritical. In this range, the slope of the straight line is 1. As the chain length increases, the chains interact more strongly and thus become less mobile, resulting in an increasing zero viscosity. The slope now assumes larger values. In most cases, it is approximately 3.4, as previously mentioned.

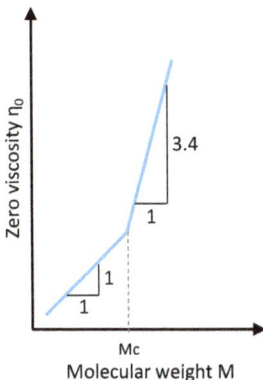

Figure 7.35
Qualitative relationship between the zero viscosity and the molecular weight

If the molecular weight shift factor a_M is used, then the relationship with the molecular weight is described by the equation:

$$a_M = k_A \cdot M^{3.4} \tag{7.21}$$

where
k_A is a constant related to a_M:

$$k_A = \frac{1}{M_{reference}^{3.4}} \tag{7.22}$$

where

$M_{reference}$ is the molecular weight of the reference curve. The master curve is given by the function $\tau = f\,(\dot{\gamma} \cdot a_M)$.

To assess the influence of the molecular weight on the zero viscosity, a rotational measurement was carried out on three isotactic PP (iPP) grades. With the aid of rotational rheometry, the zero viscosities of the individual materials were determined and correlated with the molecular weight.

In Figure 7.36, the viscosity and the shear stress of the iPP grades are plotted against the angular frequency. The measurements were carried out at a temperature of 180 °C.

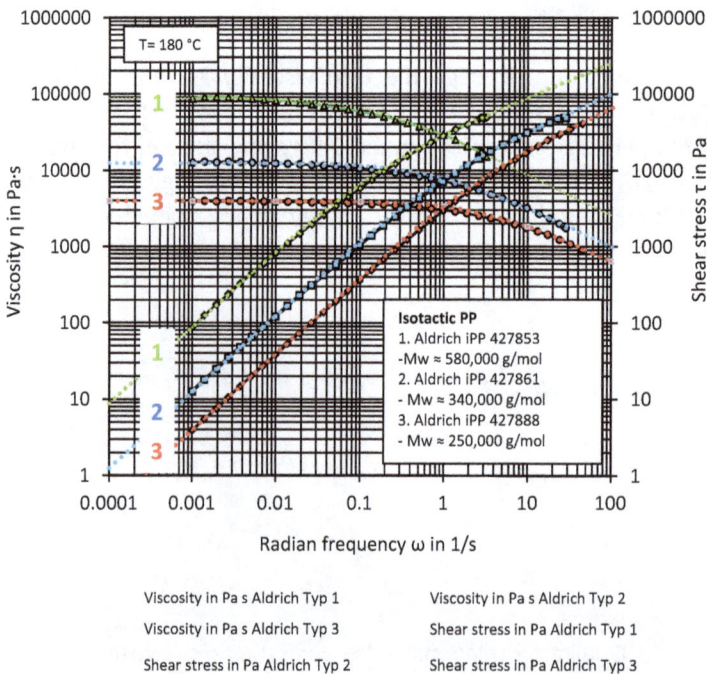

Figure 7.36 Viscosity and shear stress of different iPP grades (Aldrich)

The diagram illustrates the different viscosities of the iPP grades at 180 °C depending on the molecular weight as a function of the shear rate. With increasing molecular weight—that is, with increasing chain length—the zero viscosity, in particular, increases sharply.

A regression equation was then established and the zero viscosity was determined as a function of the molecular weight of the three iPP grades. The following applies:

$$\eta_0 = K \cdot Mw^{exponent} \tag{7.23}$$

where
K is a material constant
Mw is the molecular weight

Figure 7.37 shows the relationship between the zero viscosity measured with the rotational viscometer and the molecular weight, for three identical iPP grades with different molecular weights.

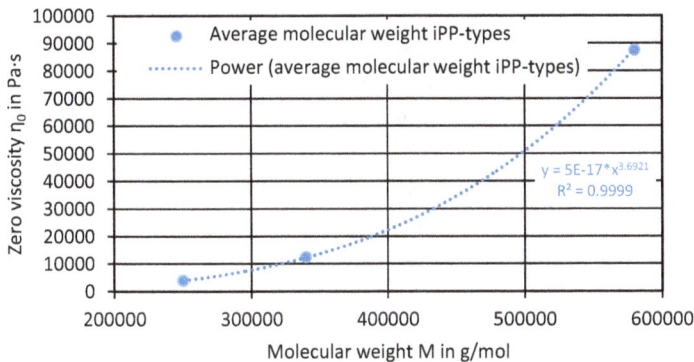

Figure 7.37 Relationship between the zero viscosity and the molecular weight

It can be seen clearly that the zero viscosity increases exponentially as a function of the molecular weight. The exponent, in this case, is 3.7. The material constant in this example is $K = 5.0 \cdot 10^{-17}$. This value is also given in the literature [13] for some polypropylene types. The equation to describe the molecular-weight-dependent zero viscosity would be as follows:

$$\eta_0 = 5.0 \cdot 10^{-17} \cdot Mw^{3.7} \tag{7.24}$$

7.3.5 Molecular Weight Distribution

The molecular weight distribution also has an influence on the flow curve. Figure 7.38 shows two flow curves with the same average molecular weight but a different molecular weight distribution. In the range of very low shear rates, the zero viscosities are identical.

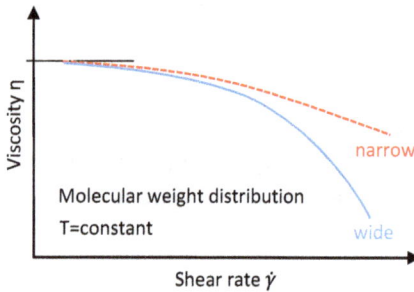

narrow

Molecular weight distribution

T=constant

wide

Shear rate $\dot{\gamma}$

Figure 7.38

Qualitative viscosity curve for polymers with different molecular weight distributions [12]

As Figure 7.38 illustrates, a wide molecular weight distribution results in an early transition from Newtonian to shear-thinning viscous flow behavior. Furthermore, the shear-thinning viscosity is more pronounced. In contrast, the behavior of polymers with a narrow molecular weight distribution is less shear-thinning viscous and they only enter the shear-thinning viscosity range at higher shear rates. As a result, the two curves are not congruent, since both the transition point and the slope of the flow curve change depending on the molecular weight distribution.

The transition point of the flow curve from the Newtonian to the shear-thinning viscosity range is described in the Carreau approach (Chapter 8, Figure 8.4) with the quantity $B = \frac{1}{\dot{\gamma}_0}$, which is designated as the reciprocal transition shear rate. The narrower the molecular weight distribution is, the further the reciprocal transition shear rate shifts to the right.

The same applies to the molecular weight. As the molecular weight increases, the zero viscosity increases, and the transition point of the flow curve shifts to the left. The slope does not change in this case—that is, the curves are congruent.

Figure 7.39 shows the viscosity curve for two types of polypropylene with different molecular weights and different molecular weight distributions. It is clear that the PP with the higher molecular weight (bottom image) has a higher zero viscosity (approximately 10,000 Pa · s) than the PP with the lower molecular weight (top image) (approximately 500 Pa · s). Furthermore, the transition point from the Newtonian to the shear-thinning viscosity range occurs at lower shear rates for the PP with the higher molecular weight than for the PP with the lower molecular weight.

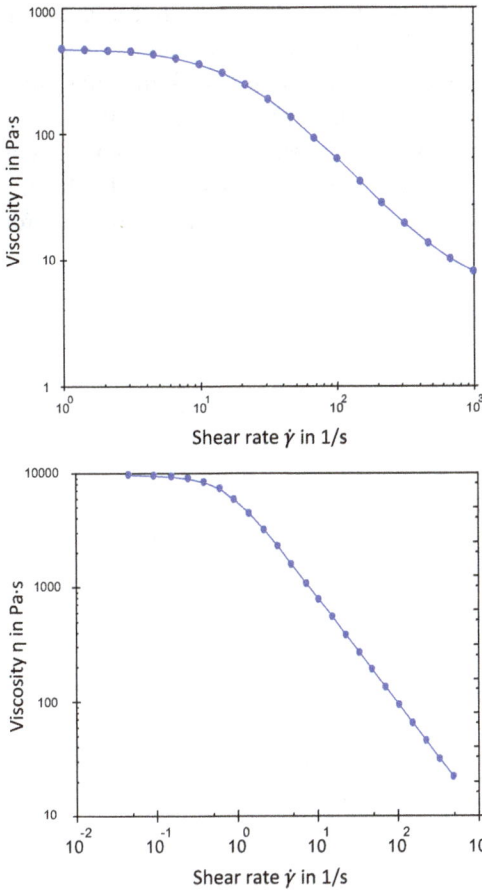

Figure 7.39
Viscosity as a function of shear rate for two different types of PP (top: lower molecular weight, bottom: higher molecular weight)

Figure 7.40 shows qualitatively how the zero viscosity is plotted versus the reciprocal transition shear rate B.

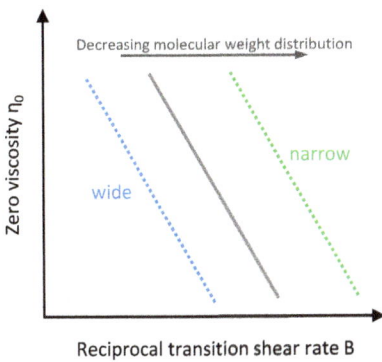

Figure 7.40
Qualitative relationship between the molecular weight distribution and the reciprocal transition shear rate B as a function of the zero viscosity [12]

Figure 7.41 shows the strand extension measured in the high-pressure capillary viscometer, for polypropylenes with different molecular weight distributions, as a function of the shear rate. It is clear that the polypropylene PP1 with a narrow molecular weight distribution, that is, with higher-molecular-weight components, exhibits a significantly stronger swelling. This clearly shows that the elastic phenomena, such as the swelling effect, during processing are influenced much more strongly by the molecular weight distribution than by the viscous properties. In profile extrusion, for example, such differences must be taken into account.

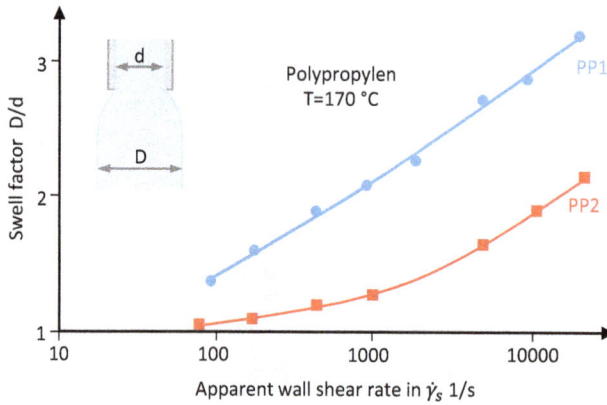

Figure 7.41 Strand extension as a function of the apparent wall shear rate for two polypropylene types with different molecular weight distributions

7.3.6 Influence of Molecular Weight and Molecular Weight Distribution on Storage and Loss Moduli During Oscillation

To visualize polymer types with a narrow and a wide molecular weight distribution and also to illustrate different average molecular weights, an oscillation rheometer (the so-called frequency sweep) can be used [14, 15]. For a viscoelastic fluid, this is shown in Figure 7.42.

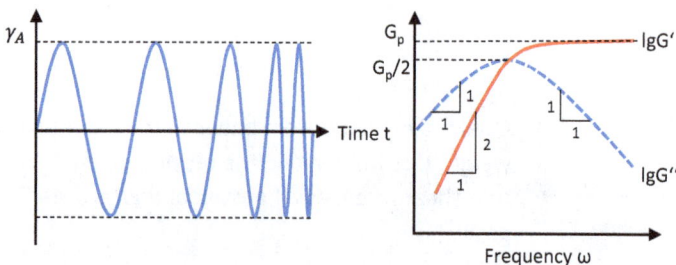

Figure 7.42 Frequency sweep and storage and loss moduli for a Maxwell fluid at constant amplitude

Figure 7.43 shows the result of an oscillation measurement for a PE-HD. After an amplitude sweep, the frequency sweep was carried out. The diagram shows the storage modulus G' and the loss modulus G'' as a function of the frequency. It is clear that the two curves intersect. This intersection point is referred to as the crossover point (COP). To the left of the intersection point, the loss modulus is above the storage modulus. For this reason, the polymer shows predominantly viscous properties in this frequency range, while to the right of the intersection, at higher frequencies, the elastic properties predominate.

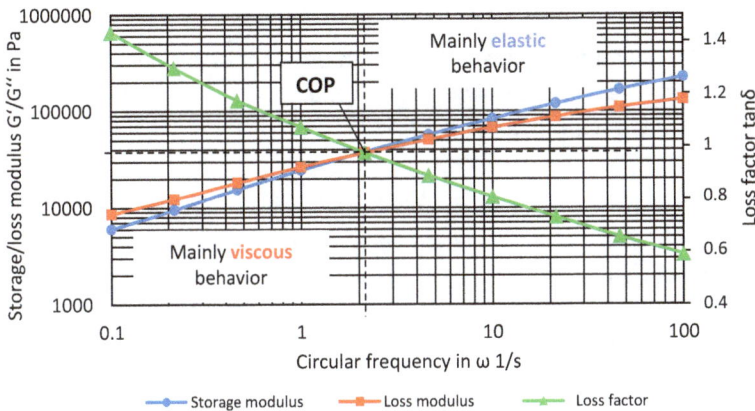

Figure 7.43 Loss factor as a function of storage and loss moduli at 180 °C

In the high shear rate range (angular frequencies), the molecular chains can no longer move in this way, as they become entangled or block each other. In this state, the molecular chains can only vibrate and no longer follow the oscillating movement.

7.3.6.1 Comparison of Polymer Types with Different Molecular Weight Distributions

$G'(\omega)$ is often used to investigate the molecular weight distribution (MWD) (Figure 7.44). If polymers show two plateaus in the G' curve, then the upper plateau is referred to as G_P and the lower one as the rubber-elastic plateau GP_R. The curve is then divided into four areas [14, 15].

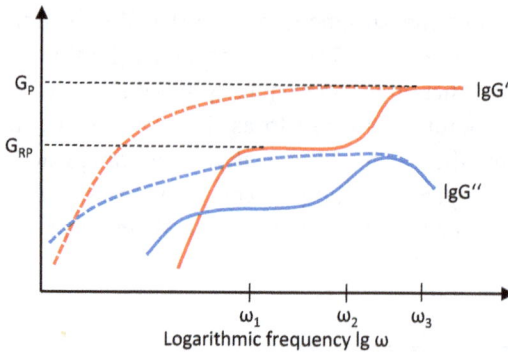

Figure 7.44 $G'(\omega)$ and $G''(\omega)$ for two polymer types with different molecular weight distributions. The dashed line shows a polymer with a wide MWD and the solid line a polymer with two fractions, each with a narrow MWD

1. **Flow zone 1: $\omega < \omega_1$**

   At low frequencies, the behavior of the polymer types can be described with a simple Maxwell model. Here, $G'(\omega)$ and $G''(\omega)$ have a slope of 2 and 1, respectively. The polymer exhibits viscoelastic behavior with a dominant viscous component. The following applies: $G'' > G'$.

2. **Rubber-elastic area**

   If there is a plateau GP_R between ω_1 and ω_2 in the medium frequency range, this is the rubber-elastic range. The longer chains no longer slide past each other, a kind of network is formed, and the polymer has viscoelastic behavior. The elastic components predominate. The following applies: $G' > G''$.

3. **Transition zone or flow zone 2**

   In the region between ω_2 and ω_3, G' and G'' continue to increase. In this range, only short and very mobile molecular chains can be deformed. The elastic part becomes more and more dominant.

4. **Glassy state with the plateau G_P at $\omega > \omega_3$**

   For a narrow MWD, $G'(\omega)$ reaches its constant maximum and $G''(\omega)$ decreases. In this range, movement between the molecular chains is no longer possible. This state can be described as a kind of frozen state. The molecular chains can vibrate a little but are not able to follow the fast oscillatory movements. The polymer now shows the behavior of a solid with a clear dominant elastic component.Polymers with a wide MWD consist of a mixture of long and short molecular chains. The first rubber-elastic plateau GP_R (Figure 7.44) no longer exists in this case. Both G' and G'' increase with increasing frequency ω up to G_P and G''_{max}, respectively.

Figure 7.45 shows the storage and loss moduli for two types of polypropylene with different molecular weights determined with the amplitude sweep. For the PP with the lower molecular weight (Figure 7.45, top), the storage modulus is considerably smaller than the loss modulus, that is, the values are very different and lie below 10,000 Pa. In contrast, the values for the storage and loss moduli for the PP types with the higher molecular weight (Figure 7.45, bottom) are over 10,000 Pa and at an almost identical level.

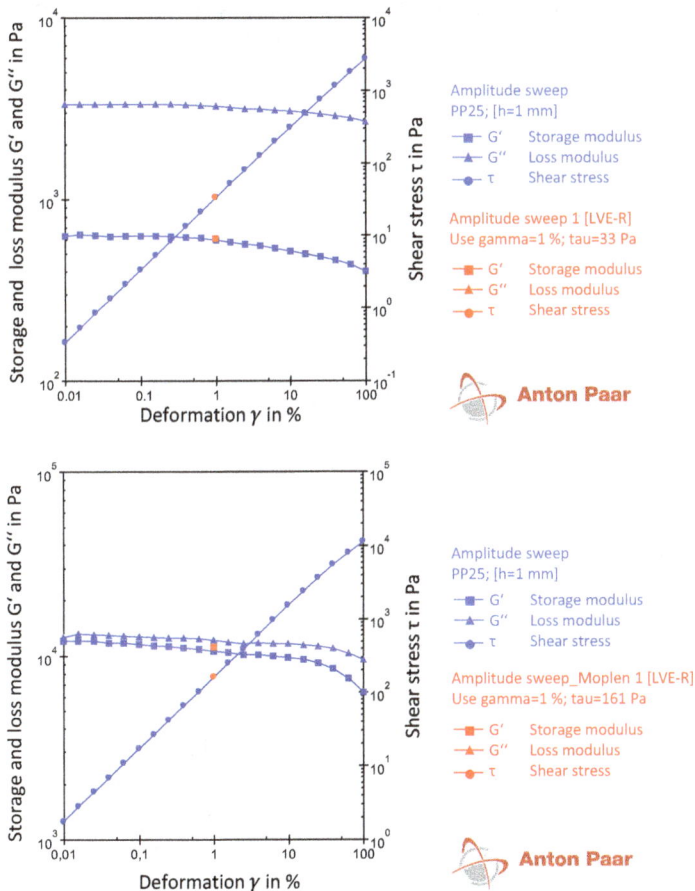

Figure 7.45 Storage and loss moduli and shear stress for two PP types with different molecular weights determined using the amplitude sweep

7.3.6.2 Comparison of Polymer Types with Different Molecular Weights and Molecular Weight Distributions

If we compare two polymer types with the same molecular weight distribution but different average molecular weights, we can see that the crossover point (COP) between the storage modulus G' and the loss modulus G'' occurs at lower frequencies for types with a higher molecular weight. This is because the molecules are less flexible and therefore less mobile. The lower the molecular weight—that is, the shorter the chains—the further the intersection point shifts to the right, as the short chains remain mobile even at higher frequencies.

This transition point can also be used to investigate the molecular weight distribution of different polymers (Figure 7.46). If we have two polymer types with the same average molecular weight, but different MWDs, it can be seen that the intersection/transition point between G' and G'' occurs at lower G values for types with a wide MWD. This means that a wide MWD shifts the intersection point of the curves downwards, while a narrow MWD causes an upward shift.

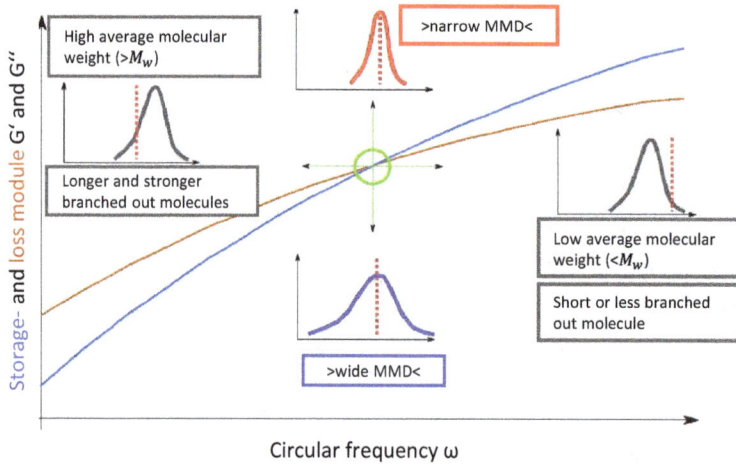

Figure 7.46 Transition point of G' and G'' as a function of the molecular weight and the molecular weight distribution

How the crossover point shifts as a function of the molecular weight and the molecular weight distribution is again shown by the measurement of two types of polypropylene with different molecular weights and different molecular weight distributions. The top image in Figure 7.47 shows the curves of the storage and loss moduli as a function of the angular frequency for a PP type with a rather low molecular weight and a broader molecular weight distribution. Compared to the PP type with the higher molecular weight and the rather narrower molecular weight distribution (Figure 7.47, bottom), the COP shifts to the right and downwards. In contrast, the elastic properties (storage modulus) predominate in the PP with the higher molecular weight and narrower molecular weight distribution (Figure 7.47, bottom) even at lower angular frequencies.

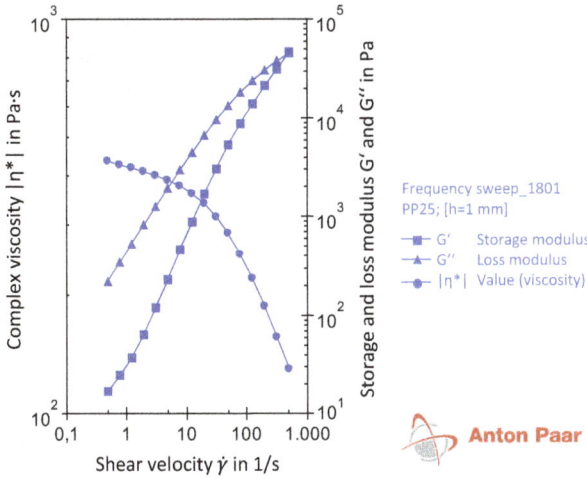

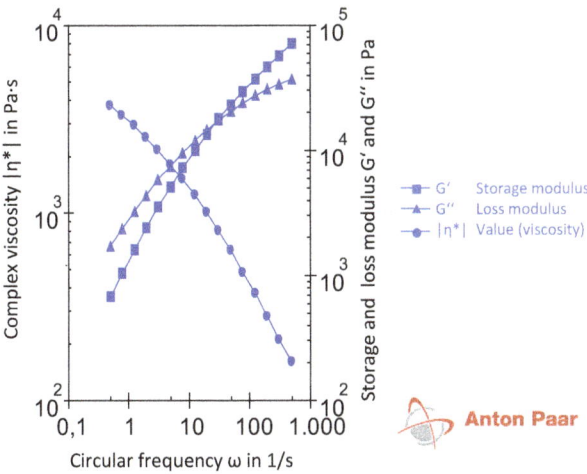

Figure 7.47 Crossover point (COP) of G' and G'' as a function of the molecular weight and the molecular weight distribution

7.4 Influence of Residual Moisture on Shear Viscosity

As a general rule, for plastics that have hygroscopic properties it must be ensured that they are sufficiently dry before a measurement is taken. The residual moisture can lead to decomposition in plastics that are sensitive to hydrolysis. Furthermore, the evaporating water also influences the measurement result, as it reduces the viscosity like a blowing agent. Figure 7.48 shows the measurement of viscosity as a function of residual moisture and time for a polyethylene terephthalate (PET) using rotational rheometry. PET is sensitive to hydrolysis and tends to absorb water quickly. It becomes clear that the viscosity increases with a residual moisture content of more than 40 ppm to 50 ppm. In this respect, the material must be sufficiently dried beforehand. A polyamide, for example, would behave in a similar way. For this reason, the viscosity of these plastics is usually measured in solution (e.g., with an Ubbelohde capillary viscometer).

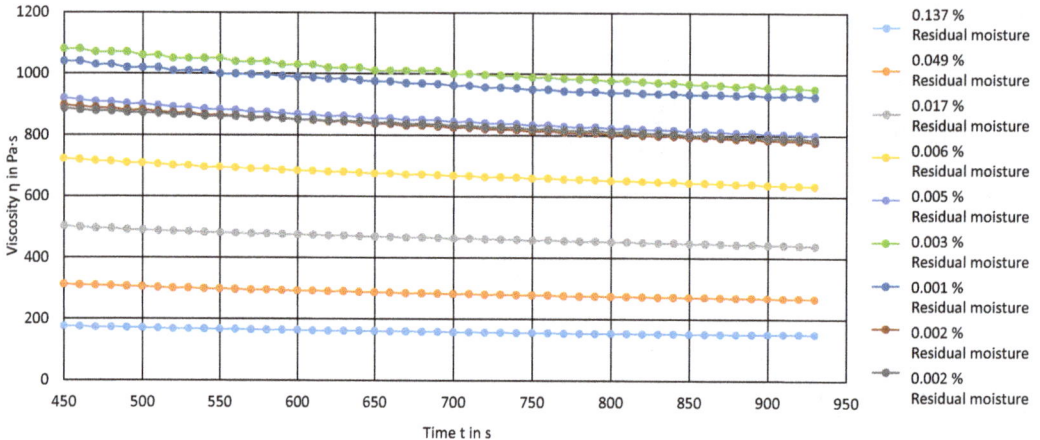

Figure 7.48 Viscosity as a function of time for different residual moisture contents, for a PET

7.5 *Task*: Description of the Flow Behavior with a Master Curve

The support values for three viscosity–shear rate curves of PMMA (type Plexiglas 6N; Röhm GmbH) are *given*; see Table 7.5.

- Determine the temperature shift factor a_T with a calculation using the appropriate shift approach for the temperatures 200 °C and 240 °C. Assume the reference temperature $\vartheta = 220$ °C.

- Draw the temperature shift in the $\tau = f(\dot{\gamma})$ and the $\eta = f(\dot{\gamma})$ representations.

- Determine the temperature shift factor graphically and compare the values.

See Figure 7.49 and Figure 7.50.

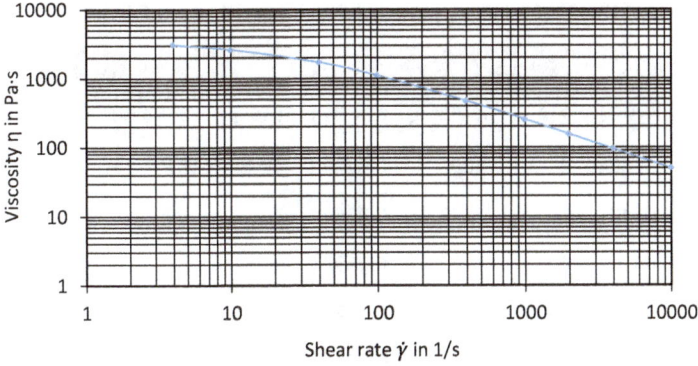

Figure 7.49 Viscosity as a function of shear rate for a PMMA, for $\vartheta = 220\,°C$

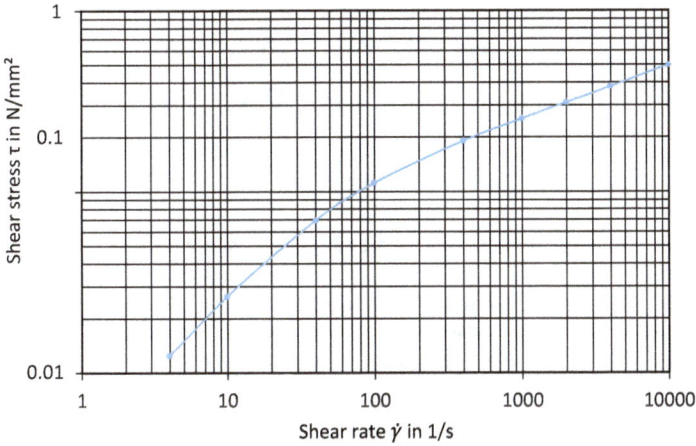

Figure 7.50 Wall shear stress as a function of the shear rate for a PMMA, for $\vartheta = 220\,°C$

Table 7.5 Support Values of the Three Flow Curves of Plexiglas 6N (from CAMPUS)

No.	$\vartheta_1 = 200\,°C$			$\vartheta_2 = 220\,°C$			$\vartheta_3 = 240\,°C$		
	$\dot{\gamma}\,[s^{-1}]$	$\eta\,[Pa\cdot s]$	$\tau\,[\frac{N}{mm^2}]$	$\dot{\gamma}\,[s^{-1}]$	$\eta\,[Pa\cdot s]$	$\tau\,[\frac{N}{mm^2}]$	$\dot{\gamma}\,[s^{-1}]$	$\eta\,[Pa\cdot s]$	$\tau\,[\frac{N}{mm^2}]$
0	0.4	16,600	0.00664	4	3090	0.0124	10	871	0.00871
1	1	14,800	0.0148	10	2640	0.0264	40	695	0.0278
2	4	10,900	0.0436	40	1730	0.0692	100	537	0.0537
3	10	7800	0.078	100	1120	0.112	400	289	0.116
4	40	3730	0.149	400	478	0.191	1000	167	0.167
5	100	2040	0.204	1000	253	0.253	2000	105	0.21
6	400	771	0.308	2000	155	0.31	4000	64	0.256
7	1000	407	0.407	4000	95	0.38	100,000	34	0.34
8	2000	247	0.494	10,000	50	0.5	40,000	13	0.52

References

[1] Menges, G.; Haberstroh, E.; Michaeli, W.; Schmachtenberg, E.: *Menges Werkstoffkunde Kunststoffe*. Hanser, Munich, 2011

[2] Menges, G. Geisbüsch, P.: *Verbesserung der Fliessfähigkeit hochgefüllter Formmassen beim Spritzgießen, Extrudieren und Beschichten mittels Rakel*. Westdeutscher Verlag, Opladen, 1979

[3] Pahl, M.; Gleißle, W.; Laun, H.-M.: *Praktische Rheologie der Kunststoffe und Elastomere*. VDI-Verlag GmbH, Düsseldorf, 1995

[4] Menges, G.; Geisbüsch, P.: The glass fiber orientation and its influence on the mechanical properties of thermoplastic polymer melts – An estimation method. *Colloid & Polymer Science*, 1982, pp. 73–81

[5] Baur, E.; Drummer, D.; Osswald, T. A.; Rudolph, N.: *Saechtling Kunststoff Taschenbuch*. Hanser, Munich, 2013

[6] Bonten, C.: *Plastics Technology: Introduction and Basics*. Hanser, Munich, 2014

[7] Braun, D.; Becker, G.; Carlowitz, B.: *Die Kunststoffe: Chemie, Physik, Technologie – Kunststoff-Handbuch 1*. Hanser, Munich, 1990

[8] Ehrenstein, G. W.: *Polymer Werkstoffe: Struktur – Eigenschaften – Anwendung*. Hanser, Munich, 2011

[9] Baur, E.; Harsch, G.; Moneke, M.: *Werkstoff-Führer Kunststoffe: Eigenschaften, Prüfungen, Kennwerte*. Hanser, Munich, 2019

[10] Schwarzl, F. R.: *Polymer Mechanics: Structure and Mechanical Behavior of Polymers*. Springer, Berlin · Heidelberg, 1990

[11] Kaiser, W.: *Plastics Chemistry for Engineers: From Synthesis to Application*. Hanser, Munich, 2015

[12] Frick, A.; Stern, C.: *Introduction to Plastics Testing*. Hanser, Munich, 2017

[13] Dealy, J. M.; Larson, R. G.: *Structure and Rheology of Molten Polymers: From Structure to Flow Behavior and Back Again*. Hanser, Munich, 2006

[14] Mezger, T. G.: *The Rheology Handbook*. Vincentz Network, Hanover, 2016

[15] Mezger, T. G.: *Applied Rheology: With Joe Flow on the Rheology Road*. Anton Paar GmbH, Graz, 2014

8 Viscometry—Mathematical Description of the Flow Curve

Today, computer-aided simulation programs such as Cadmould®, Moldex®, Moldflow®, and Sigmasoft®, among others, are often used for the design of molded parts and injection molds. These programs can be employed on the basis of the finite element method to calculate and graphically display

- The filling behavior
- The printing requirements
- Temperature curves
- Shear stresses
- Shear rates
- Orientations
- Shrinkage/distortion
- Other relevant variables for a molded part

This allows the molded part to be optimized in advance, that is, before mold production. By varying the number and positions of the injection points, air inclusions and weld lines can be predicted. Furthermore, changes to the molded part geometry (ribs, openings, etc.) can be implemented without great effort.

For these calculations, however, in addition to the material data sheets and process parameters, the programs also require mathematical approaches that can be used to describe the shear-thinning viscous and possibly the viscoelastic flow behavior of the plastic melts. These mostly empirically found approaches for describing the viscoelastic flow behavior of plastic melts will be listed and explained in this chapter.

In contrast to Newtonian fluids, plastic melts flow in a shear-thinning viscous manner. This means that the shear stress and the viscosity are not only a function of temperature and pressure but also depend on the shear rate. The shear stress is not a linear relationship but an increasing, concave-down curve and the viscosity decreases with increasing shear rate.

Figure 8.1 illustrates that the shear-thinning viscous flow behavior of plastic melts must be described by equations that include the shear rate as well as the temperature dependence. These approaches are also known as material laws. Looking at the graph on the right in Figure 8.1, it can be seen that this flow curve consists of two almost linear areas when plotted logarithmically. This has the advantage that this flow curve can be described mathematically in a relatively simple way.

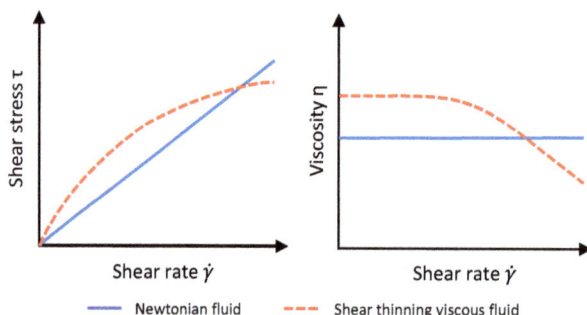

Figure 8.1 Flow curves for Newtonian and shear-thinning viscous fluids

8.1 Ostwald and de Waele's Power Approach (Power-Law Model)

If one plots the flow curves of different polymers on a double-logarithmic scale, curves that consist of two approximately linear sections (Newtonian zero-viscosity range, shear-thinning viscous range) and a transition range are obtained. In many cases, one only moves in one of the two areas, so that a function of the following form is suitable for the mathematical description of the curve section:

$$\dot\gamma = \phi \cdot \tau^m \tag{8.1}$$

In this equation, known as the power approach of Ostwald and de Waele, in addition to the shear rate and the shear stress, the fluidity ϕ and a flow exponent m are included. The unit of the fluidity ϕ is $Pa^{-m} \cdot s^{-1}$. For Newtonian media, the fluidity is the reciprocal of the dynamic viscosity. The flow exponent m describes the deviation from the Newtonian behavior (see Figure 8.2) and thus takes into account the characteristic shear-thinning viscous flow behavior of the plastic melts. The greater m is, the more shear-thinning viscous is the melt behavior; for $m = 1$, Newtonian flow is present. The following applies:

$$m = \frac{\Delta (\log \dot{\gamma})}{\Delta (\log \tau)} = \frac{\log \dot{\gamma}_2 - \log \dot{\gamma}_1}{\log \tau_2 - \log \tau_1} \tag{8.2}$$

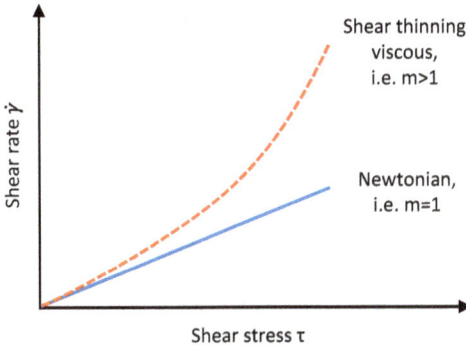

Shear thinning
viscous,
i.e. m>1

Newtonian,
i.e. m=1

Shear rate $\dot{\gamma}$

Shear stress τ

Figure 8.2
Relationship between shear stress
and shear rate [1]

Thus, m is the gradient of the flow curve $\dot{\gamma} = f(\tau)$ (see Figure 8.2) in the area under consideration, plotted in the double-logarithmic diagram. The exponent usually assumes values between 1 and 6 for plastic melts. Shear rates of $10^0 \, s^{-1}$ to $10^4 \, s^{-1}$ are relevant for the design of dies and molds in plastics processing. In these ranges, m assumes a value between 1 and 4. For $m = 1$, $\phi = 1/\eta$, that is, the power approach becomes Newton's law of friction. With Newton's law of friction, $\eta = \frac{\tau}{\dot{\gamma}}$, the following viscosity function is obtained from the power approach:

$$\eta = \phi^{-1} \cdot \tau^{1-m}$$

or

$$\eta = \phi^{-\frac{1}{m}} \cdot \dot{\gamma}^{\frac{1-m}{m}}$$

Introducing K and n as follows:

$$K = \phi^{-\frac{1}{m}} \; ; \; n = \frac{1}{m} = \frac{\Delta (\log \tau)}{\Delta (\log \dot{\gamma})}$$

the usual representation of the viscosity function as a function of the shear rate is obtained:

$$\eta = K \cdot \dot{\gamma}^{n-1}$$

The factor K is called the consistency factor or unity viscosity, as it indicates the viscosity at a shear rate of $\dot{\gamma} = 1 \, s^{-1}$ (see Figure 8.6). The unit of the consistency factor K is $Pa \cdot s^n$. The viscosity exponent n describes the gradient of the first flow curve in the $\tau = f(\dot{\gamma})$ representation. In the second flow curve, $\eta = f(\dot{\gamma})$, $n-1$ corresponds to the slope in the shear-thinning viscous range (see Figure 8.6). For Newtonian media, the viscosity exponent n assumes the value 1. For the shear-thinning viscous plastic melts, $0 < n < 1$ applies.

As a result, there are two different versions of the power approach. The first one is:

$$\eta = \phi^{-\frac{1}{m}} \cdot \dot{\gamma}^{\frac{1-m}{m}} \tag{8.3}$$

Because it is easier (ϕ has very small values, for example), the other one is also frequently used:

$$\eta = K \cdot \dot{\gamma}^{n-1} \tag{8.4}$$

The following applies for conversions:

$$K = \phi^{-\frac{1}{m}} \tag{8.5}$$

and

$$n = \frac{1}{m}$$

The empirically found power approach is often used in practice as it is mathematically quite easy to handle. The disadvantage, however, is that the flow and viscosity curves can only be described well in some areas using this approach. The flow exponent depends on the shear rate and therefore m and n are not constant values. The specification for the flow and viscosity exponents is only valid for a certain range of the flow curve. Furthermore, for mathematical reasons, when the shear rate approaches zero, the viscosity approaches infinity.

For this reason, an approximation range, that is, a validity range for the shear rate, is always specified for the material data of the power approach. The viscosity may only be calculated using the power approach within this range.

It should also be noted that the flow exponent m is included in the unit of fluidity and the viscosity exponent n in the unit of the consistency factor.

Task: Graphical Determination of the Constants of the Power Approach

The constants m, n, ϕ, and K of the power approach are to be determined graphically using the diagram in Figure 8.3. It should be noted that the flow exponent m is only ever determined for one point (shear rate and shear stress), as the gradient of the flow curve changes with shear-thinning viscous fluids.

The following parameters must be determined for the power approach:

- The flow exponent m ($1 < m < 6$)
- The viscosity exponent n ($0 < n < 1$)
- The fluidity ϕ (calculate in Pa, not in MPa)
- The consistency factor K for $\dot{\gamma} = 10^2\,\mathrm{s}^{-1}$ (calculate in Pa, not in MPa)
- The viscosity for different shear rates: $\dot{\gamma} = 10^{-1}\,\mathrm{s}^{-1}$, $\dot{\gamma} = 10^0\,\mathrm{s}^{-1}$, $\dot{\gamma} = 10^2\,\mathrm{s}^{-1}$, and $\dot{\gamma} = 10^3\,\mathrm{s}^{-1}$

Discuss the calculated viscosities by comparing them with the values from the flow curve. Pay particular attention to the units when calculating the consistency factor and when determining the fluidity!

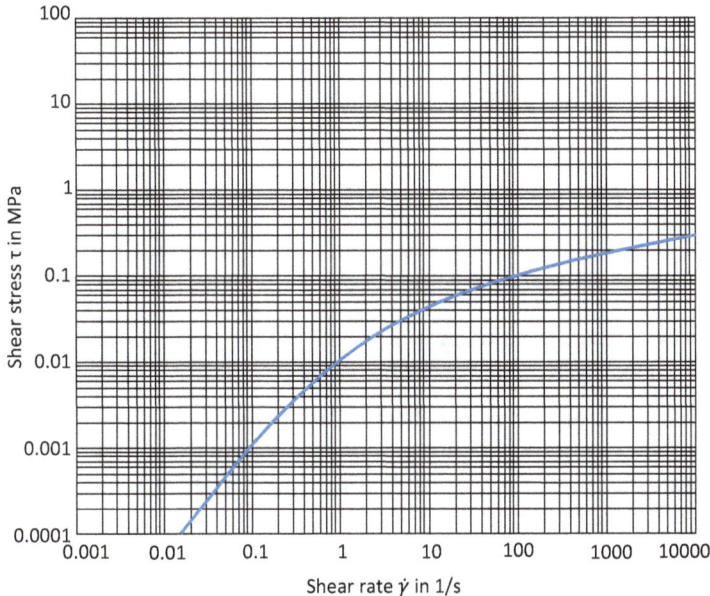

Figure 8.3 Determining the constants for the power approach from the flow curve

8.2 The Carreau Approach

The Carreau approach describes the curve of the viscosity function with a so-called three-parameter model:

$$\eta = \frac{A}{(1 + B \cdot \dot{\gamma})^C} \tag{8.6}$$

Using Newton's law of friction, $\tau = \eta \cdot \dot{\gamma}$, it follows for the Carreau approach:

$$\tau = \frac{A \cdot \dot{\gamma}}{(1 + B \cdot \dot{\gamma})^C} \tag{8.7}$$

In this approach, the three parameters are as follows (Figure 8.4):

- A is the zero viscosity.
- B is the so-called reciprocal transition shear rate.

- C is the slope of the viscosity curve in the shear-thinning viscous range. It assumes values between 0 and 1: $0 < C < 1$. The closer C approaches the value 1, the more shear-thinning viscous the material behavior is. For $C = 0$, Newtonian flow behavior is present.

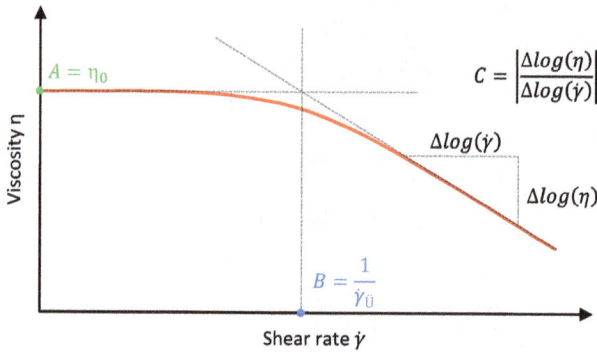

$A = \eta_0$

$C = \left|\dfrac{\Delta log(\eta)}{\Delta log(\dot\gamma)}\right|$

$\Delta log(\dot\gamma)$

$\Delta log(\eta)$

$B = \dfrac{1}{\dot\gamma_0}$

Viscosity η

Shear rate $\dot\gamma$

Figure 8.4 Description of the viscosity curve with the Carreau model

The three parameters P_1, P_2, and P_3 can also be found in the literature, where P_1 describes the zero viscosity, P_2 corresponds to the reciprocal transition shear rate, and P_3 indicates the slope of the flow curve in the shear-thinning viscous range.

The Carreau model has the advantage that it correctly reproduces the actual material behavior within a wide shear rate range. The approach also provides reasonable viscosity values for low shear rates ($\dot\gamma \to 0$). The following applies:

$$\lim_{\dot\gamma \to 0} \eta = \lim_{\dot\gamma \to 0} \frac{A}{(1 + B \cdot \dot\gamma)^C} = A = \eta_0 \tag{8.8}$$

For Newtonian flow behavior, the flow curve in the $\eta = f(\dot\gamma)$ representation is horizontal. This means that the gradient C assumes the value zero. As a result, the Carreau equation is reduced to:

$$\eta = \frac{A}{(1 + B \cdot \dot\gamma)^0} = A = \eta_0 \tag{8.9}$$

This approach therefore describes the flow curve in a broader range than that of the power approach.

8.2.1 *Task*: Graphical Determination of the Constants of the Carreau Approach

In the following, the three constants A, B, and C are determined using Figure 8.5.

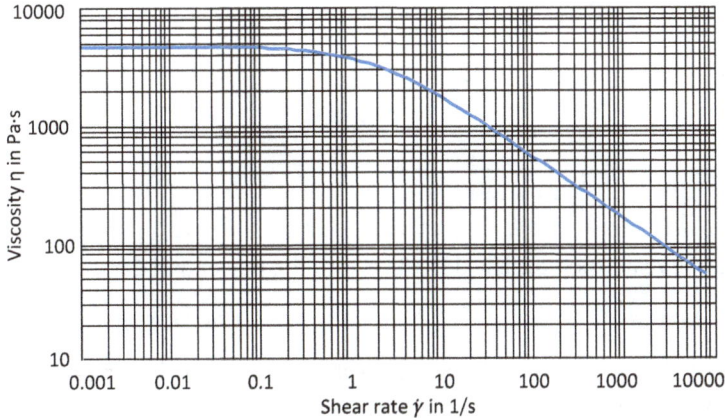

Figure 8.5 Determining the constants for the Carreau approach

The constants for the Carreau approach are determined as follows:

1. The parameter A corresponds to the zero viscosity, i.e., the viscosity for very low shear rates. The value can be read off the ordinate (horizontal segment of the flow curve).

2. The parameter B corresponds to the reciprocal transition shear rate $1/\dot{\gamma}_U$. To determine this, a tangent is added to the flow curve in the shear-thinning viscous flow range and another one is added in the zero-viscosity range. The shear rate can be read on the x axis at the intersection of the tangents.

3. The constant C is determined from the gradient of the flow curve in the shear-thinning viscous range. The following applies:

$$C = \left| \frac{\Delta \log (\eta)}{\Delta \log (\dot{\gamma})} \right| = \left| \frac{\log (\eta_2) - \log (\eta_1)}{\log (\dot{\gamma}_2) - \log (\dot{\gamma}_1)} \right| \tag{8.10}$$

With the same logarithmic scale of the axes, C can also be read from the gradient triangle using a ruler.

Figure 8.6 shows an example of the superposition of the flow curve descriptions by the power approach and the Carreau approach. It is clear that the power approach represents a linear equation. In the shear-thinning viscous range, this straight line coincides with the curve of the Carreau approach. Only in this range is the viscosity of the plastic melt correctly described by the power approach.

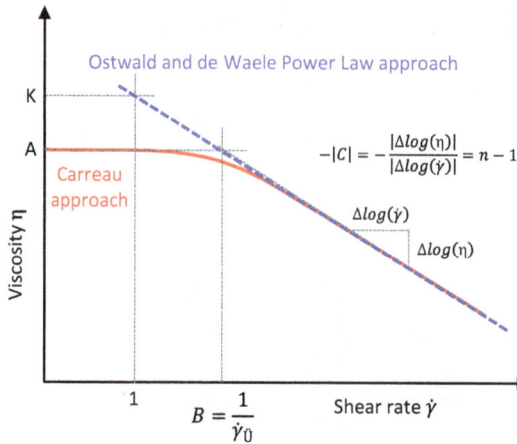

Figure 8.6 Viscosity curve mathematically described with the power and Carreau approaches

For this range, the gradient C of the Carreau approach corresponds to the gradient $n-1$ of the power approach. Thus:

$$-|C| = n - 1 \tag{8.11}$$

As soon as the two curves no longer coincide, the power approach no longer applies. In this case, we are outside the validity or approximation range. This range is always specified when determining the material data. As soon as the plastic is no longer processed within the specified validity range, the approach with the given material parameters may no longer be used.

It is also clear how the value of the consistency factor K can be determined. To do this, extend the straight line of the power function up to a shear rate of $1\,\text{s}^{-1}$ and read off the value for this point on the y axis. Pay attention to the units here again, as the viscosity has the unit $\text{Pa} \cdot \text{s}$, while the unit of the consistency factor K is $\text{Pa} \cdot \text{s}^{n}$.

8.2.2 Consideration of Temperature Dependence in the Carreau Approach

The parameters A and B are temperature-dependent in the Carreau approach, as the flow curve shifts to the lower right for an increasing temperature and to the upper left for a decreasing temperature. The gradient remains constant. This means that the values A and B are always specified for a given reference temperature. The gradient C is not a function of the temperature. The temperature dependence of the variables A and B can be taken into account by the temperature shift factor. According to Section 7.2.2, the following relationships apply to viscosity and shear rate:

$$\eta_{\text{wanted}} = a_T \cdot \eta_{\text{reference}} \tag{8.12}$$

and

$$\dot{\gamma}_{\text{wanted}} = \frac{1}{a_T} \cdot \dot{\gamma}_{\text{reference}}$$

If these equations are used in Carreau's approach:

$$\eta_{\text{reference}} = \frac{A}{\left(1 + B \cdot \dot{\gamma}_{\text{reference}}\right)^C} \tag{8.13}$$

the equation obtained is:

$$\eta_{\text{wanted}} = \frac{A \cdot a_T}{\left(1 + B \cdot \dot{\gamma}_{\text{wanted}} \cdot a_T\right)^C} \tag{8.14}$$

This equation can be used to describe viscosity as a function of shear rate and temperature.

The same procedure—that is, implementing the temperature dependence— can also be carried out using the power approach. In the power approach, the variables K and ϕ are temperature-dependent. The equations for this can be found at the beginning of this book, under "Important Rheological Formulas".

8.3 The Cross–WLF Approach

Another mathematical approach to describing the flow curve that is often used in computer-aided simulation programs (Cadmould®, Moldflow®, Moldex®, etc.) is the Cross–WLF approach. It is similar to the Carreau approach but relates to the glass transition temperature of the plastic. Here, n is the viscosity exponent. The shear stress τ^* and the zero viscosity D_1 refer to the glass temperature D_2 of the material to be described. T^* is the pressure-dependent glass temperature of the plastic and the variables A_1 and A_2 or A_3 are material constants that characterize the temperature dependence of the material. Furthermore, the Cross–WLF approach contains the parameter D_3, which describes the pressure dependence of the viscosity (see Section 7.3.3).

The basic equation of the Cross–WLF approach is as follows:

$$\frac{\eta_{\dot{\gamma}} - \eta_\infty}{\eta_0 - \eta_\infty} = \frac{1}{\left[1 + K \cdot \dot{\gamma}\right]^{1-n}} \tag{8.15}$$

With $\eta_\infty = 0$ and $K = \frac{\eta_0}{\tau^*}$, it follows:

$$\eta\left(\dot{\gamma}\right) = \frac{\eta_0}{1 + \left(\frac{\eta_0 \dot{\gamma}}{\tau^*}\right)^{1-n}}$$

(8.16)

The following equation for calculating the zero viscosity describes the temperature dependence of this quantity:

$$\eta_0\left(T\right) = D_1 \cdot e^{\left[\frac{-A_1\left(T-T^*\right)}{A_2 + \left(T-T^*\right)}\right]}$$

(8.17)

In addition to the temperature dependence, the approach also takes into account the pressure dependence of the viscosity via the variable D_3, which has the unit K/Pa. With $A_2 = A_3 + D_3 \cdot p$ and $T^* = D_2 + D_3\, p$:

$$\eta_0\left(T, p\right) = D_1 \cdot e^{\left[\frac{-A_1\left(T-D_2-D_3 p\right)}{A_3 + D_3 \cdot p + T - D_2 - D_3 \cdot p}\right]}$$

(8.18)

If the pressure coefficient D_3 is not given ($D_3 = 0$ K/Pa), it follows that $T^* = D_2$ and $A_2 = A_3$.

The parameter D_3 is generally considered important in the following cases:

- The flow path to wall thickness ratios are greater than 100.

- The injection pressure is above 100 MPa.

- The wall thickness of the molded part is less than 2 mm.

- Certain materials such as polycarbonate, i.e., amorphous plastics, are being used.

As an **example**, the following parameters are used to calculate the viscosity as a function of the shear rate for four different temperatures. The material constants for the Cross–WLF approach are:

- $n = 0.2907$

- $\tau^* = 23{,}500$ Pa

- $D_1 = 2.64 \cdot 10^{19}$ Pa $\cdot$ s

- $D_2 = 233.15$ K

- $D_3 = 0$ K/Pa

- $A_1 = 44.335$

- $A_3 = 51.6$ K, as $D_3 = 0$ K/Pa

- $T^* = 233.15$ K, as $D_3 = 0$ K/Pa

As it turns out, the parameter D_3 is not included in this material data sheet. Unfortunately, this applies to most types of material in the simulation programs. Furthermore, a constant value is assumed for D_3 when considering the pressure-dependent viscosity. However, as shown in Section 7.3.3, D_3 in principle also depends on the shear rate and is therefore not a constant value. This could also be proven by comparing the simulation with reality using a flow spiral.

In the literature, for example, the information listed in Table 8.1 is given for D_3. It is clear that the influence of pressure on viscosity is greater for amorphous plastics than for semicrystalline plastics.

Table 8.1 D_3 Data

PS	$D_3 = 0.1$ K/bar
PC	$D_3 = 0.03$ K/bar
PMMA	$D_3 = 0.023$ K/bar
PE-HD	$D_3 = 0.012$ K/bar
PP	$D_3 = 0.014$ K/bar

If the value D_3 is not given, the pressure-dependent glass temperature D_2 can be read from the pvT diagram of the plastic. This shifts to higher temperatures with increasing pressure.

If the variables listed above are now used in the Cross–WLF approach without taking D_3 into account, the viscosity curve shown in Figure 8.7 results.

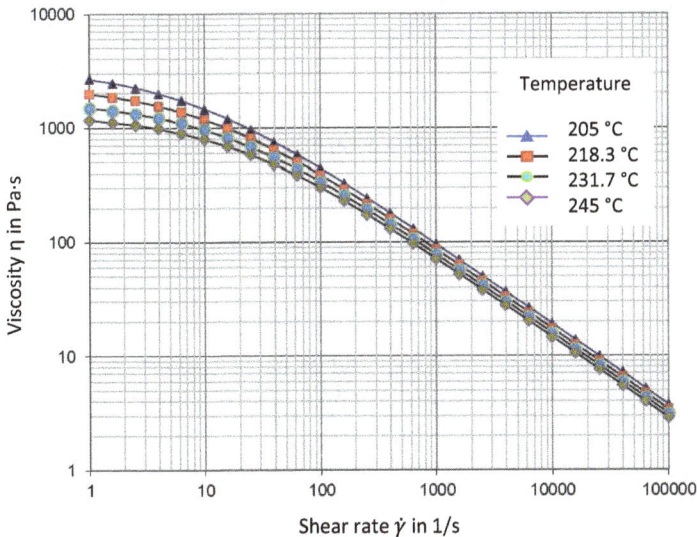

Figure 8.7 Viscosity curve calculated with the Cross–WLF model for different temperatures without the pressure coefficient D_3

8.4 Polynomial Approaches

8.4.1 Polynomial Approach According to Münstedt

An almost congruent agreement with the measured flow curve $\eta = f\,(\dot\gamma)$ is provided by this polynomial approach introduced by Münstedt [2]:

$$\log(\eta) = A_1 + A_2 \cdot \log(\dot\gamma) + A_3 \cdot (\log(\dot\gamma))^2 + A_4 \cdot (\log(\dot\gamma))^3 + A_5 \cdot (\log(\dot\gamma))^4 \tag{8.19}$$

with the constants A_1 to A_5. If we insert $\tau = \eta \cdot \dot\gamma$, we get the corresponding polynomial for the flow curve $\tau = f\,(\dot\gamma)$:

$$\log(\tau) = A_1 + (A_2 + 1) \cdot \log(\dot\gamma) + A_3 \cdot (\log(\dot\gamma))^2 + A_4 \cdot (\log(\dot\gamma))^3 + A_5 \cdot (\log(\dot\gamma))^4 \tag{8.20}$$

The flow curves can be described with this fourth-order polynomial over a shear rate range of four decades with deviations of less than 2 % between the measured curve and the curve given by the equation. For the representation of the master curve, the equation is supplemented by inserting the displacement factor A_T:

$$\log(\eta) = \log(A_T) + A_1 + A_2 \cdot \log(\dot\gamma \cdot A_T) + A_3 \cdot (\log(\dot\gamma \cdot A_T))^2 + A_4 \cdot (\log(\dot\gamma \cdot A_T))^3 + A_5 \cdot (\log(\dot\gamma \cdot A_T))^4 \tag{8.21}$$

The coefficients A_1 to A_5 are determined using the familiar method of linear regression via support values taken from the measured curve.

The derivative of the equation can also be used to determine the flow exponent of the power approach:

$$\frac{1}{m} = 1 + A_1 + 2 \cdot A_2 \cdot \log(\dot\gamma \cdot A_T) + 3 \cdot A_3 \cdot (\log(\dot\gamma \cdot A_T))^2 + 4 \cdot A_4 \cdot (\log(\dot\gamma \cdot A_T))^3 \tag{8.22}$$

When using the polynomials described here, it should be noted that extrapolation—that is, a calculation with curve values outside the support values range—can lead to large errors.

8.4.2 Biquadratic Polynomial Approach

Instead of calculating the temperature shift separately, the following biquadratic polynomial approach can also be used [2]:

$$\log(\eta) = A_1 + A_2 \cdot \log(\dot{\gamma}) + A_3 \cdot T + A_4 \cdot (\log(\dot{\gamma}))^2 + A_5 \cdot \log(\dot{\gamma}) \cdot T + A_6 \cdot T^2$$

(8.23)

with the constants A_1 to A_6 and the melt temperature T.

8.4.3 Polynomial Approaches for Complex Flow Behavior

For more complicated temperature shift laws than those described so far (e.g., for in-homogeneous melts such as filled thermoplastics or liquid crystal polymers (LCPs)), more complex polynomial approaches must be used. For example, the following poly-nomial approach with 15 coefficients is used [2]:

$$\log(\eta) = S_1 + S_2 \qquad (8.24)$$

where

$$S_1 = A_1 + A_2 \cdot \log(\dot{\gamma}) + A_3 \cdot T + A_4 \cdot (\log(\dot{\gamma}))^2$$
$$+ A_5 \cdot \log(\dot{\gamma}) \cdot T + A_6 \cdot T^2 + A_7 \cdot (\log(\dot{\gamma}))^2 \cdot T + A_8 \cdot \log(\dot{\gamma}) \cdot T^2$$

$$S_2 = A_9 \cdot (\log(\dot{\gamma}))^2 \cdot T^2 + A_{10} \cdot (\log(\dot{\gamma}))^2 + A_{11} \cdot (\log(\dot{\gamma}))^3 \cdot T^2$$
$$+ A_{12} \cdot \log(\dot{\gamma}) \cdot T^3 + A_{13} \cdot T^3 + A_{14} \cdot (\log(\dot{\gamma}))^4 \cdot T + A_{15} \cdot T^4$$

A_1 to A_{15} are constants (coefficients of the polynomial).

8.5 *Task*: Determination of the Consistency Factor and the Viscosity Exponent

The viscosity curve for a plastic is given (Figure 8.8). *Calculate* the consistency factor K and the viscosity exponent n of the Ostwald and de Waele power-law approach.

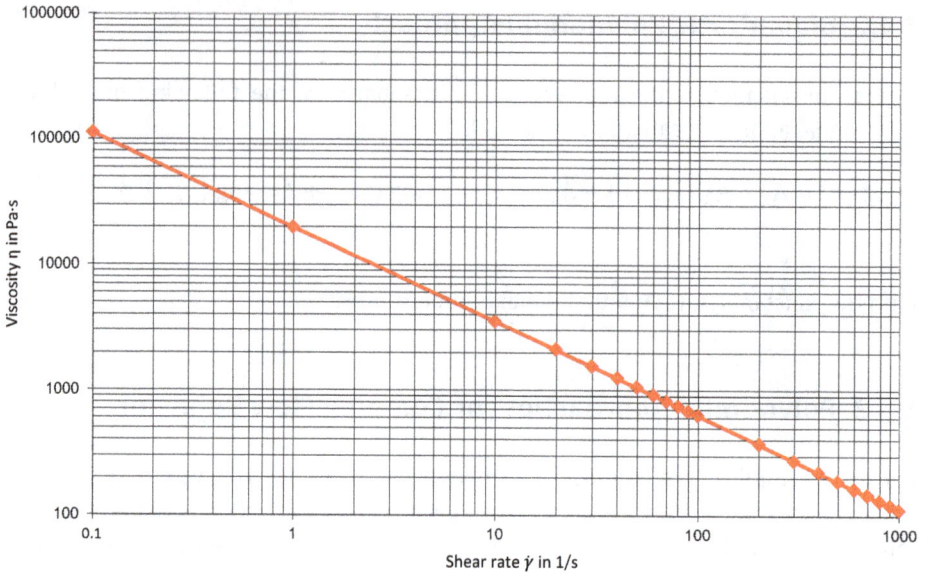

Figure 8.8 Viscosity as a function of shear rate

The power-law approach is:

$$\eta = K \cdot \dot{\gamma}^{n-1} \tag{8.25}$$

By applying the logarithm we get:

$$\log(\eta) = \log(K) + (n-1) \cdot \log(\dot{\gamma}) \tag{8.26}$$

$(n - 1)$ is therefore the gradient in the $\log(\eta)$, $\log(\dot{\gamma})$ plot of the viscosity function.

With the unit of K, it must be ensured that it depends on the value of the viscosity exponent n.

For the same material, the zero viscosity is 1000 Pa·s.

Question: What are the parameters A, B, and C of the Carreau approach?

8.6 *Task*: Comparison of the Material Laws (Power Approach and Carreau Approach)

This task is intended to compare the flow curves of the Ostwald and de Waele power approach and the Carreau approach.

- *Calculate* the viscosities for the shear rates specified in Table 8.2 and a temperature of 230 °C.

- *Draw* the flow curves on a logarithmic plot.

- *Discuss* the differences for low shear rates and also for high shear rates. Give reasons for the differences.

- *Determine* the approximation range for the power approach.

The material data of a polystyrene from BASF for a temperature of 230 °C is *given* in Table 8.3 and Table 8.4.

Table 8.2 Given Shear Rates and Calculated Viscosities with the Carreau and the Power Approaches

$\dot{\gamma}$ [s^{-1}]	η [Pa·s] Carreau and power approaches	$\dot{\gamma}$ [s^{-1}]	η [Pa·s] Carreau and power approaches
0.000001		1	
0.00001		10	
0.0001		100	
0.001		1000	
0.01		10,000	
0.1		100,000	

Table 8.3 Material Constants for the Carreau Function

Material	P_1 [Pa·s]	P_2 [s]	P_3 [-]
PS 168 N	$0.357 \cdot 10^4$	0.07441	0.8162

Table 8.4 Material Constants for the Power Approach

Material	Flow exponent [-]	Fluidity [Pa$^{-m} \cdot$ s^{-1}]
PS 168 N	4.22	$7.64 \cdot 10^{-19}$

References

[1] Menges, G.; Haberstroh, E.; Michaeli, W.; Schmachtenberg, E.: *Werkstoffkunde Kunststoffe*. Hanser, Munich, 2011

[2] Pahl, M.; Gleißle, W.; Laun, H.-M.: *Praktische Rheologie der Kunststoffe und Elastomere*. VDI-Verlag GmbH, Düsseldorf, 1995

9 Calculation of Flow Processes

The flow properties of different media have been described in the previous chapters. Viscometry/rheometry can be used to determine the material properties of the media and display the corresponding flow curves. Using mathematical approaches such as the Ostwald and de Waele power approach, the Carreau approach, or the Cross–WLF approach, these flow curves can also be described as a function of temperature.

The actual flow processes of the plastic melt play a major role in the design of injection molding and extrusion molds or hot runner systems. The relationship between the pressure and the volume flow rate of the melt is the main focus here and is required for the mathematical design. The same applies to the balancing of melt systems in extrusion and injection molding.

For this reason, the basic equations required to calculate the pressure, velocity, and volume flow rate are derived in this chapter. Specifically, it deals with the equations for:

- Shear stress and shear rate
- Velocity distribution over the flow cross section
- Volume flow rate and pressure function

First, the derivation is based on Newtonian flow behavior. The flow behavior of the plastic melt can then be taken into account by implementing the Ostwald and de Waele power approach or the Carreau approach.

9.1 Calculation of the Volume Flow Rate and Pressure Function for Newtonian Fluids

9.1.1 Assumptions for Simplifying the Equations

The basic principles for calculating flow processes are the conservation equations for mass, momentum, and energy. These equations are simplified with the help of a few assumptions. The assumptions are as follows:

- *Steady-state flow:* No change in flow over time at any point in the flow channel!

- *Laminar layer flow (creeping flow):* $Re < 2300$, always applies to plastic melt, as the inertial forces are negligible compared to the frictional forces.

- *Isothermal flow:* All fluid particles have the same temperature, i.e., cooling effects and dissipation are not taken into account!

- *Fully developed hydrodynamic flow:* No inlet and outlet effects.

- *Incompressible fluid:* Density ρ is constant. This assumption is not entirely correct, as the density of plastics depends on the pressure and temperature. This becomes clear in the *pvT* diagram, for example.

- *Neglect of elastic effects* such as normal stresses.

- *No external forces:* Gravity is neglected!

With these by no means completely unrealistic assumptions, it is possible to derive the following simple basic equations.

9.1.2 Flow Channel with Rectangular Cross Section

To derive the relationship between pressure and volume flow rate, a momentum balance on a volume element is carried out (Figure 9.1). In this case, all the forces acting on a volume element in the direction of flow are considered. Inlet and outlet effects are neglected.

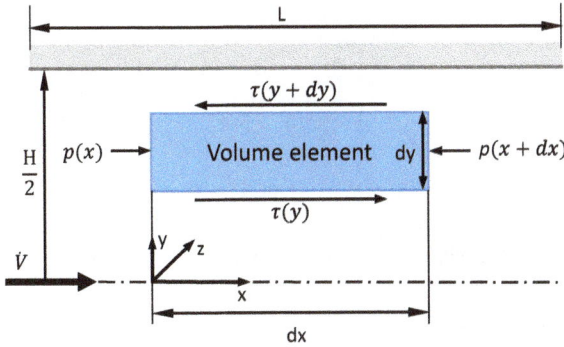

Figure 9.1 Volume element in rectangular channel

From the equilibrium of the forces on the volume element in the direction of flow, it follows:

$$p(x)\,dydz - p(x+dx)\,dydz + \tau(y)\,dxdz - \tau(y+dy)\,dxdz = 0 \qquad (9.1)$$

Development into a Taylor series—and a first-order approximation—gives:

$$p(x+dx) = p(x) + \frac{\partial p}{\partial x}dx \qquad (9.2)$$

and for

$$\tau(y+dy) = \tau(y) + \frac{\partial \tau}{\partial y}dy$$

It follows:

$$p(x)\,dy - p(x)\,dy - \frac{\partial p}{\partial x}dxdy + \tau(y)\,dx - \tau(y)\,dx - \frac{\partial \tau}{\partial y}dydx = 0 \qquad (9.3)$$

The following therefore applies:

$$-\frac{\partial p}{\partial x} - \frac{\partial \tau}{\partial y} = 0 \qquad (9.4)$$

With the assumption of fully developed incompressible flow (see Figure 9.2), the following is valid for a volume element:

$$\frac{\partial p}{\partial x} = -\frac{\Delta p}{L} \qquad (9.5)$$

As the pressure decreases in the direction of flow, the pressure difference can be assumed to be negative.

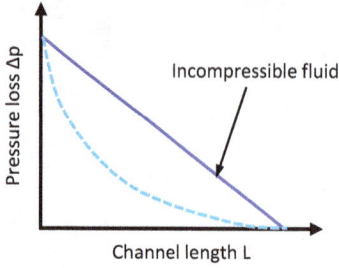

Figure 9.2

Pressure curve along the flow path for a compressible and an incompressible medium

If Equation 9.5 is inserted into Equation 9.4, then:

$$\frac{\Delta p}{L} - \frac{d\tau}{dy} = 0 \rightarrow \frac{d\tau}{dy} = \frac{\Delta p}{L} \tag{9.6}$$

The first indefinite integration results in:

$$\tau(y) = \frac{\Delta p}{L} \cdot y + c_1 \tag{9.7}$$

The integration constant c_1 results from the indefinite integration and it is determined using boundary conditions. The following can be used as boundary conditions:

- $\tau\left(y = \frac{H}{2}\right) = \tau_{max}$
- $\tau(y = 0) = 0$ as all forces must disappear for $y = 0$

With the boundary condition $\tau(y = 0) = 0$, the first constant of integration is: $c_1 = 0$. Therefore, the following results from Equation 9.7:

$$\tau(y) = \frac{\Delta p}{L} \cdot y \tag{9.8}$$

This equation makes it clear that the shear stress does not depend on the fluid. Only geometric parameters and the pressure are included in the equation, not material variables. It can also be seen that this is a linear equation. In the center of the flow channel ($y = 0$), the shear stress is zero, and at the flow channel wall ($y = H/2$), the wall shear stress is maximum (Figure 9.3).

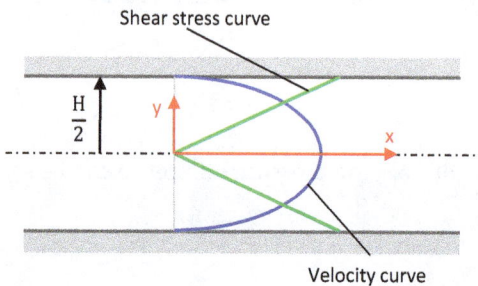

Figure 9.3

Shear stress and velocity curve for a slot flow

> The shear stress does not depend on a material quantity (viscosity), but only on the driving pressure and the flow cross section.

A material law can now be introduced into this relationship. The following applies to the simplest case of Newtonian flow behavior:

$$\tau = \eta \cdot \dot{\gamma} = -\eta \cdot \frac{dv}{dy} \tag{9.9}$$

The shear rate is replaced by the velocity gradient. This is negative, as the velocity decreases towards the wall.

If Equation 9.9 is inserted into Equation 9.8, then the following applies:

$$-\eta \frac{dv}{dy} = \frac{\Delta p}{L} \cdot y \tag{9.10}$$

After further integration, we get:

$$v(y) = -\frac{1}{\eta} \cdot \frac{\Delta p}{L} \cdot \frac{y^2}{2} + c_2 \tag{9.11}$$

The second integration constant results from a further boundary condition. Maximum velocity prevails in the center of the channel, while the velocity at the channel wall is zero, assuming wall adhesion. The following therefore applies:

- $v(y = 0) = v_{max}$
- $v\left(y = \frac{H}{2}\right) = 0$

This results in the second integration constant c_2:

$$c_2 = \frac{1}{\eta} \cdot \frac{\Delta p}{L} \cdot \frac{H^2}{8} \tag{9.12}$$

Then, the following is valid for the velocity distribution over the channel cross section:

$$v(y) = \frac{1}{\eta} \cdot \frac{\Delta p}{L} \cdot \left(\frac{H^2}{8} - \frac{y^2}{2}\right) \tag{9.13}$$

This equation states that the velocity curve has the form of a quadratic parabola. However, this only applies to Newtonian fluids, as Newton's law of friction was used in the derivation. For shear-thinning viscous fluids, the shape of the velocity curve will change as a function of the shear-thinning viscosity (m). When deriving the velocity curve for plastic melts, Newton's law of friction is not used, but a material law that takes the shear-thinning viscosity of the plastic melt into account.

The maximum velocity is, for $y = 0$:

$$v_{max} = \frac{H^2 \cdot \Delta p}{8 \cdot \eta \cdot L} \tag{9.14}$$

For the average velocity $\bar{v}$, $\dot{V} = A \cdot \bar{v}$. The volume flow rate results from the integration of the velocity curve, with $\dot{V} = \int v(y)\, dA$. Thus, it follows:

$$\bar{v} = \frac{1}{A} \int_{-\frac{H}{2}}^{+\frac{H}{2}} v(y)\, dA \tag{9.15}$$

with $dA = B \cdot dy$

If we now introduce Equation 9.13 into Equation 9.15 and integrate this with the integration limits $-H/2$ to $+H/2$, the result is the average velocity

$$\bar{v} = \frac{H^2 \cdot \Delta p}{12 \cdot \eta \cdot L} \tag{9.16}$$

For the volume flow rate $\dot{V}$, as stated above, $\dot{V} = A \cdot \bar{v}$, and from this, it follows:

$$\dot{V} = \frac{H^3 \cdot B \cdot \Delta p}{12 \cdot \eta \cdot L} \tag{9.17}$$

This equation is also known as the Hagen–Poiseuille law.

The wall shear rate can be derived from the equation for the wall shear stress and from the Hagen–Poiseuille equation:

$$\dot{\gamma}_W = \frac{6\dot{V}}{B \cdot H^2} \tag{9.18}$$

The Hagen–Poiseuille equation can be used to calculate the pressure loss if the volume flow rate is known, and vice versa. In principle, this equation only applies if the condition $B \gg H$ is fulfilled. As a good approximation, the limit value $B/H = 10$ can be used. Only if this is the case, the boundary influences can be neglected and then a one-dimensional slot flow can be assumed. If the ratio $B/H \geq 10$ is not fulfilled, the calculated volume flow rate must be corrected using a correction factor. The influence of the side walls can be taken into account using this correction factor. The correction is made as follows:

$$\dot{V}_{actual} = \dot{V}_{calculated} \cdot F_{PN} \tag{9.19}$$

The correction factor (recommended by Squire [1, 2]) F_{PN} (N: Newtonian fluid, i.e., $m = 1$) can be taken from Figure 9.4. The curve shown can be calculated for Newtonian fluids using the approximation:

$$F_{PN} = 1.008 - 0.7474 \cdot \frac{H}{B} + 0.1638 \cdot \left(\frac{H}{B}\right)^2 \tag{9.20}$$

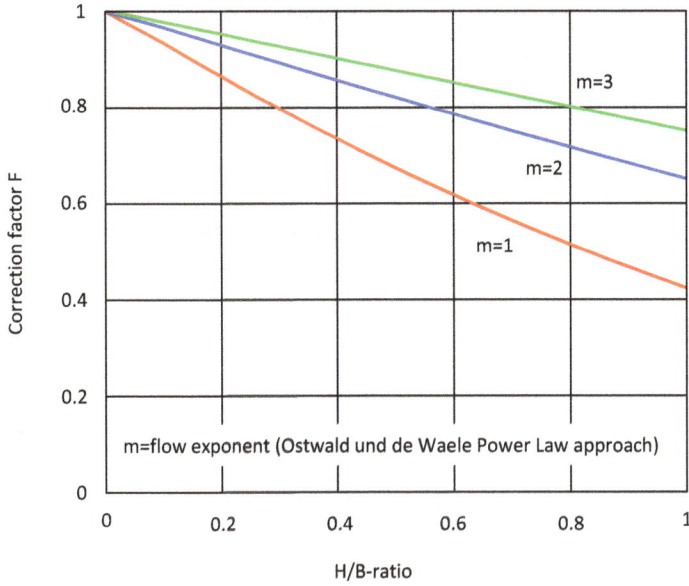

Figure 9.4 Correction factor F for the volume flow rate calculations for rectangular channels

In Figure 9.4, the correction factor F is also shown for different flow exponents. For fluids with shear-thinning viscous flow behavior ($m > 1$), the correction factor F is greater than for Newtonian fluids.

9.1.3 Flow Channel with Circular Cross Section

Using the same boundary conditions that have already been listed for the flow channel with rectangular cross section, the relationship between pressure and volume flow rate for the circular cross section(Figure 9.5) is derived.

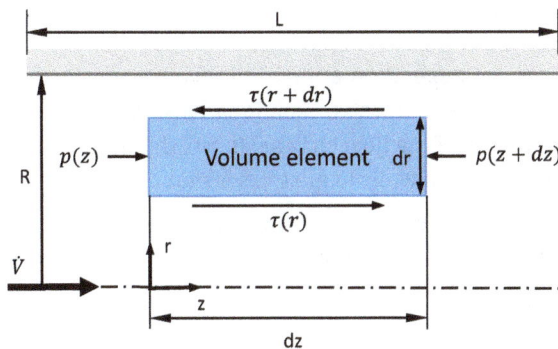

Figure 9.5 Volume element in the pipe section

From the equilibrium of the forces on the volume element in the direction of flow, it follows:

$$2\pi r \cdot dr \left(p\left(z\right) - p\left(z + dz\right)\right) + 2\pi r\tau \cdot dz - \tau\left(r + dr\right) \cdot 2\pi\left(r + dr\right) \cdot dz = 0$$

(9.21)

Development into a Taylor series—and a first-order approximation— provides the following:

$$p\left(z + dz\right) = p\left(z\right) + \frac{\partial p}{\partial z}dz$$

(9.22)

$$\tau\left(r + dr\right) = \tau\left(r\right) + \frac{\partial\tau}{\partial r}dr$$

(9.23)

Assuming a fully developed incompressible flow, the following applies:

$$\frac{\partial p}{\partial z} = -\frac{\Delta p}{L}$$

(9.24)

This is also negative, as the pressure decreases in the direction of flow. If all higher order elements are neglected, the following differential equation is obtained:

$$\frac{\Delta p}{L} = \frac{\tau}{r} + \frac{d\tau}{dr} = \frac{1}{r}\frac{\partial}{\partial r}\left(\tau r\right)$$

(9.25)

After integration:

$$\tau(r) = \frac{\Delta p r}{2L} + \frac{1}{r}c_1$$

(9.26)

Boundary conditions are formulated again to determine the integration constant c_1. For $r = 0$, all forces must disappear, so the first boundary condition is $\tau(r = 0) = 0 \implies c_1 = 0$. Thus:

$$\tau(r) = \frac{\Delta p r}{2L}$$

(9.27)

This means that the shear stress does not depend on a material quantity (viscosity), but only on the driving pressure and the flow cross section. It can also be seen that this is again a linear equation. In the center of the flow channel ($r = 0$), the shear stress is zero, and at the flow channel wall ($r = R$), the wall shear stress is maximum.

A material law can now be introduced into the relationship for the shear stress. The following applies to the simplest case of Newtonian flow behavior:

$$\tau = \eta \cdot \dot{\gamma} = -\eta\frac{dv_z}{dr}$$

(9.28)

This is negative, as the velocity decreases towards the channel wall.

If the relationship from Equation 9.28 is substituted into Equation 9.27, the result is

$$-\eta \frac{dv_z}{dr} = \frac{\Delta p r}{2L} \tag{9.29}$$

It follows:

$$\frac{dv_z}{dr} = -\frac{\Delta p r}{2 \cdot \eta \cdot L} \tag{9.30}$$

After another indefinite integration and taking into account the boundary condition $v_z = 0$ for $r = R$ (wall adhesion), the velocity curve over the channel cross section is as follows:

$$v_z\,(r) = \frac{R^2 \cdot \Delta p}{4 \cdot \eta \cdot L} \left[1 - \left(\frac{r}{R}\right)^2 \right] \tag{9.31}$$

Here, too, a velocity profile of a quadratic parabola is obtained. Again, this only applies to Newtonian fluids. In the case of shear-thinning viscous fluids, the velocity profile also changes depending on the shear-thinning viscosity (m).

The maximum flow velocity is, for $r = 0$:

$$v_{z,\text{max}} = \frac{R^2 \cdot \Delta p}{4 \cdot \eta \cdot L} \tag{9.32}$$

It can also be used for writing the parabolic velocity curve for Newtonian fluids (Figure 9.6) as:

$$v_z\,(r) = v_{z,\text{max}} \left[1 - \left(\frac{r}{R}\right)^2 \right] \tag{9.33}$$

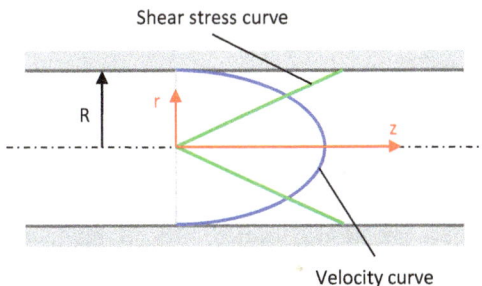

Shear stress curve

Velocity curve

Figure 9.6
Shear stress and velocity curve over the channel cross section

For the average velocity, the following applies:

$$\overline{v_z} = \frac{1}{A} \int\limits_{0}^{R} v\,(r)\,dA \tag{9.34}$$

with $dA = 2\pi r \cdot dr$

This means that the average velocity $\overline{v_z}$ is:

$$\overline{v_z} = \frac{R^2 \cdot \Delta p}{8 \cdot \eta \cdot L} \tag{9.35}$$

A comparison of the average velocity with the maximum velocity results in:

$$v_{average} = \frac{v_{maximum}}{2} \tag{9.36}$$

However, this relationship only applies to Newtonian fluids, as will be shown later. For the volume flow rate $\dot{V}$, $\dot{V} = A \cdot \overline{v}$, and from that we obtain:

$$\dot{V} = \frac{\pi \cdot R^4 \cdot \Delta p}{8 \cdot \eta \cdot L} \tag{9.37}$$

This equation is also known as the Hagen–Poiseuille law.

The wall shear rate can be derived from the equation for the wall shear stress and from the Hagen–Poiseuille equation:

$$\dot{\gamma}_W = \frac{4\dot{V}}{\pi \cdot R^3} \tag{9.38}$$

The Hagen–Poiseuille equation can be used to calculate the pressure loss if the volume flow rate is known, and vice versa. As can be seen (Equation 9.37), this equation includes the radius to the fourth power. This means that even a slight change in the radius results in a significant change in the volume flow rate. This effect is even more pronounced with shear-thinning viscous materials.

9.1.4 Channel with Circular Ring Cross Section

Circular ring and rectangular cross sections are closely related. The circular ring cross section is derived from the rectangular cross section if you imagine it bent together from a rectangle. This is permissible if the height of the rectangular gap is small compared to the mean diameter of the circular ring. For this condition, a channel with a one-dimensional flow is again obtained. It is neglected that the local area $dA = 2\pi r\,dr$ of the annulus increases from the inside to the outside. The error caused by this is very small for the usual dimensions of circular ring surfaces (e.g., plastic pipes). The permissible limit value can be assumed here: $s/R_m < 0.1$ (see relationships used below).

The equations for the shear stress, the volume flow rate, and the shear rate are obtained from the equations for the rectangular gap. In these equations, the following relationships are used (Figure 9.7):

- $H = R_a - R_i = s$
- $B = (R_a + R_i) \cdot \pi = 2 \cdot R_m \cdot \pi$

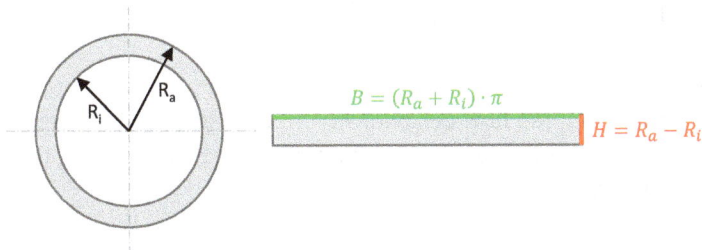

Figure 9.7 Circular ring cross section and unwinding

The complete equation for calculating the pressure loss for a circular ring is as follows:

$$\frac{\Delta p}{z} = \frac{8 \cdot \dot{V} \cdot \eta}{\pi \cdot R_a^4 \left[1 - \left(\frac{R_i}{R_a}\right)^4 - \frac{\left(1 - \left(\frac{R_i}{R_a}\right)^2\right)^2}{\ln\left(\frac{1}{\frac{R_i}{R_a}}\right)} \right]} \tag{9.39}$$

9.2 Calculation of the Volume Flow Rate and Pressure Function for Shear-Thinning Viscous Fluids

9.2.1 Consideration of the Shear-Thinning Viscosity Using the Power Approach

To derive the equations for the shear-thinning viscous fluid, the mathematical description of the flow curve is used instead of Newton's law of friction for the calculations in Section 9.1. Otherwise, the derivation of the equations for the velocity and the volume flow rate is the same as for the Newtonian fluid. The equations for pipe flow with a circular cross section are listed below as an example. The same relationship applies to the shear stress curve as for the Newtonian fluid, since no material variables are included in this equation.

$$\tau\,(r) = \frac{\Delta p \cdot r}{2L} \tag{9.40}$$

To calculate the velocity, the equation derived from the momentum balance uses the power-law approach of Ostwald and de Waele ($\dot{\gamma} = \phi \cdot \tau^m$) instead of Newton's law of friction ($\tau = \eta \cdot \dot{\gamma}$). This provides the local velocity:

$$v\,(r) = v_{max} \left(1 - \left(\frac{r}{R} \right)^{m+1} \right) \tag{9.41}$$

It is immediately clear that the velocity curve is now a function of the shear-thinning viscosity of the plastic. For Newtonian flow behavior, $m = 1$. This means that the velocity curve has the shape of a quadratic parabola. For plastics, $1 < m < 4$, meaning that the velocity curve changes as a function of m.

With

$$v_{max} = \frac{R^{m+1} \cdot \phi \cdot \Delta p^m}{2^m \cdot (m+1) \cdot L^m} \tag{9.42}$$

the average velocity is calculated:

$$v_{average} = v_{max} \cdot \frac{m+1}{m+3} \tag{9.43}$$

This equation indicates that the average velocity no longer corresponds to half of the maximum velocity.

The volume flow rate results from the continuity relationship:

$$\dot{V} = \frac{\pi \cdot R^{m+3} \cdot \phi \cdot \Delta p^m}{2^m \cdot (m+3) \cdot L^m} = v_{max} \cdot \frac{m+1}{m+3} \cdot \pi \cdot R^2 \tag{9.44}$$

This equation shows that the radius is now included in the calculation of the volume flow rate with an exponent greater than four ($m + 3$). In this regard, a change in the radius for shear-thinning viscous fluids has an even greater effect on the pressure loss or the volume flow rate than for Newtonian fluids.

For the wall shear rate, the following results:

$$\dot{\gamma}_W = \frac{(m+3) \cdot \dot{V}}{\pi \cdot R^3} \tag{9.45}$$

This equation shows that the wall shear rate of shear-thinning viscous media is always greater than that of Newtonian media.

All equations that apply to shear-thinning viscous fluids become the corresponding equations for Newtonian fluids if $m = 1$ and $\phi = 1/\eta$.

When calculating flows in the rectangular channel, a correction factor must be introduced if the condition $B/H > 10$ is not fulfilled. The calculated volume flow rate must be converted into the actual volume flow rate using the following equation:

$$\dot{V}_{actual} = \dot{V}_{calculated} \cdot F_{PS} \tag{9.46}$$

The correction factor F_{PS} (S: shear-thinning viscous fluids) is greater than the correction factor F_{PN} (N: Newtonian fluids) because the local volume flow rate shifts more towards the center of the pipe as the shear-thinning viscosity increases. The wall influence thus decreases. The following is valid for the relationship between F_{PN} and F_{PS}:

$$F_{PS} = F_{PN}^{\frac{1}{m}} \tag{9.47}$$

The curve of the correction factors for $m = 2$ and $m = 3$ has already been shown graphically in a diagram (Figure 9.4).

Another possible correction is to introduce a substitute channel height instead of the specified channel height. The equation is:

$$H_{substitute} = \left(1 - 0.35\frac{H}{B}\right)H \tag{9.48}$$

This correction has approximately the same effect as $F_{PN}^{1/2}$.

9.2.2 Consideration of the Shear-Thinning Viscosity Using the Carreau Approach

When using the Carreau approach, the equation for the volume-flow-rate–pressure function and the material law are structured similarly. If the volume flow rate is specified, the pressure requirement can be calculated using

$$\Delta p = \frac{L\dot{V}\overline{a}}{\left(1 + \overline{b}\dot{v}\right)^{C}} \tag{9.49}$$

For the circular nozzle, the following applies for $\overline{a}$ and $\overline{b}$:

$$\overline{a} = \frac{8A}{\pi R^4} \tag{9.50}$$

$$\overline{b} = \left(\frac{4 - 4C}{4 - 3C}\right)^{\frac{1}{C-1}} \cdot \left(\frac{4B}{\pi R^3}\right) \tag{9.51}$$

If, at a given pressure, the volume flow rate is required, an iteration calculation is necessary, since $\dot{V}$ is on both sides of the equation:

$$\dot{V} = \frac{\Delta p \left(1 + \bar{b}\dot{v}\right)^{C}}{\bar{a}L} \tag{9.52}$$

9.3 Normalized Velocity and Shear Rate Curves

In this section, the curves of the velocity and the shear rate profiles of the Newtonian and the shear-thinning viscous fluids are compared with each other. The flow behavior of the shear-thinning viscous fluid is described using the power approach. As a simplification, it is assumed that the flow exponent m does not depend on the shear rate.

For the graphical representation of the velocity profiles, the equation from Section 9.1 and Section 9.2 is used in a dimensionless, that is, normalized, form. The representation can then be made independently of the individual example. The local velocity can be related to the maximum velocity. Thus, the following is valid:

$$\frac{v(r)}{v_{max}} = 1 - \left(\frac{r}{R}\right)^{m+1} \tag{9.53}$$

with

$$v_{max} = \phi \cdot \frac{2}{m+1} \cdot \left(\frac{R}{2}\right)^{m+1} \cdot \left(\frac{\Delta p}{L}\right)^{m} \tag{9.54}$$

Figure 9.8 shows the dimensionless representation of the relative velocity as a function of the relative pipe radius for different flow exponents. In this example, the local velocity $v(r)$ was related to the average velocity $v_{average}$, which was plotted on the y axis. The relative pipe radius r/R is plotted on the x axis. For $r/R = 0$, we are in the middle of the channel, and for $r/R = 1$, we are looking at the channel wall.

$$\frac{v(r)}{v_{average}} = \frac{m+3}{m+1} \cdot \left[1 - \left(\frac{r}{R}\right)^{m+1}\right] \tag{9.55}$$

with

$$v_{average} = v_{max}\frac{m+1}{m+3} \tag{9.56}$$

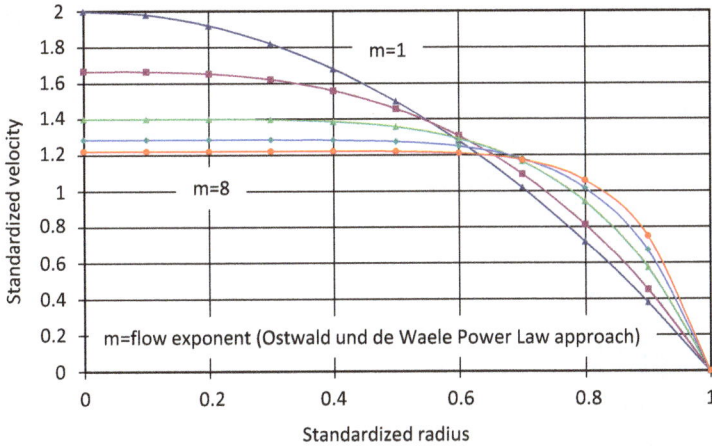

Figure 9.8 Normalized velocity as a function of the normalized pipe radius

The profiles in Figure 9.8 correspond to the flow exponents $m = 1$ (Newtonian fluid), $m = 2$, $m = 4$, $m = 6$, and $m = 8$. A parabolic profile is displayed for the Newtonian flow. With increasing shear-thinning viscosity (m increases), the curve becomes flatter, that is, the flow profile becomes "blunter". It can be seen that as m increases, the profiles increasingly approach the plug flow. Additionally, it can be observed that the velocity gradient on the channel wall increases as m increases. This means that the shear rate at the channel wall increases as a function of m. It is also clear that the relationship $v_{average} = v_{max}/2$ only applies to Newtonian fluids in the center of the channel.

The equations for the description of the shear rate profiles are obtained by differentiating the equation for the velocity profile (Equation 9.53) with respect to r:

$$\frac{dv(r)}{dr} = \dot{\gamma}(r) = \frac{(m+1)\, v_{max}}{R} \left(\frac{r}{R}\right)^m \tag{9.57}$$

The shear rate profiles are also described in a dimensionless way. For this purpose, the local shear rate is related to the maximum wall shear rate of the shear-thinning viscous fluid. This is obtained from the equation for $r = R$. The following applies:

$$\frac{\dot{\gamma}(r)}{\dot{\gamma}_w} = \left(\frac{r}{R}\right)^m \tag{9.58}$$

This equation shows that, for $m = 1$, the shear rate is linear. The larger the value m, the more parabolic the shear rate curve.

Figure 9.9 shows schematically the velocity, shear rate, and shear stress profiles and, as a supplement, the viscosity profiles for Newtonian flow, shear-thinning viscous flow, and—as a special case—plug flow. The shear stress profiles are the same in all three cases, as this equation does not contain any material parameters, regardless of the type of fluid. Boundary influences (freezing effects, for example) are neglected.

The plug flow occurs with very pronounced shear-thinning viscosity or with highly filled polymers.

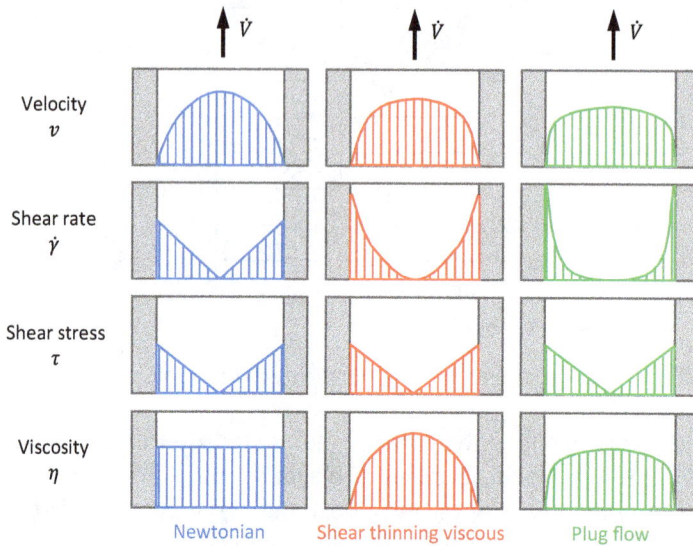

Figure 9.9 Velocity, shear rate, shear stress, and viscosity curves for different fluids

If the maximum velocity of a fluid with shear-thinning viscosity is compared with that of a Newtonian fluid, it is initially assumed that the volume flow rate is the same. Equal volume flow rate means that, taking into account the continuity relationship, the average velocities of Newtonian (N) and shear-thinning viscous fluids (S) are the same for the same pipe radius. From the equations

$$\overline{v}_{Newtonian} = \frac{v_{max,Newtonian}}{2} \tag{9.59}$$

and

$$\overline{v}_{shear-thinning} = v_{max,shear-thinning}\,\frac{m+1}{m+3} \tag{9.60}$$

we obtain:

$$\frac{v_{max,Newtonian}}{2} = v_{max,shear-thinning}\,\frac{m+1}{m+3} \tag{9.61}$$

It follows from this:

$$\frac{v_{max,shear-thinning}}{v_{max,Newtonian}} = \frac{m+3}{2\,(m+1)} \tag{9.62}$$

Flow Exponent	1	2	3	4	5
$V_{max,shear-thinning}$ / $V_{max,Newtonian}$	1	0.83	0.75	0.7	0.66

The values in the table above show that the maximum velocity in the center of the flow channel decreases with increasing shear-thinning viscosity (increasing m). Since the average velocity remains the same, the velocity profile must change depending on the shear-thinning viscosity.

9.4 *Task*: Effect of the Flow Channel on the Melt Volume Flow Rate

The effect of the manufacturing accuracy of the flow channel diameter on the melt volume flow rate will now be investigated.

The following data is given for an injection mold with two cavities. Starting from the bar gate, the melt flow is divided and directed to the respective cavities. L_1 and L_2 are the two flow paths to the cavities. R_1 and R_2 are the radii of the two flow paths. The radius of the second flow path is only 0.01 mm larger than the radius of the first one. The power approach is used to take the shear-thinning viscosity into account.

Given data:

- $R_1 = 0.5$ mm
- $R_2 = 0.51$ mm
- $L = 80$ mm
- $\Delta p = 400$ bar
- $m = 2$
- $\phi = 2 \times 10^5 \, mm^{2m}/(N^m s)$

We are looking for the volume flow rates $\dot{V}_1 = f(R_1)$ and $\dot{V}_2 = f(R_2)$ and then the percentage difference.

a) First calculate the volume flow rates using the coupled (implemented) approach (Equation 9.44).

b) Afterwards, calculate the volume flow rates using the Hagen–Poiseuille approach without the integrated power approach. To do this, you must first calculate the viscosity. Use the given pressure, and also use it to determine the wall shear stresses (Equation 9.27). Then, calculate the viscosities using the power approach of Ostwald and de Waele. Finally, the volume flow rates can be determined using the Hagen–Poiseuille equation.

c) Compare the results and discuss them.

References

[1] Drazin, P. G.; Reid, W. H.: *Hydrodynamic Stability*, Cambridge University Press, 1981

[2] Squire, H. B.: On the Stability of Three-Dimensional Disturbances in Viscous Flow. *Proceedings of the Royal Society A*, 1933, Vol. 142, pp. 621–628

10 The Representative Shear Rate Method

The application of the Hagen–Poiseuille equations can only be successful if the shear-thinning viscous behavior of the plastics and the temperature dependence of the viscosity are taken into account when determining the viscosity. To simplify matters, it is initially assumed that the viscosity is constant across the cross section. Average or representative values [1–3] are therefore used to determine the viscosity.

With regard to the shear rate, the following model is used for this purpose. If a mean value is defined for the flow rate, the shear rate distribution associated with this mean value can be calculated. For Newtonian fluids, the shear rate increases linearly from the center of the channel to the wall of the channel, since the velocity curve corresponds to a parabola and the shear rate is the velocity gradient. The shear-thinning viscous behavior of plastics can be described by the flow exponent m (power approach of Ostwald and de Waele). The greater m is, the more shear-thinning viscous the plastic behavior is. The velocity curve can also be described as a function of m. The greater m is, the "blunter" (flatter) the velocity profile becomes. As a result, the curve of the shear rate also changes as a function of m. For $m = 2$, a parabola is obtained for the shear rate. The greater m is, the "blunter" the curve of the shear rate becomes. This is illustrated by the following equation:

$$\frac{dv(r)}{dr} = \dot{\gamma}(r) = \frac{(m+1)\,v_{max}}{R}\left(\frac{r}{R}\right)^m \tag{10.1}$$

In the center, the shear rate assumes the value zero for all media. At the wall, the shear rate increases with increasing m due to the increasing velocity gradient. As a result, the shear rate curves of the Newtonian and the shear-thinning viscous media must intersect. This intersection point is referred to as the representative point, which is used for the calculation (Figure 10.1). For a pipe cross section, this representative point is approximately 0.815 when viewed from the center and 0.772 for a rectangular cross section.

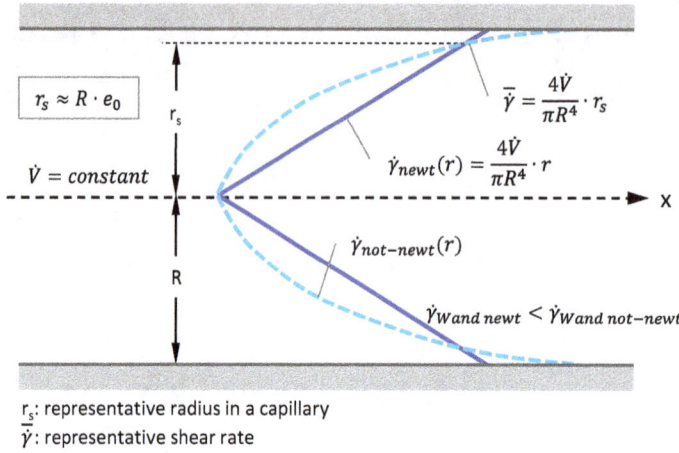

$r_s \approx R \cdot e_0$

$\dot{V} = constant$

r_s

R

$$\overline{\dot{\gamma}} = \frac{4\dot{V}}{\pi R^4} \cdot r_s$$

$$\dot{\gamma}_{newt}(r) = \frac{4\dot{V}}{\pi R^4} \cdot r$$

$\dot{\gamma}_{not-newt}(r)$

$\dot{\gamma}_{Wand\,newt} < \dot{\gamma}_{Wand\,not-newt}$

r_s: representative radius in a capillary
$\overline{\dot{\gamma}}$: representative shear rate

Figure 10.1 Method for determining the "representative" shear rate

If the shear rate curves are plotted as a function of the shear-thinning viscosity (flow exponent m), it can be seen that the point of intersection with the curve of the Newtonian material ($m = 1$) shifts towards the channel wall for an increasing m and therefore the representative point is not constant. This is also shown in Figure 10.2. However, for the typical range of shear-thinning viscosity of plastics, the intersection points at $e_{circle} = 0.815$ and $e_{rectangle} = 0.772$ can be assumed to be sufficiently accurate.

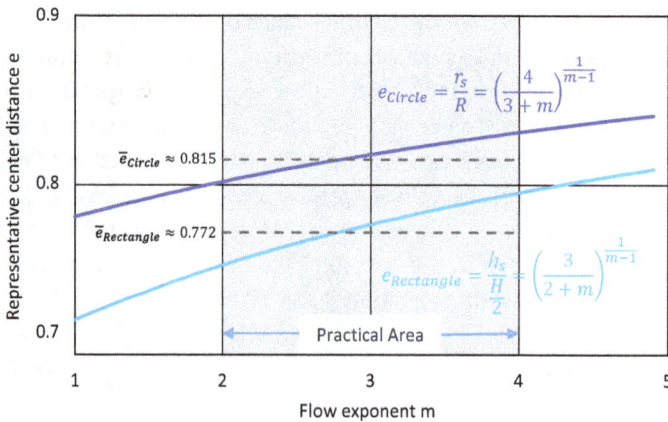

$$e_{Circle} = \frac{r_s}{R} = \left(\frac{4}{3+m}\right)^{\frac{1}{m-1}}$$

$\overline{e}_{Circle} \approx 0.815$

$\overline{e}_{Rectangle} \approx 0.772$

$$e_{Rectangle} = \frac{h_s}{\frac{H}{2}} = \left(\frac{3}{2+m}\right)^{\frac{1}{m-1}}$$

Practical Area

Representative center distance e

Flow exponent m

Figure 10.2 Representative distance e from the channel center for plastic melts with different flow exponents

The equations of the representative shear rate for the three basic geometries are as follows:

Disk:

$$\bar{\dot{\gamma}} = e_{\text{rectangle}} \frac{6 \cdot \bar{v}_r}{H} = e_{\text{rectangle}} \frac{3 \cdot \dot{V}}{\pi \cdot r \cdot H^2} \tag{10.2}$$

Plate:

$$\bar{\dot{\gamma}} = e_{\text{rectangle}} \frac{6 \cdot \bar{v}_x}{H} = e_{\text{rectangle}} \frac{6 \cdot \dot{V}}{B \cdot H^2} \tag{10.3}$$

Pipe:

$$\bar{\dot{\gamma}} = e_{\text{circle}} \frac{8 \cdot \bar{v}_z}{D} = e_{\text{circle}} \frac{32 \cdot \dot{V}}{\pi \cdot D^3} \tag{10.4}$$

where
$e_{\text{rectangle}} = 0.772$
$e_{\text{circle}} = 0.815$

References

[1] Schümmer, P.: Rheologie 1 Umdruck zur Vorlesung, Institut für mechanische Verfahrenstechnik, RWTH Aachen, 1992

[2] Giesekus, H.; Langer, G.: Die Bestimmung der wahren Fließkurven nicht-newtonscher Flüssigkeiten und plastischer Stoffe mit der Methode der repräsentativen Viskosität, *Rheologica Acta*, January 1977, Vol. 16, Issue 1, pp. 1–22

[3] Menges, G.; Haberstroh, E.; Michaeli, W.; Schmachtenberg, E.: *Werkstoffkunde der Kunststoffe*. Hanser, Munich, 2011

11 Calculation of Flow Processes during Injection Molding

For the design and construction of injection molds, it is necessary to know the injection pressure required to fill the mold. This enables the mold designer to estimate the necessary clamping force, for example, and thus to determine the machine size or obtain information about the maximum number of cavities that can be filled. Furthermore, the pressure requirement calculation is a prerequisite for the design of sprue systems that are not naturally balanced [1]. In this chapter, some simple calculation strategies that allow the injection pressure requirement to be estimated during the filling process will be shown.

11.1 Model Presentation

The bases for calculating the mold filling process are the conservation equations for mass, momentum, and energy known from thermodynamics and fluid mechanics. To formulate these equations in practical cases, it is useful to break down complicated molded parts into so-called basic geometries, as shown in Figure 11.1. For these basic geometries, mainly circular disks, plate-shaped, and cylindrical molded part areas, the balance equations are established.

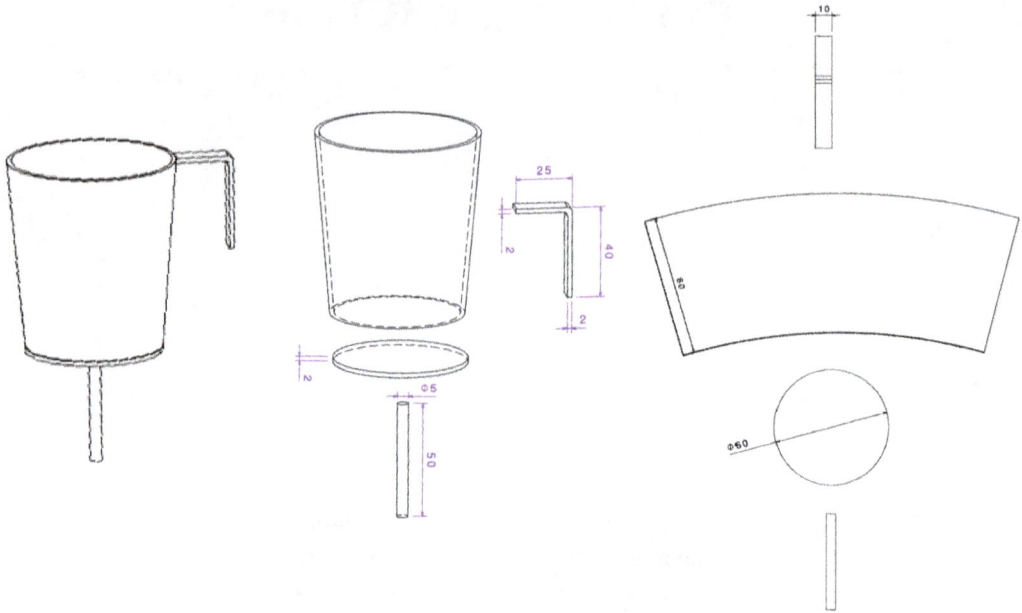

Figure 11.1 Breakdown of a molded part into basic geometries [1]

The following simplifications are employed:

- Incompressibility of the melt
- Laminar layered flow
- Wall adhesion
- Neglect of acceleration and gravitational forces
- Isothermal conditions
- Temporally and spatially averaged conservation equations

On this basis, the Hagen–Poiseuille law can be derived from the conservation equations for mass and momentum for the three basic geometries. (See these basic geometries in Figure 11.2.)

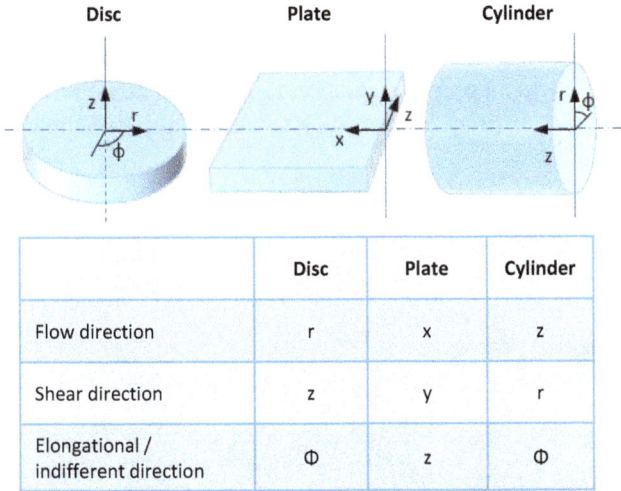

	Disc	Plate	Cylinder
Flow direction	r	x	z
Shear direction	z	y	r
Elongational / indifferent direction	Φ	z	Φ

Figure 11.2 Representation of the three basic geometries

Disk:

$$\frac{\Delta p}{r} = \frac{12 \cdot \bar{v}_r \cdot \eta}{H^2}$$

(11.1)

Plate:

$$\frac{\Delta p}{x} = \frac{12 \cdot \bar{v}_x \cdot \eta}{H^2}$$

(11.2)

Cylinder:

$$\frac{\Delta p}{z} = \frac{32 \cdot \bar{v}_z \cdot \eta}{D^2}$$

(11.3)

From the continuity equation $\dot{V} = \bar{v} \cdot A$, it follows:

Disk:

$$\frac{\Delta p}{r} = \frac{6 \cdot \dot{V} \cdot \eta}{\pi \cdot R \cdot H^3}$$

(11.4)

Plate:

$$\frac{\Delta p}{x} = \frac{12 \cdot \dot{V} \cdot \eta}{B \cdot H^3}$$

(11.5)

Cylinder:

$$\frac{\Delta p}{z} = \frac{128 \cdot \dot{V} \cdot \eta}{\pi \cdot D^4}$$

(11.6)

It is noticeable that in Equation 11.4 the pressure gradient in the disk geometry for a given volume flow rate is a function of the radius. For $r = R$, the radius is simplified from the equation. Furthermore, when applying Equation 11.1, it must be noted that the velocity in the disk equation is a function of r. As r increases, the velocity decreases at a constant volume flow rate.

The viscosity that occurs at a representative shear rate is used in these equations. The following then applies to the three basic geometries:

Disk:

$$\bar{\dot{\gamma}} = e_{\text{rectangle}} \cdot \frac{6 \cdot \bar{v}_r}{H} = e_{\text{rectangle}} \cdot \frac{3 \cdot \dot{V}}{\pi \cdot r \cdot H^2} \tag{11.7}$$

Plate:

$$\bar{\dot{\gamma}} = e_{\text{rectangle}} \cdot \frac{6 \cdot \bar{v}_x}{H} = e_{\text{rectangle}} \cdot \frac{6 \cdot \dot{V}}{B \cdot H^2} \tag{11.8}$$

Cylinder:

$$\bar{\dot{\gamma}} = e_{\text{circle}} \cdot \frac{8 \cdot \bar{v}_z}{D} = e_{\text{circle}} \cdot \frac{32 \cdot \dot{V}}{\pi \cdot D^3} \tag{11.9}$$

where
$e_{\text{rectangle}} = 0.772$
$e_{\text{circle}} = 0.815$

The equations presented assume isothermal conditions. In the real injection molding process, however, the melt temperature often changes considerably, as dissipation caused by shearing adds energy to the material, on the one hand, and energy is removed by the cold mold wall, on the other hand. In order to capture this influence, it would be necessary to be able to solve the energy equation and the momentum equation together. However, this is only possible with great numerical effort and it is therefore not recommended for initial rough calculations. Consequently, first of all, we will assume isothermal conditions for the sake of simplicity.

A simple way of calculating the viscosity as a function of the shear rate and the temperature is the power-law approach in the following form:

$$\eta = K \cdot a_T{}^n \cdot \dot{\gamma}^{n-1} \tag{11.10}$$

Another form of the power-law approach is

$$\eta = a_T{}^{\frac{1}{m}} \cdot \phi^{-\frac{1}{m}} \cdot \dot{\gamma}^{\frac{1-m}{m}} \tag{11.11}$$

The temperature influence a_T can be calculated using an Arrhenius approach in the form

$$a_T = e^{\left[\frac{E_0}{R} \left(\frac{1}{T} - \frac{1}{T_{\text{reference}}} \right) \right]} \tag{11.12}$$

or with the William–Landel–Ferry approach (WLF approach):

$$\log{(a_T)} = \frac{8.86\,(T_{\text{reference}} - T_s)}{101.6K + (T_{\text{reference}} - T_s)} - \frac{8.86\,(T - T_s)}{101.6K + (T - T_s)} \tag{11.13}$$

The power-law approach approximates the viscosity function on a double-logarithmic scale using a straight line. In the range of higher shear rates, which are usually present during the mold filling process, the power-law approach provides a good approximation of the real viscosity function.

The values for the parameters of the material laws, which can be determined with the aid of regression analyses, are given for some thermoplastics in the materials table in Chapter 17.

The flattening of the viscosity function at low shear rates (Newtonian flow behavior) cannot be captured with the power-law approach. For this reason, the specified range of validity (shear rate range) must always be strictly observed. This is because at low shear rates, it is possible that a too-high viscosity is determined using the power law, which leads to incorrect calculations, especially for materials with a Newtonian flow behavior over a wide range of shear rates in the viscosity function (see Figure 8.6).

A constitutive law that approximates the viscosity function in the low shear rate range better than the power law is the Carreau–WLF approach:

$$\eta = \frac{a_T \cdot P_1}{(1 + a_T \cdot \dot{\gamma} \cdot P_2)^{P_3}} \tag{11.14}$$

Other approaches used to describe the viscosity of plastic melts include the following:

- The Cross–Williamson model
- The four-parameter Carreau model
- The De Kee model
- The Carreau–Yasuda model
- The Bingham model
- The Casson model
- The Herschel–Bulkley model

As explained in Chapter 9, in principle, the equations of Hagen–Poiseuille only apply if the condition $B \gg H$ is fulfilled. As a good approximation, the limit value $B/H = 10$ can be used, the boundary influences can be neglected, and a one-dimensional slot flow can be assumed. If $B/H \geq 10$ is not fulfilled, the calculated volume flow rate must be corrected using a correction factor, which allows the influence of the side walls to be taken into account:

$$\dot{V}_{\text{actual}} = \dot{V}_{\text{calculated}} \cdot F_{\text{PS}} \tag{11.15}$$

The relationship between the correction factors F_{PS} (S: shear-thinning viscous fluids) and F_{PN} (N: Newtonian fluids)—where F_{PS} is greater than F_{PN} because the local volume flow rate shifts more towards the center of the pipe as the shear-thinning viscosity increases (the wall influence thus decreases)—is the following:

$$F_{PS} = F_{PN}^{\frac{1}{m}} \tag{11.16}$$

Figure 11.3 shows the correction factor F for different flow exponents: for fluids with shear-thinning viscous flow behavior ($m > 1$), F is greater than for Newtonian fluids. The correction factor (recommended by Squire) F_{PN} ($m = 1$) can be taken from this figure, where the curve shown can be calculated for Newtonian fluids using the approximation

$$F_{PN} = 1.008 - 0.7474 \cdot \frac{H}{B} + 0.1638 \cdot \left(\frac{H}{B}\right)^2 \tag{11.17}$$

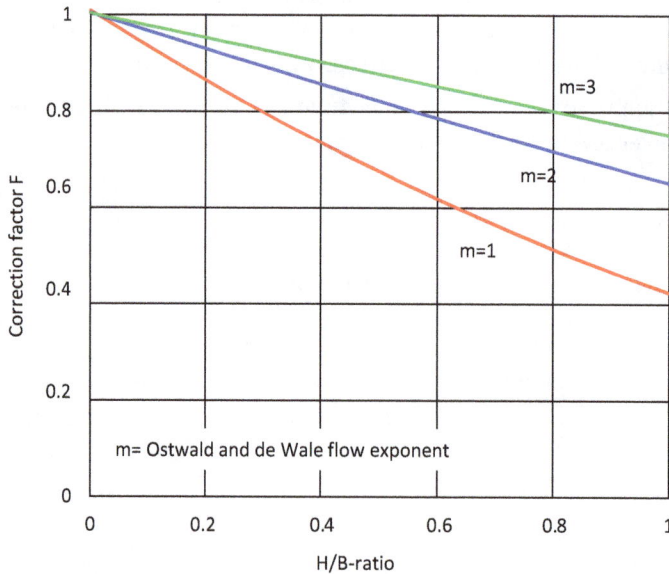

Figure 11.3 Correction factor F for the volume flow rate calculations for rectangular channels

Another possible correction (which has approximately the same effect as the correction factor $F_{PN}^{1/2}$) is to introduce a substitute channel height instead of the specified channel height:

$$H_{substitute} = \left(1 - 0.35 \cdot \frac{H}{B}\right) \cdot H \tag{11.18}$$

11.2 General Procedure for Calculating Pressure Loss

The pressure loss is calculated in steps. The following variables must be determined:

1. Average flow front rate or velocity (volume flow rate)

2. Representative shear rate

3. Temperature shift factor a_T

4. Viscosity ($\eta = f(\dot{\gamma}, \vartheta_{melt})$)

5. Pressure loss with Hagen–Poiseuille

11.2.1 *Tasks*: Calculation Examples

11.2.1.1 Pressure Loss: Plate Geometry

Consider (*given*) a plate-shaped molded part as shown in Figure 11.4 (see Table 11.1). The molded part is gated via a strip gate with a manifold channel on one end face. The pressure losses in the gate/gating system are to be neglected in the calculation. Furthermore, isothermal conditions are assumed. This means that dissipation and cooling effects are neglected.

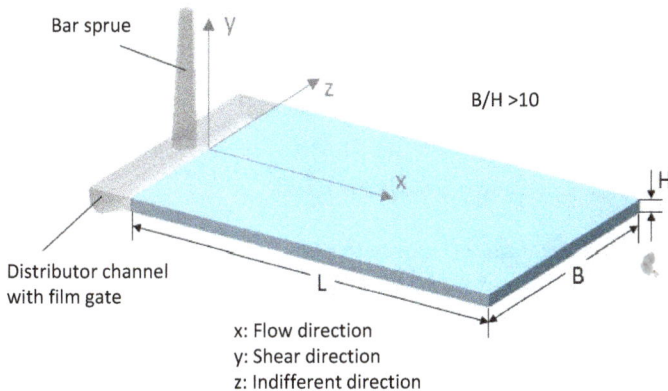

x: Flow direction
y: Shear direction
z: Indifferent direction

Figure 11.4 Molded panel with lateral gating

Table 11.1 Calculation Data

Geometric data	Length L	100 mm
	Width B	30 mm
	Wall thickness H	3 mm
Processing data	Filling time	1.8 s
	Processing temperature	240 °C
	Cavity temperature	50 °C
Material data	Type	Polypropylene P 1320 L

Question: How high is the pressure required to fill the molded part?

Calculate the pressure first with the average flow front rate (velocity) and then with the volume flow rate.

11.2.1.2 Pressure Loss: Disk Geometry

Figure 11.5
Display of a DVD

Consider (*given*) a DVD with the characteristics listed in Table 11.2 (Figure 11.5). The molded part is injected centrally via a bar gate and a screen gate. Isothermal conditions can be assumed for the calculations. This means that dissipation and cooling effects are neglected.

Table 11.2 Calculation Data

Geometric data	Bar sprue diameter D	3 mm
	Bar sprue length L	25 mm
	DVD diameter D_D	120 mm
	DVD wall thickness H	0.5 mm
Processing data	Injection time	1.2 s
	Processing temperature	300 °C
	Mold temperature	120 °C
Material data	Type	PC Makrolon 2800
	Glass temperature T_G	120 °C

As mentioned above, the material data for the power-law approach can be found in Chapter 17.

Question: How high is the pressure required to fill the molded part including the bar sprue?

Calculate the pressure first with the average flow front rate (velocity) and then with the volume flow rate.

11.2.2 Influence of Material Properties on the Manufacturing Process

Two different materials are available for the production of a plate-shaped injection-molded part. The rheological behavior of these materials can be fully described by the Carreau approach and the WLF approach. The known constants are listed in Table 11.3.

Table 11.3 Material Parameters for the Carreau Approach

Constant	Material 1	Material 2	Unit
Zero viscosity A	12,000	7000	Pa·s
Reciprocal transition rate B	0.4	0.1	s
Slope of the viscosity function C	0.7	0.8	–
Glass temperature T_G	90	–20	°C
Reference temperature	230	230	°C

The geometric dimensions of the molded part are

- Height $H = 2\,\text{mm}$

- Length $L = 100\,\text{mm}$

- Width $B = 25\,\text{mm}$

Gating takes place via a film gate on the narrow side of the plate, whose pressure losses are negligible.

Answer the Following *Questions*:

Question 1:

Both materials are processed at a temperature $T = 230\,°\text{C}$ and a volume flow rate $\dot{V} = 20\,\text{cm}^3/\text{s}$.

- Which material requires the lower pressure? How high is it?

Assume isothermal conditions for your calculations.

Question 2:

In the next step, the volume flow rate is reduced to $\dot{V} = 10\,\text{cm}^3/\text{s}$.

- Which material causes the lower pressure loss? How high is it?

Question 3:

- Justify the result of Question 2 on the basis of the given material constants (sketch of the flow curves).

Question 4:

Now the processing temperature is increased to $T = 260\,°\text{C}$. The volume flow rate is $\dot{V} = 20\,\text{cm}^3/\text{s}$.

- Which material has the higher viscosity for this operating point?

Interpret and discuss the results.

11.2.3 *Task*: Pressure Losses during Injection Molding and the Resulting Real Locking Force

Calculate the necessary locking force for the locking mechanism shown in Figure 11.6.

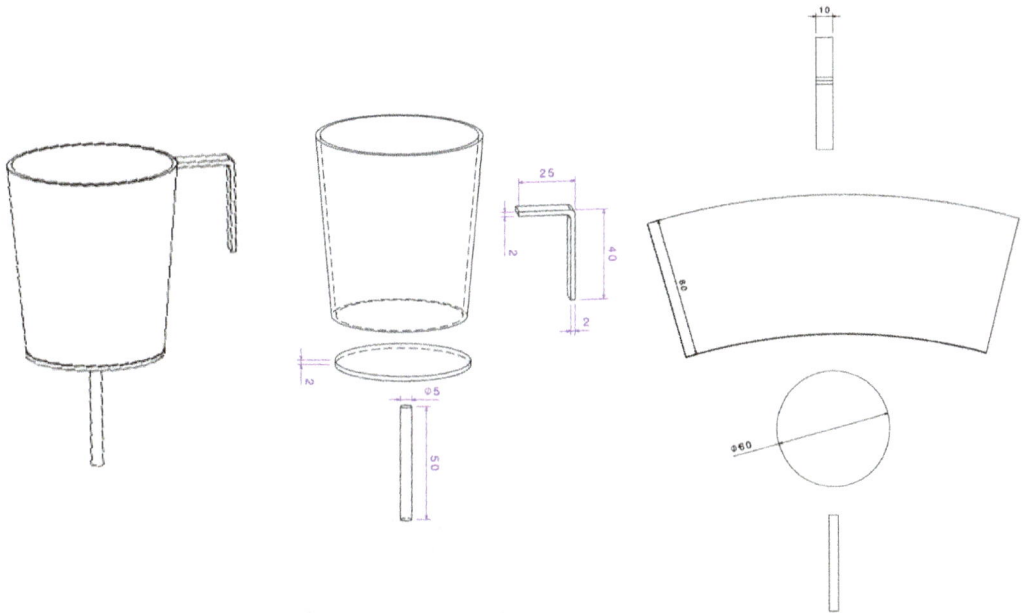

Figure 11.6 Cup broken down into basic geometries

The wall thickness of the handle is 2 mm. The handle width is 10 mm. The demolding taper of the side wall is 5°.

- Molded part data:
  - $m = 38\,g$
- Processing data:
  - $t_{inj} = 1.2\,s$
  - $\vartheta_M = 220\,°C$
- Material data:
  - PS 475K (see materials table)
  - $\rho = 1.1\,g/cm^3$

Note: When calculating the locking force, take into account the actual pressure acting at this point. The floor, the handle, and the side wall must be taken into account.

11.2.4 *Task*: Consideration of Dissipation and Cooling Effects (Non-Isothermal Flow)

During injection molding, there are dissipation effects that occur due to friction be-tween the polymer chains (shear flow) and cooling effects, which occur due to contact with the cold mold wall. Both effects normally must be taken into account in the calculation. The following example is intended to include both the dissipative and the cooling effects in a simple way when calculating pressure losses.

Consider (*given*) a plate-shaped molded part as shown in Figure 11.4 (see Table 11.4). The molded part is gated via a film gate on one end face. The pressure losses in the sprue and gate system are to be neglected in the calculation.

Table 11.4 Information for the Calculation

Geometric data	Panel height $H = 3$ mm Panel width $B = 30$ mm Flow path $L = 100$ mm
Processing data	Material temperature: 220 °C Mold temperature: 50 °C Filling time: 0.4 s
Material data	Material type: Polypropylene PP 1320 L Consistency factor $K = 8.05 \cdot 10^4$ kg/(ms^{2-n}) Viscosity exponent $n = 0.24$ Constant for the Arrhenius approach $\alpha = 6.07 \cdot 10^{-3}$ 1/°C Validity range for the power-law approach: 10^2 s^{-1} to 10^4 s^{-1} Effective thermal diffusivity $a_{\text{eff}} = 0.065$ mm^2/s Thermal conductivity $\lambda = 0.17$ W/mK

a) *Calculate* the pressure required for filling. The power-law approach applies:

$$\eta = K \cdot e^{-\alpha \cdot v_{\text{M}}} \cdot \dot{\gamma}^{n-1} \tag{11.19}$$

During filling, the temperature of the melt is changed by heat conduction and dissipation (Figure 11.7). The following is valid:

$$\Delta \bar{\vartheta}_{\text{diss}} = \frac{a_{\text{eff}}}{\lambda} \cdot \eta \cdot \dot{\gamma}^2 \cdot \Delta t \tag{11.20}$$

The average temperature, due to the cooling of the melt as a result of mold contact, can be calculated using the cooling equation:

$$t_{\text{k}} = \frac{H^2}{4 \cdot a_{\text{eff}} \cdot \mu_1{}^2} \cdot \ln\left(\frac{2}{\mu_1{}^2} \cdot \frac{\vartheta_{\text{M}} - \vartheta_{\text{W}}}{\vartheta_{\text{E}} - \vartheta_{\text{W}}}\right) \tag{11.21}$$

with $\mu_1 = \frac{\pi}{2}$

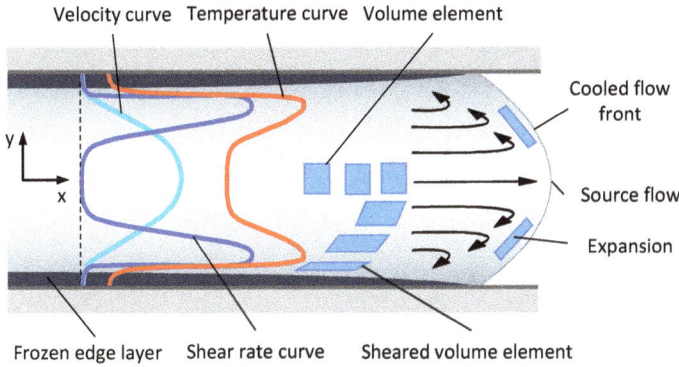

Figure 11.7 Laminar pressure flow of a plastic melt

b) *Calculate* the viscosity for the actual average temperature.

c) Assuming that the average melt temperature prevails throughout the entire molded part, how large is the frozen surface layer at the time of complete filling? The yield point temperature is 130 °C.

d) *Calculate* the local temperature profile over the cross section of the molded part as a function of time for the following time steps: $t = 0.1/0.2/0.3/0.4/0.5/0.6/0.8/1/2/4/6/8/10$ s.

Use the following equation to display the temperature profiles:

$$\frac{\partial \vartheta}{\partial t} = a \cdot \frac{\partial^2 \vartheta}{\partial x^2} \tag{11.22}$$

The equation should be solved either analytically or numerically. Solving the problem using predefined solutions should be avoided. Set the result as shown in Figure 11.8, as a function of time and location for the molded part.

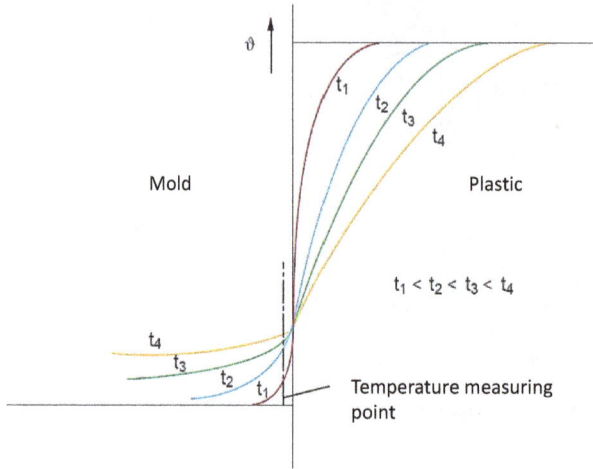

Figure 11.8 Temperature curves over the molded part cross section for different time steps

11.3 Calculation of the Optimum Filling Time (Injection Rate) during Injection Molding Using the Brinkman Number

In reality, the injection molder determines the optimum mold filling time or injection rate, for example, by changing the injection rate on the machine and reading the necessary filling pressure on the machine control system. As soon as the filling pressure has reached its minimum, the filling time can be considered optimal. Naturally, other aspects must also be taken into account at the same time, such as free jet formation, diesel effects, or ribs that need to be filled. These effects can also influence the injection rate and also lead to a certain velocity profile when injecting the melt into the mold.

When the hot melt is injected into the comparatively cold injection mold, two effects that are responsible for the necessary filling pressure are superimposed. Figure 11.9 is used to explain these effects.

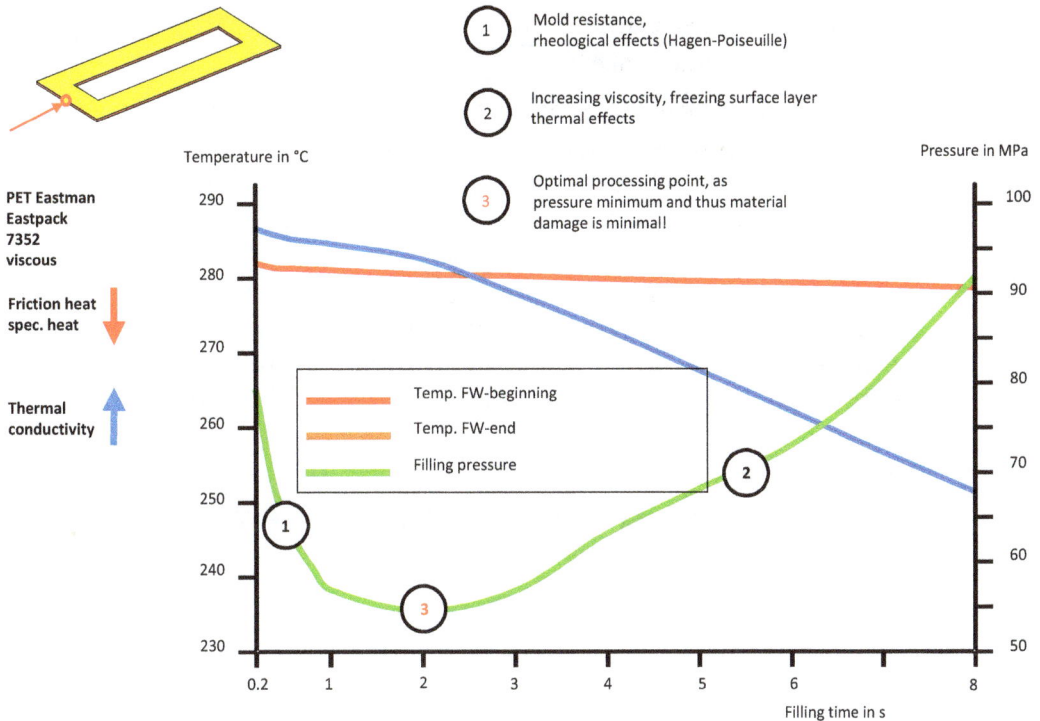

Figure 11.9 Pressure and temperature curves as a function of filling time

If the filling process is very fast, that is, the injection rate is very high, the necessary filling pressure is correspondingly high. The reason for the high filling pressure required can be explained using the Hagen–Poiseuille equation. In this equation, the injection rate is in the numerator: the greater the injection rate, the greater the necessary filling pressure. The increase (1) (Figure 11.9) for decreasing filling times is therefore a purely rheological effect.

For increasing filling times, the required filling pressure decreases until it subsequently increases again. The increase (2) has thermal causes. As the filling time increases, the melt cools increasingly during the filling process. This means that the layer thickness of the frozen surface layers (top and bottom) increases and at the same time the viscosity of the melt increases during the filling process. As a result, the pressure must increase.

The optimum injection time is reached at point (3)—that is, the minimum pressure is reached. This point can be calculated using a dimensionless indicator, the so-called Brinkman number. It represents the ratio of the dissipated energy to the energy lost by heat conduction. The rheological and thermal effects are therefore included in the Brinkman number:

$$Br = \frac{\eta \cdot v_m^2}{\lambda \cdot \Delta\vartheta} = \frac{\text{rheological effects}}{\text{thermal effects}} \tag{11.23}$$

In the numerator of Equation 11.23 we find the rheological effects (Figure 11.9 (1)), while the denominator contains the thermal, namely cooling, effects (Figure 11.9 (2)). If the Brinkman number assumes the value 1, the thermal and rheological effects balance each other out, which means that the optimum filling time has been achieved. We are at point (3) on the curve. This assumes isothermal conditions in the melt.

The viscosity is calculated using the power-law approach. For the solution, the power-law approach is used in the equation of the Brinkman number. The same applies to the representative shear rate, which is included in the power-law approach. If necessary, the temperature shift factor a_T must also be included.

Task: Optimum Filling Time

The geometry of the part is *given* as shown in Figure 11.10; see Table 11.5.

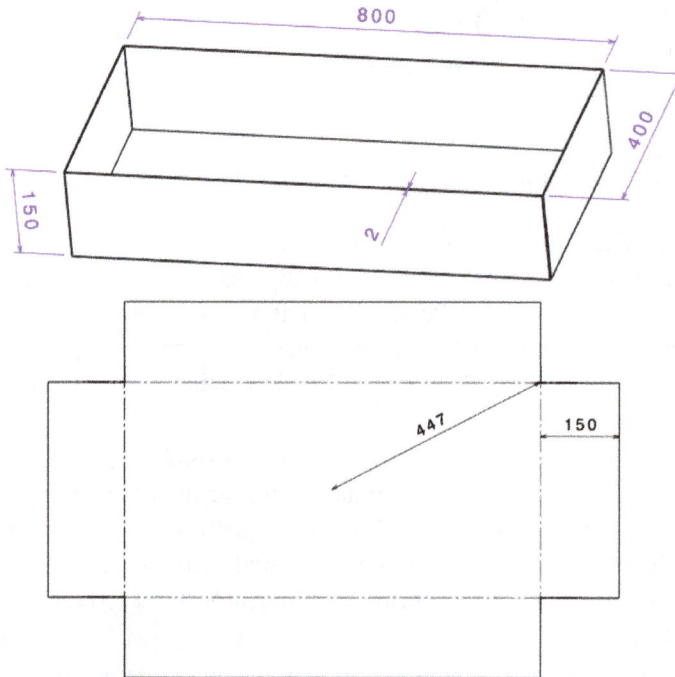

Figure 11.10 Molded part geometry for a calculation example

Table 11.5 Processing and Material Data

Material	ABS Terluran from BASF
Specific heat capacity	2.4 kJ/kg·K
Density at melting temperature	1.01 g/cm³
Effective thermal diffusivity	0.076 mm²/s
Processing temperature	250 °C
Mold temperature	60 °C/40 °C
Glass temperature	100 °C
Reference temperature	210 °C
Consistency factor K	6,306·10⁵ Pa·sⁿ
Viscosity exponent n	0.236

Determine the optimum filling time for the molded part shown in Figure 11.10. Then change the mold temperature to 40 °C and determine the optimum injection time for this value. Discuss the results!

References

[1] Lichius, U.; Schmidt, L.: *Rechnergestütztes Konstruieren von Spritzgießwerkzeugen: systematisches Entwickeln von Betriebsmitteln, Aufbau und Funktion von Spritzgießwerkzeugen*. Vogel, Würzburg, 1986

12 Calculation of Flow Processes in Hot Runner Systems and Extrusion Molds

12.1 Basics of the Pressure Curve over the Length of Composite Channel Systems

In hot runner systems [1] and extrusion molds, the flow channels are usually made up of a sequence of individual channels with different geometries. The different flow channels can be arranged in series or in parallel. For this reason, this chapter will deal with the calculation of the volume flow rate and the pressure loss for this arrangement.

The calculations are based on the following assumptions:

- The pressure curve $\Delta p/L$ over the flow path is a linear relationship (incompressible medium).

- The power-law approach is used:
    - The flow exponent m and the fluidity ϕ in the power-law approach are assumed to be constant.

- The flow is in a steady state.

- This is a laminar layered flow.

- Pressure losses at cross-sectional transitions (extension/inlet pressure losses) are not taken into account. The same applies to additional pressure losses that occur at deflections. Only the so-called shear pressure losses are taken into account.

- The temperature in the flowing fluid is assumed to be constant (isothermal flow).

- For hot runner systems, pipes with a circular cross section are considered.

When a melt flows, energy provided in the form of pressure is required to overcome the flow resistance (fluid friction). Part of this energy is converted into heat and molecular transformation (orientation). A pressure drop can therefore always be observed in the direction of flow. An increase in pressure cannot take place.

12.1.1 Pressure Curve in Parallel Pipes

Figure 12.1 shows two parallel pipes with different diameters and different flow path lengths. In this diagram, the relative pressure p/p_{max} was plotted as a function of the relative pipe length z/L. In the following analysis, it is assumed that for both flow paths the pressure at the start of the flow path, p_{max}, is the same. Ambient pressure prevails at the end of the flow path for both channels. Therefore, the following applies:

$$\Delta p_{tot} = \Delta p_1 = \Delta p_2 \tag{12.1}$$

As a result, the total volume flow rate must be divided according to the resistance (geometry) of the individual channels:

$$\dot{V}_{tot} = \dot{V}_1 + \dot{V}_2 \tag{12.2}$$

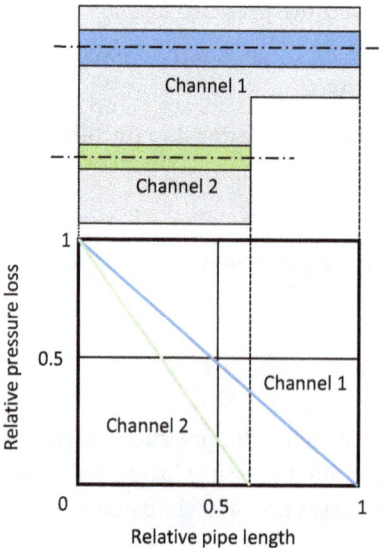

Figure 12.1
Parallel connection of two pipes with representation of the pressure curve over the channel length

12.1.2 Pressure Curve in Pipes Arranged in Series

Another arrangement is the series-connected flow channels. For example, pipes with different diameters can be connected in series; see Figure 12.2.

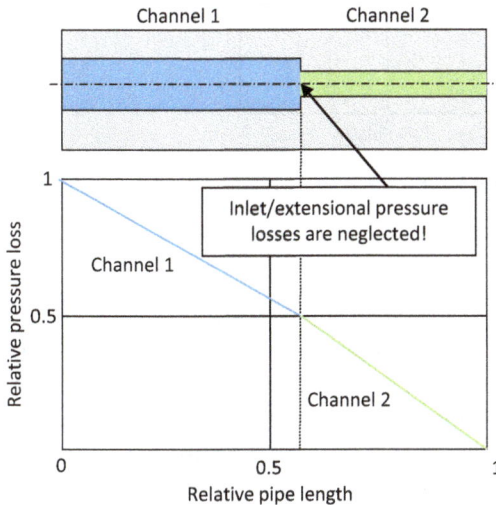

Figure 12.2
Series connection of two pipes with representation of the pressure curve over the channel length

At the point where the diameter changes, the pressure curve has a kink. The volume flow rate is the same in the individual channel sections. Therefore, the following applies:

$$\dot{V}_{tot} = \dot{V}_1 = \dot{V}_2 \tag{12.3}$$

Depending on the flow resistance, the pressure requirement (pressure gradient) in the channel sections changes. The total pressure requirement over the length L is made up of the individual pressure losses. Therefore:

$$\Delta p_{tot} = \Delta p_1 + \Delta p_2 \tag{12.4}$$

Figure 12.3 shows again the pressure curve in pipes connected in series with a stepped diameter. The pressure curve has a kink at the point of the cross-sectional jump. Inlet pressure losses are neglected here. Otherwise, the pressure curve would have a jump at the change in cross section. Depending on the diameter of the flow channel, the gradient of the pressure curve differs for the respective section. The smaller the channel diameter, the steeper the gradient.

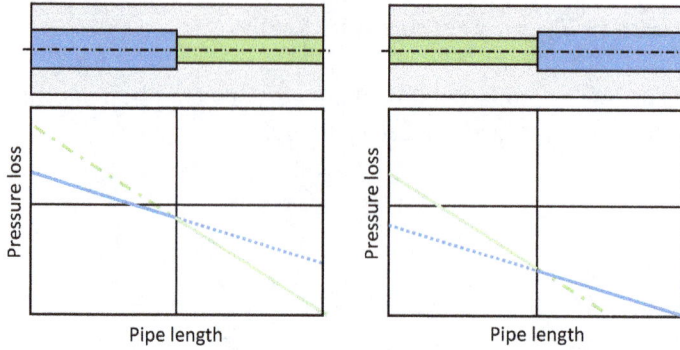

Figure 12.3 Pressure curve in pipes with stepped diameter

12.1.3 Conical Flow Channels

A conical change in the flow channel over the channel length is replaced in the calculation by a series connection of cylindrical channel sections. Figure 12.4 shows such an arrangement. With the converging channel, the differently steep curve sections of the pressure curve, which have kinks at transitions, merge into a smooth curve if infinitely small cylindrical pipe sections are assumed at the boundary transition.

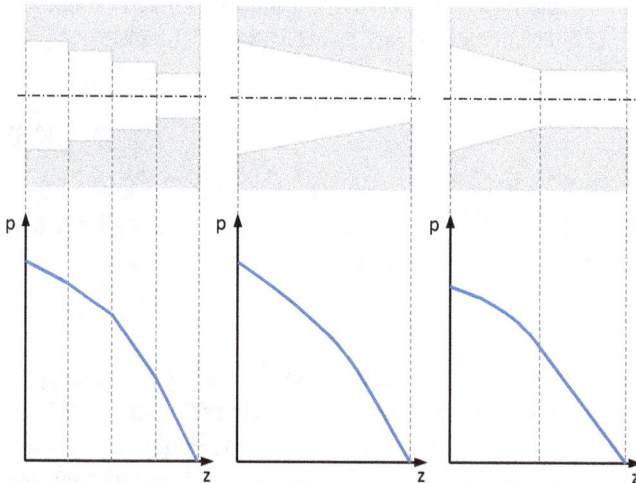

Figure 12.4 Pressure curve in conical pipes (qualitative)

If a cylindrical channel is connected to a channel with a conical cross section, the pressure curve has a continuous transition where the channel has a kink in its course. The slopes of the tangents of the two pressure curves are the same at the transition point.

12.1.4 Pressure Curve for an Arbitrarily Composed Channel

As a further example, let us now consider a flow channel that is made up of different pipe sections. Such a series connection of channels with different geometries can be found, for example, in a mold for extrusion blow molding. The following special features apply to the pressure curve:

- If the diameters of the flow channels are the same, then the curve $p = f(z)$ has the same gradient for these areas.

- If the radius is the same in the conical and the cylindrical part of the pipe, the gradient of the curve (or pressure gradient) at this point is also the same.

Figure 12.5 shows the result of a pressure measurement taken on an extrusion blow mold. The basic characteristics of the relationship between the channel geometry and the pressure curve discussed above can be found here again.

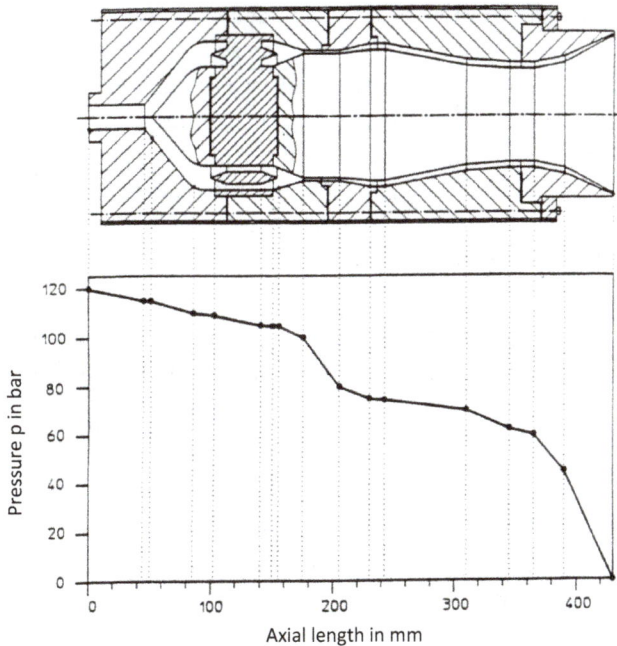

Figure 12.5 Measured pressure curve in an extrusion blow mold [2]

12.2 Rheological Design of Hot Runner Systems for Injection Molding

Hot runner systems are always designed rheologically in injection molding technology. The aim of this design is to keep the pressure loss as low as possible, since the pressure loss correlates directly with the shear stress and consequently leads to mechanical stress on the plastic. Hence, in addition to the flow path length in the hot runner, the diameter plays an important role. Larger flow diameters are therefore preferable in terms of pure pressure loss. On the other hand, many plastics (PET, PVC, blends, etc.) are also thermally sensitive. This means that they degrade to a greater or lesser extent depending on the dwell time and the temperature. For this reason, efforts are made to reduce the volume in the hot runner system and, thus, to reduce the diameter. It quickly becomes clear that a compromise needs to be found between the mechanical and the thermal load on the polymer [2–5].

In the following, a distinction is made between multi-cavity molds, such as those used for the production of caps, preforms, syringe barrels, etc., and family molds. Multi-cavity molds are molds in which the cavities (molded parts) are identical. The articles mentioned are often produced in molds with a high number of cavities (16/24/32/48/56/64/72/96/128/144/192). In family molds, the cavities are not identical. For example, the components of a remote control (front, back, and cover for the battery compartment) can be produced in one mold. In addition to color identity, dimensional accuracy is also an advantage of the use of family molds compared to production in separate molds.

The aim of balancing is generally to ensure that all cavities are filled at the same time. If this is not the case—that is, the cavities are filled successively—this has a direct impact on the quality of the molded parts (stresses, burr formation, orientation, etc.) and the load on the mold.

- *The following applies to the cavity filled initially:* High pressure buildup, overmolding (flash formation, mold damage), residual stresses, orientations, etc.

- *The following applies to the cavity filled later:* Acceleration of the flow front (markings on the surface, orientations, shearing ⇒ temperature increase, shrinkage, etc.) (Figure 12.6)

- Cavity A fills faster than cavity B
- Cavity B shows higher thermal potential by the end of filling

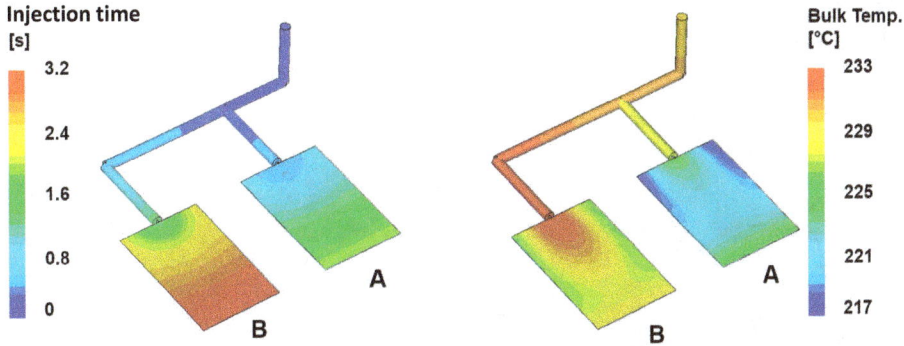

Injection time [s]

3.2
2.4
1.6
0.8
0

A

B

Bulk Temp. [°C]

233
229
225
221
217

A

B

Figure 12.6 Unbalanced distributor system (filling time and average melt temperature in the molded part)

In order to achieve the goal of simultaneous filling, the pressure loss must be the same on each flow path, that is, from the start of the flow path to the end of the flow path! This means that, for the following 72-cavity injection mold (Figure 12.7), this condition must be met:

$$\Delta p_1 = \Delta p_2 = \Delta p_3 = \Delta p_4$$

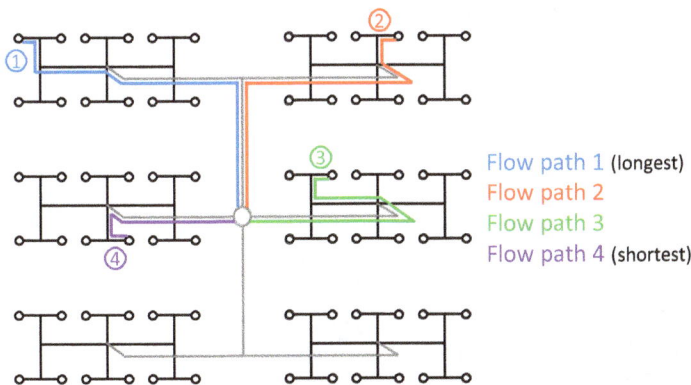

Flow path 1 (longest)
Flow path 2
Flow path 3
Flow path 4 (shortest)

Figure 12.7 Distributor system for a 72-cavity mold

Basically, there are two options for rheological balancing. The usual way is the **natural rheological balancing**. With natural rheological balancing, the flow path length is identical on all paths. The diameters are also the same on the respective distribution levels. This design guarantees an identical pressure loss for each flow path. Figure 12.8 shows a natural rheological balancing of a six-cavity mold. Balancing can take place on one (left) or two (right) distributor levels. From a purely rheological point of view, the design on two levels has advantages, as the flow paths are absolutely identical and sheared surface layers have no influence on the filling process. This is not necessarily the case with the left-hand version, with one distribution level.

one distribution layer   two distribution layers

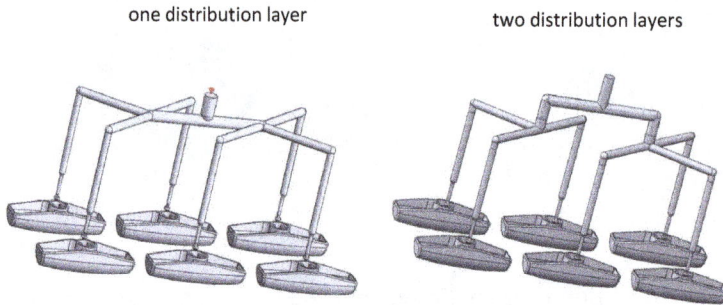

Figure 12.8 Balancing of a six-cavity mold

Figure 12.9 shows the natural rheological balancing of a 96-cavity mold for the production of PET preforms with the corresponding distributor levels. Starting from the sprue bushing, the melt is distributed to two sub-distributors in the first level and then to two further sub-distributors in the second level. This system continues until a quadruple and then the last "y" distributor is reached, and from there the melt passes through the nozzle into the actual cavity. As a rule, the diameters are reduced from the sprue bushing to the cavity, as the volume flow rate is divided in the distributor levels. This reduces the volume in the hot runner while maintaining a low pressure loss.

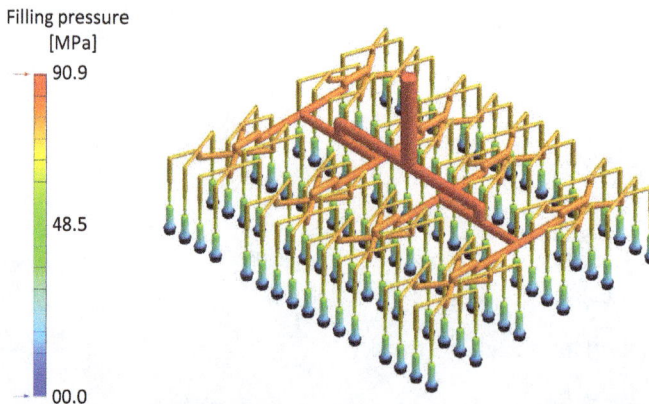

Filling pressure [MPa]

90.9

48.5

00.0

Figure 12.9 Natural rheological balancing of a 96-cavity mold for the production of preforms made of polyethylene terephthalate (PET)

Another option for balancing a multi-cavity mold is **mathematical rheological balancing**. In this variant, the diameters are adapted to the different flow paths. For a given material and defined volume flow rate, the diameters can be calculated as a function of the flow path using the Hagen–Poiseuille equation, assuming that the pressure loss should be the same on each flow path. This balancing can often be used to reduce the number of manifold levels and thus also the melt volume in the hot runner.

A disadvantage of this balancing is the operating-point dependence of the system. This means that the mathematical balancing is carried out for one material (viscosity) and one volume flow rate. If these boundary conditions change, the system is no longer balanced, that is, the cavities are no longer filled simultaneously. As a rule, this does not play a major role in reality, as the boundary conditions (material, injection time, etc.) do not usually change during the production of caps or preforms.

Another theoretical disadvantage is that the cavity with the longest flow path requires the largest flow channel diameter and the cavity with the shortest flow path has the smallest channel diameter. This could lead to different part qualities due to the different dwell times. However, if we look at the Hagen–Poiseuille equation for shear-thinning viscous media, we quickly realize that the diameter of the flow channel is included in the pressure calculation with an exponent greater than 4. For this reason, the diameter only needs to be changed marginally during balancing.

Nevertheless, when designing the hot runner system for multi-cavity molds, natural rheological balancing has prevailed. Mathematical rheological balancing is to be preferred for family molds, as the cavities are not identical (Figure 12.10) and therefore natural rheological balancing is not expedient. Here, too, it must be ensured that all cavities are filled at the same time.

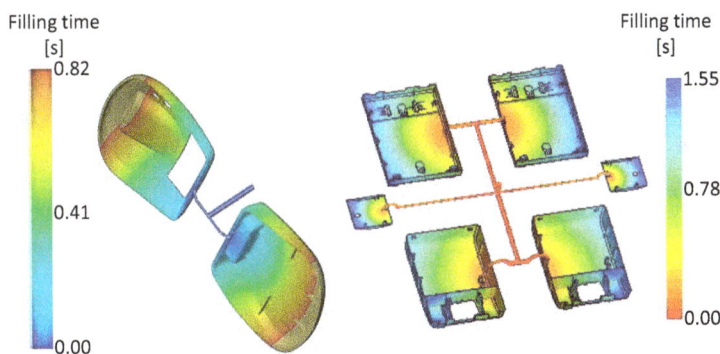

Figure 12.10 Example of family molds

As shown in Figure 12.10, balancing is achieved by adjusting the diameter of the flow paths. Regardless of whether it is a hot or a cold runner system, the pressure losses on each path from the start of the runner system to the end of the flow path are always considered. Assuming that the pressure loss must be identical on each flow path, the diameters are adjusted accordingly for otherwise specified conditions (material, volume flow rate, etc.). In this case, too, the design of the distribution system depends on the operating point. This may need to be taken into account when changing the material.

In principle, the sheared surface layers must also be taken into account when designing a hot runner manifold system. The shearing of the surface layers leads to an increase in temperature and thus to a reduction in viscosity in these surface layers. If the melt flows are now simply diverted without a sub-distributor, such as in the six-fold hot runner manifold in Figure 12.8 (left side), one flow path is preferred due to the lower viscosity. This is also shown as an example in Figure 12.11. The simulation shows that the temperature in the sheared surface layers increases, which means that the left cavity is filled earlier than the right cavity.

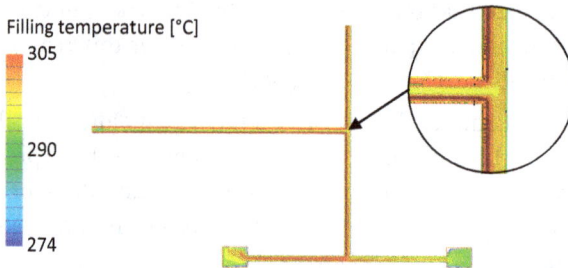

Figure 12.11 Simulation of the temperature in the sheared surface layer

The same behavior also occurs in the case of the 48-cavity mold shown in Figure 12.12. With a simple melt distribution without a sub-distributor and with a "y" design in the hot runner, the cavities 3, 10, 17, 20, 29, 32, 39, and 46 (cavities shown in red) are preferred. This can be confirmed by partial fillings (short shots). The problem can be solved, for example, by diverting the melt in a sub-distributor, as shown in Figure 12.9 for the 96-cavity mold.

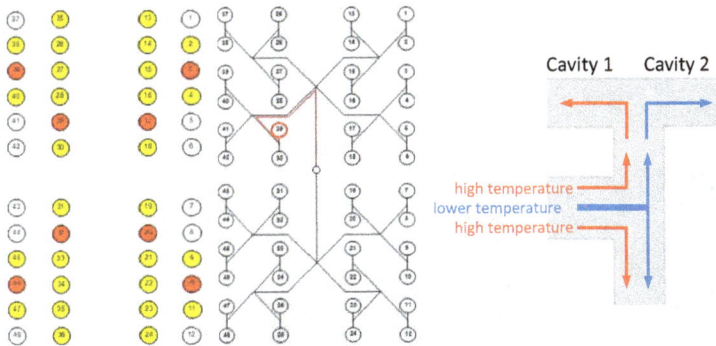

Figure 12.12 Influence of the sheared surface layers on the flow behavior in the hot runner of a 48-cavity mold

Another important point that must be taken into account when balancing hot runner systems is the pressure losses at cross-sectional jumps and deflections. In addition to shear pressure losses, so-called extensional or inlet pressure losses also occur at cross-sectional jumps, and they must not be neglected. As the material parameters required for the calculation (extensional viscosity, storage and loss moduli, normal stress difference, etc.) are not usually included in the material data sheets of the simulation programs, the pressure losses at cross-sectional discontinuities cannot be calculated correctly. When designing hot runner systems, this can lead to the pressure loss calculated being too low. These extensional pressure losses and the possible calculation models are discussed in more detail in Chapter 13.

In fluid mechanics, pressure losses that occur at deflections are taken into account with the pressure loss number ζ (zeta number). Plastics processing involves comparatively slow flows. Furthermore, the flowing medium exhibits shear-thinning viscous flow behavior. These two circumstances mean that no additional pressure losses, as known from fluid mechanics, occur at deflectors. Studies have shown that deflectors can even reduce pressure losses in hot runner systems.

For the experimental investigation of pressure losses at deflectors, a specially developed test mold was produced (Figure 12.13). This test mold allows the angle of the deflection (180°, 90°, and 45°) and the flow channel diameter ($d = 4$, 6, and 8 mm) to be varied. Two pressure transducers are installed in front of and behind the deflector. The entire test setup is temperature-controlled and thus corresponds in principle to a hot runner system.

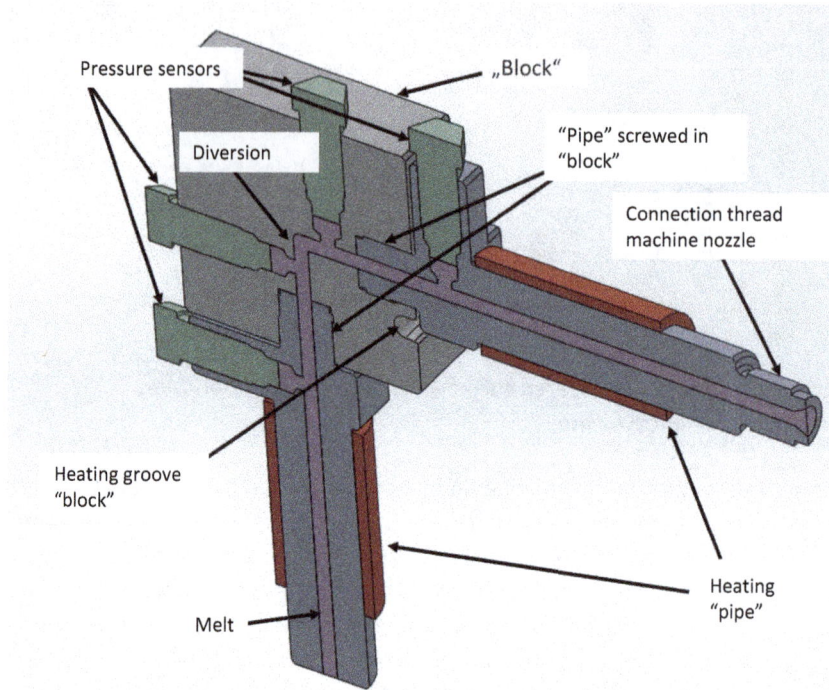

Figure 12.13 Test setup for investigating pressure losses at deflectors in the hot runner

As the diagram in Figure 12.14 illustrates, the pressure losses at deflectors are a function of the flow channel diameter and the angle of the deflector, among other things. An angle of 180° corresponds to a flow channel without deflection. The smaller the angle, the more acute the deflection. It is clear that the pressure loss decreases as the angle becomes more acute. Furthermore, the influence of the angle on the pressure loss at the deflection decreases as the diameter increases.

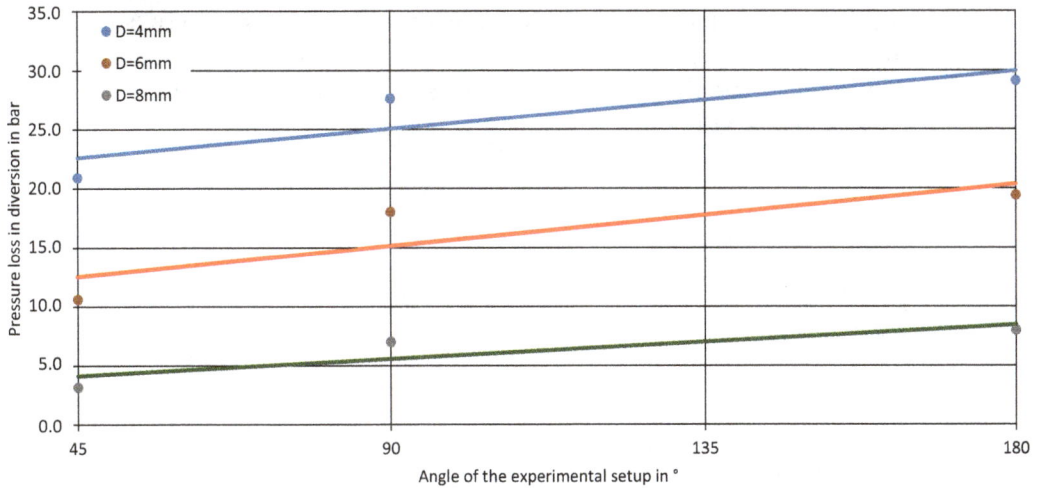

Figure 12.14 Pressure losses at deflectors in the hot runner system

The simulation depicted in Figure 12.15 confirms these results from the test series. The simulation was carried out on the same test geometries as the practical tests.

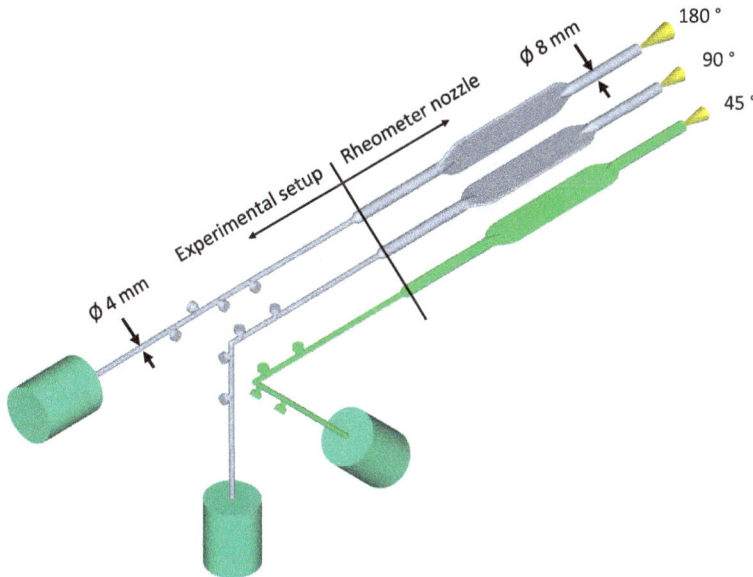

Figure 12.15 Test setup for simulating pressure losses at deflections in the hot runner

The fact that the pressure losses decrease because of deflections may also be due to the shear-thinning viscosity of the plastic. The deflection leads to additional shearing of the plastic on the inside. As a result, the temperature increases locally and the viscosity decreases, which in turn leads to a reduction in the pressure loss. Figure 12.16 shows the shear rate, temperature, and viscosity for a flow channel with a diameter of 4 mm and geometries with angles of 180°, 90°, and 45°. The more acute the deflection (smaller angle), the greater the shear of the plastic in the deflection.

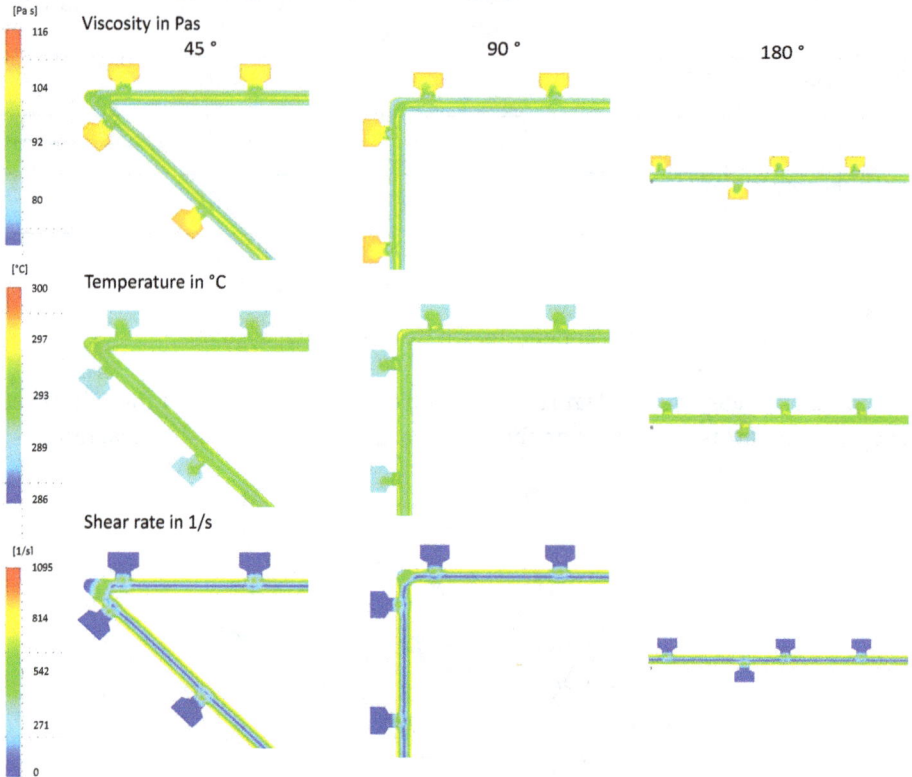

Figure 12.16 Simulation results of viscosity, temperature, and shear rate in deflections

12.3 *Tasks*: Mathematical Rheological Balancing of Hot Runner Systems

12.3.1 Double Mold with Different Melt Distribution Systems

The hot runner system shown in Figure 12.17 is to be designed for two identical molded parts but different distribution systems. The geometric data is as follows: $L = 50$ mm, $L_A = 2$ mm, $D_D = 5$ mm.

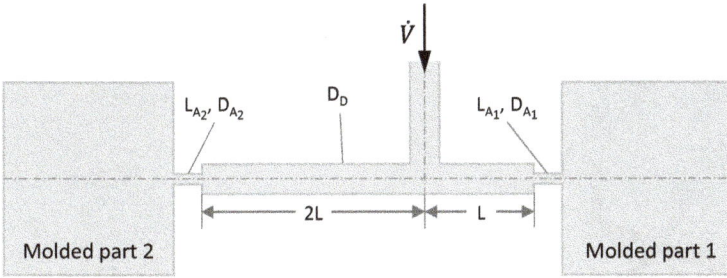

Figure 12.17 Geometry of a hot runner system for two molded parts

Care must be taken to ensure that the two identical molded parts are filled under the same conditions (simultaneous entry of the compound into the molded parts, simultaneous reaching of the flow path ends).

To ensure that extensional pressure losses can still be neglected, the following diameter ratio must be satisfied:

$$\frac{D_A}{D_D} \geq \frac{1}{3} \tag{12.5}$$

Question: How large must the gate diameter D_A be selected for the respective cavity?

a) First assume Newtonian flow behavior.

b) Then, calculate the diameter D_A for a shear-thinning viscous material with a viscosity exponent of $n = 0.7$ and a consistency factor $K = 7.8 \cdot 10^4 \left[\frac{kg}{m \cdot s^{2-n}}\right]$.

c) Discuss the case in which the viscosity exponent changes to $n = 0.4$ and the consistency factor assumes the value $K = 7.8 \cdot 10^4 \left[\frac{kg}{m \cdot s^{2-n}}\right]$.

Note: The sprue system is designed as a hot runner. Cooling and dissipation effects can be neglected.

12.3.2 Eight-Cavity Mold with Different Melt Distribution Systems

The mold in Figure 12.18 is to be mathematically rheologically balanced (see Table 12.1). Due to the symmetry, it is sufficient to consider only one segment during balancing, as Figure 12.18 illustrates.

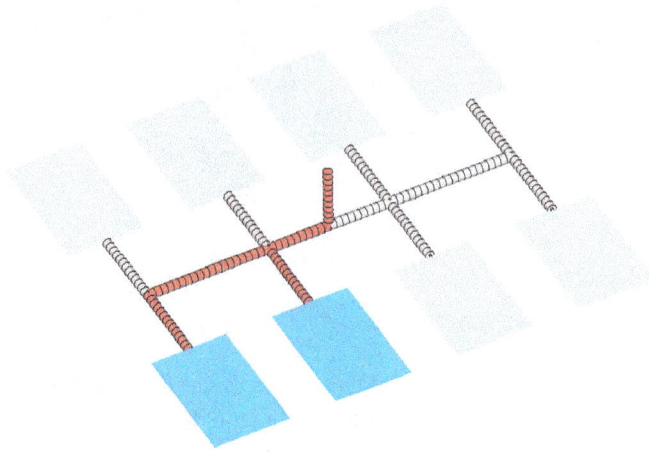

Figure 12.18 Hot runner system for an eight-cavity mold

Table 12.1 Inlet Volume Flow Rate and Material Data

Inlet volume flow rate into the sprue bushing (Figure 12.19)	$\dot{V} = 400 \dfrac{cm^3}{s}$
Manufacturer	BASF
Material type	PS 168N (see materials table in Chapter 17)

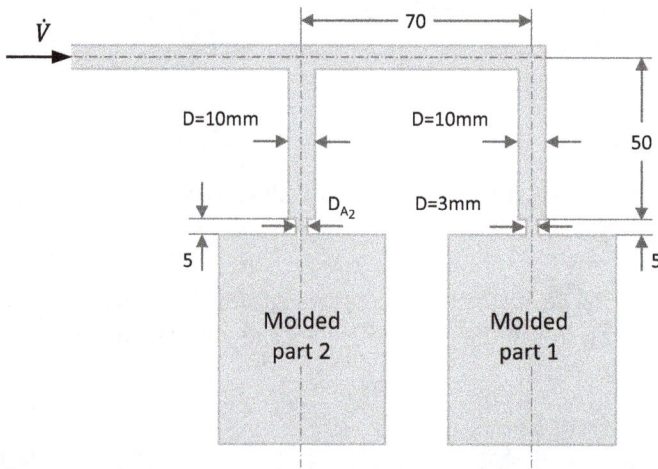

Figure 12.19 Section of an eight-way distributor

a) *Calculate* the gate diameter D_{A2} for molded part 2 with the condition that the two molded parts are filled under the same conditions. Since the molded parts are identical, in this case this means that the melt enters the molded parts at the same time and it reaches the flow path ends at the same time.

b) *Check* whether the critical shear rate is exceeded and whether the constants for the power-law approach are valid.

Note: The sprue system is designed as a hot runner. Cooling and dissipation effects can be neglected.

12.3.3 Six-Cavity Mold with Different Melt Distribution Systems

The series distributor illustrated in Figure 12.20 is to be mathematically rheologically balanced for six identical molded parts (see Table 12.2 and Table 12.3).

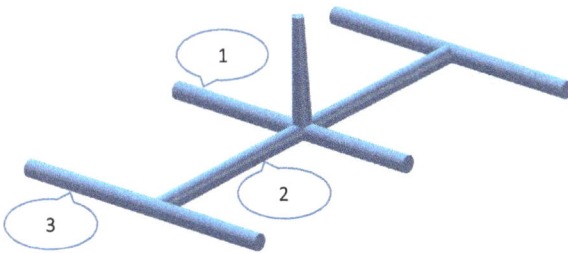

Figure 12.20 Series distributor for six molded parts

Table 12.2 Calculation Data

Geometric data	$l_1 = l_2 = l_3 = 50\,mm$, $d_2 = 12\,mm$
Pressure loss on flow path 1	$\Delta p_1 = 60\,bar$
Volume flow rate per molded part	$\dot{V}_{molded\ part} = 100\,cm^3/s$
The melting temperature can be assumed to be 230 °C.	
Material	BASF polystyrene 475 K

The following approach can be used to calculate the viscosity:

$$\eta = K_{OT} \cdot e^{-\alpha T} \cdot \dot{\gamma}^{n-1} \tag{12.6}$$

Table 12.3 Additional Calculation Data

Material parameters	$K_{OT} = 2.2 \cdot 10^5 \frac{kg}{m \cdot s^{2-n}}, \alpha = 1.1 \cdot 10^{-2}\frac{1}{°C}, n = 0.22$
Temperature range	190 °C to 250 °C
Approximation range	$400\ s^{-1} < \dot{\gamma} < 10000\ s^{-1}$

Questions:

a) How large must the diameter d_3 be?

b) How large must the diameter d_1 be?

Note: The sprue system is designed as a hot runner. Cooling and dissipation effects can be neglected.

12.3.4 Family Mold with Two Different Cavities

In Figure 12.21, a melt distribution system for a two-cavity family mold will be designed. The sprue system is designed as a hot runner manifold. Cooling and dissipation effects can be neglected.

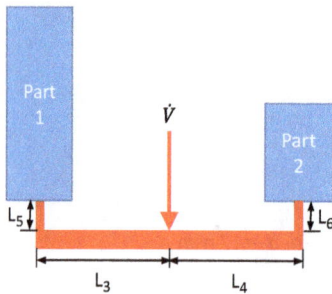

Figure 12.21
Two-cavity family mold

This is a (1 + 1)-cavity mold, that is, the cavities (molded parts) are not identical in this case!

- Both molded parts have the same diameter ($D_1 = D_2$).
- Cavity 1 has twice the flow path length as cavity 2, i.e., $L_1 = L_2$.

The following geometric data is also *given*:

- $D_1 = D_2 = 20\,mm$
- $D_3 = D_4 = 5\,mm$
- $D_5 = 3\,mm$
- $L_1 = 2L_2 = 100\,mm$

- $L_3 = L_4 = 40\,\text{mm}$
- $L_5 = L_6 = 10\,\text{mm}$

The following material and setting parameters for balancing are *given*:

- Material: ABS
- Material parameters:
    - Consistency factor $K = 25\,000\,\text{Pa s}^{0.4}$
    - Viscosity exponent $n = 0.4$
    - Temperature shift factor $a_T = 0.4$
    - Density $\rho = 1.1\,\text{g/cm}^3$
- Injection time $t_{in} = 1.2\,\text{s}$

Questions:

1. How would you carry out the balancing? Give reasons for your answer.
2. What is important for balancing (goal)?
3. Carry out suitable balancing!

12.4 Rheological Design of Extrusion Molds

In contrast to injection molding, extrusion is generally a continuous, steady-state process. This means that the melt is continuously prepared and conveyed by a screw. The melt then flows through a distribution system before the actual shaping takes place in the extrusion mold. Therefore, from a rheological point of view, the following components are of interest:

- The extruder
- The balancing and distribution system
- The extrusion mold

Today's computer-aided simulation programs make it possible to predict the flow processes in these components quite well. This allows the simulation of the mixing and homogenization processes in the screw and also the determination of the conveying capacity of the extruder in advance. The selected screw geometry can thus be optimized beforehand with regard to quality and performance without great effort.

The subsequent distribution systems serve to balance the melt. The focus here is on the relationship between pressure and volume flow rate. The pressure required for the flow through the distributor system and the subsequent die must be applied by the extruder. For example, as the conveying capacity of an extruder decreases with increasing back pressure, the pressure loss in the downstream equipment must be known. Furthermore, the distribution system must guarantee an even distribution of the volume flow rate at the outlet of the die. This means that the volume flow rate must be distributed in such a way that the outlet rate of the melt is the same at all points. This design calculation process is called rheological balancing.

If thermally sensitive materials are processed, the residence time of the melt in the flow channel must be known. The rheological approaches are also required for these calculations.

Wide slot dies are used, for example, for extruding sheets and films. One task of these dies is to distribute the usually circular strand ejected by the extruder as evenly as possible across the desired width, so that an extrudate with a uniform thickness across the width is produced. Figure 12.22 shows the most important types of melt distribution systems. These are

- The T distributor
- The fish-tail distributor
- The coat-hanger distributor

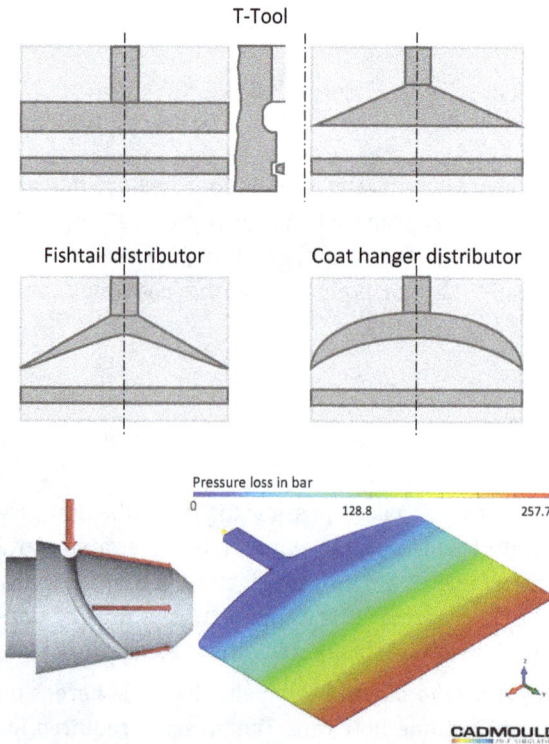

Figure 12.22 T distributor, fish-tail distributor, coat-hanger distributor, side fed mandrel [2], and simulated coat-hanger distributor

Side fed mandrels (Figure 12.22, bottom left), such as those used for extrusion blow molding, also require an appropriate distribution of the melt up to the outlet. For this reason, coat-hanger distributors or so-called heart curves are also used here. Today's simulation programs allow a rheological design of these systems, as shown in Figure 12.22 (bottom right). In the following exercise, the possibilities of manual design with the corresponding boundary conditions will be presented and explained [2, 4, 6, 7].

One disadvantage of the T distributor is that it cannot be optimally rheologically balanced. The reason for this is that the flow resistance across the width of the mold is at its lowest value in the middle. As a result, the maximum volume flow rate emerges in the center of the T distributor, while the volume flow rate in the outer areas is lower. This can be counteracted with corrective elements, such as a pressure bar or flexible lips [2, 4, 6, 7].

Coat-hanger distributors are, next to fish-tail distributors, the distribution systems that find the most applications. Coat-hanger distributors, in particular, achieve a very good distribution effect of the melt when designed accordingly. Theoretically, it is even possible to design the distributor system independently from the operating point, as will be shown later. However, this is a more or less theoretical design assumption. Figure 12.23 shows the complex geometry of a coat-hanger distribution system. It quickly becomes clear that the manufacturing effort is not insignificant and, therefore, the manufacturing costs are correspondingly high [2].

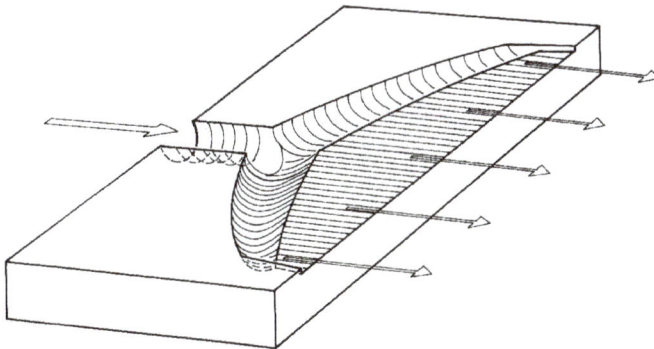

Figure 12.23 Coat-hanger distributor [2]

The content of this exercise is the presentation of different concepts for the analytical and numerical design of slot dies. The boundary conditions for the design, the derivation, and the application of the design formulas are described. Following this theoretical part, practical examples of mold design will be presented. The aim of this exercise is to teach the procedure for the theoretical design of distributor molds and the application of the design formulas determined. This will be done analytically and numerically.

In addition to the requirement of a uniform volume flow rate distribution across the width, a number of additional requirements must be taken into account in the theoretical design of wide slot distributor molds, depending on the application. These are

- Low average residence time and narrow residence-time spectrum (particularly important for thermally sensitive polymers such as rigid PVC)

- Compliance with material-dependent limit values for shear rate and wall shear stress

Different distribution concepts have been developed to meet these requirements.

All distributor systems used today consist of a distributor channel (pipe), which is located more or less at a right angle to the direction of extrusion, and a so-called island (slot) or throttle field. The individual distributor systems differ in the curve of the island length $y(x)$ and the distributor channel cross section (here, $R(x)$) over the width. The gap height in the island area is generally constant.

12.4.1 Mathematical Prerequisites for Balancing

The content of the theoretical design of a distributor is the coordination of the distributor channel cross section and the island length in such a way that the desired goal—namely, uniform melt distribution across the width at the mold outlet—is achieved. This is synonymous with a linear decrease in the volume flow rate in the distributor channel from the melt entry into the distributor to the end of the distributor channel.

Provided that the two halves of the distributor are symmetrical, that is, the incoming volume flow rate is evenly distributed, this results in the following:

$$\bar{v}_{slot} = \text{constant} = \frac{\dot{V}_{tot}}{B \cdot H} \tag{12.7}$$

$$\dot{V}_0 = \frac{\dot{V}_{tot}}{2} \tag{12.8}$$

and:

$$\dot{V}_{pipe}(x) = \dot{V}_0 \cdot \frac{2x}{B} = \dot{V}_{tot}\left(\frac{x}{B}\right) = \dot{V}_0 \cdot \frac{x}{L} \tag{12.9}$$

The following still applies:

$$\Delta p(x) = \Delta p_{pipe}(x) + \Delta p_{slot}(x) = \text{constant} \tag{12.10}$$

The flow paths are shown in Figure 12.24. The coordinate system is shown in Figure 12.25.

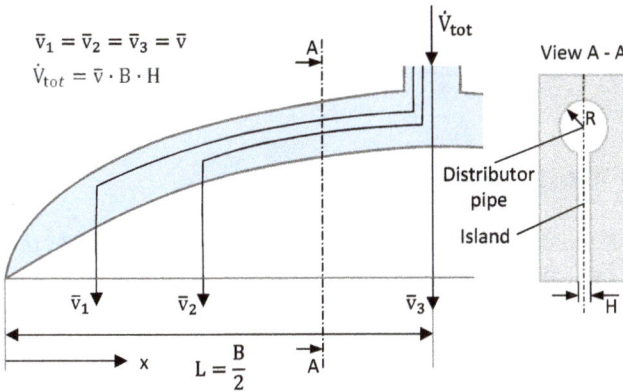

$$\bar{v}_1 = \bar{v}_2 = \bar{v}_3 = \bar{v}$$
$$\dot{V}_{tot} = \bar{v} \cdot B \cdot H$$

Figure 12.24 Flow paths in a coat-hanger distributor [2]

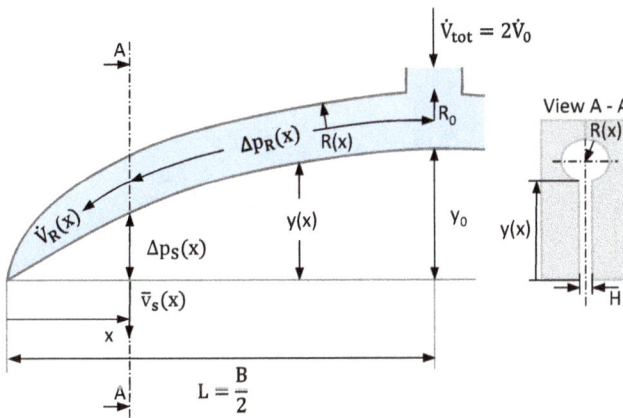

Figure 12.25 Coordinate system of a coat-hanger distributor [2]

The mathematical formulation of the requirement for the same flow resistance (Δp = constant) on all flow paths across the mold width results in the first derivative:

$$\frac{d\Delta p\,(x)}{dx} = 0 \tag{12.11}$$

The following boundary conditions are also applied to the design principles shown below:

- Isothermal, steady-state, laminar, and fully developed stratified flow
- Incompressibility of the melt
- Neglect of run-in and run-out effects
- Wall adhesion
- Viewing the distribution channel as a pipe or flat slot

- No crosscurrents in the island area
- Viewing the island area as an ideal flat slot with a constant slot height

12.4.2 Analytical Balancing: Fish-Tail Distributor

The special feature of the fish-tail distributor is that the geometric progression of the island field (slot) is assumed to be linear (Figure 12.26). This assumption results in the following curve:

$$y(x) = y_0 \cdot \frac{x}{L} \tag{12.12}$$

or

$$y(x) = y_0 \cdot \frac{2x}{B} \tag{12.13}$$

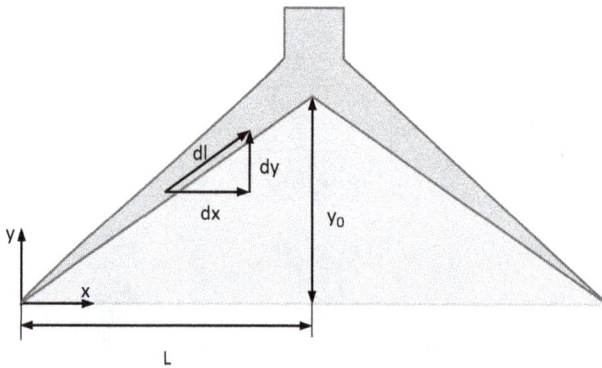

Figure 12.26 Geometry of a fish-tail distributor [2]

The pressure loss in the distributor (pipe) results from the flow path with the Hagen–Poiseuille equation and the representative variables:

$$\Delta p_{\text{pipe}}(x) = \int_x^L \frac{8 \cdot \bar{\eta}_{\text{pipe}}(x) \cdot \dot{V}_0 \cdot x}{\pi \cdot R^4(x) \cdot L} dl \tag{12.14}$$

The pressure loss in the island (slot) is also obtained using the Hagen–Poiseuille equation and the representative variables:

$$\Delta p_{\text{slot}}(x) = \frac{12 \cdot \bar{\eta}_{\text{slot}}(x) \cdot \dot{V}_0 \cdot y_0}{L^2 \cdot H^3} \cdot x \tag{12.15}$$

The local distributor length dl can be reformulated with geometric data for the mold width dx and L and the maximum island length dy and y_0 as follows:

$$dl = \frac{dx}{\cos\left(\arctan\left(\frac{y_0}{L}\right)\right)} \tag{12.16}$$

The pressure losses in the pipe and the slot are then added up to obtain $\Delta p_{tot}(x)$:

$$
\begin{aligned}
\Delta p_{tot}(x) &= \Delta p_{slot}(x) + \Delta p_{pipe}(x) \\
&= \frac{12 \cdot \bar{\eta}_{slot}(x) \cdot \dot{V}_0 \cdot y_0}{L^2 \cdot H^3} \cdot x + \int_x^L \frac{8 \cdot \bar{\eta}_{pipe}(x) \cdot \dot{V}_0 \cdot x}{\pi \cdot R^4(x) \cdot L} \frac{dx}{\cos\left(\arctan\left(\frac{y_0}{L}\right)\right)}
\end{aligned}
\tag{12.17}
$$

and the minimum of the pressure loss is determined, that is, the first derivative is established:

$$\frac{d\Delta p_{tot}(x)}{dx} = 0 \tag{12.18}$$

The result is the radius of the distributor pipe over the mold width:

$$R(x) = \left[\frac{1}{3\pi} \frac{\bar{\eta}_{pipe}(x)}{\bar{\eta}_{slot}(x)} \frac{B \cdot H^3 \cdot x}{y_0 \cdot \cos\left(\arctan\left(\frac{2 \cdot y_0}{B}\right)\right)} \right]^{\frac{1}{4}} \tag{12.19}$$

This equation can only be solved iteratively, as the viscosity in the pipe is a function of the radius $R(x)$. The procedure is described below involving Equation 12.32.

The advantages of the fish-tail distributor are [2]:

- The production costs are considerably lower than those of the coat-hanger distributor, as the distributor pipe is linear.

- The actual length of the distributor is taken into account when calculating the pressure loss in the distributor. This is particularly important for distributors with a great island length.

The main disadvantage of the fish-tail distributor is the different material loads along different flow paths. The reason for this is the inhomogeneous shear rate over the length of the distributor pipe. Furthermore, the residence time of the melt in the system varies. The product quality could be negatively affected under these conditions.

12.4.3 Analytical Balancing: Coat-Hanger Distributor

With coat-hanger distributors, the curve of the island length $y(x)$ over the width is not specified. One approach to designing the coat-hanger distributor can be to keep the melt load in the distributor system as uniform and low as possible. A uniform melt load can be achieved by assuming that the shear rate and the shear stress are the same at all points in the distributor channel.

A low melt load can be achieved, for example, by a short residence time of the melt in the distributor system. For this purpose, the melt is considered on its flow path and the average residence time of the melt is calculated. This results from the flow path in the pipe and in the island as well as from the average velocity in the pipe and in the island (slot) [2]:

$$\bar{t}_v = \int_x^L \frac{1}{\bar{v}_{pipe}(x)} dx + \frac{y(x)}{\bar{v}_{slot}(x)} \tag{12.20}$$

With

$$\bar{v}_{pipe}(x) = \frac{\dot{V}_{pipe}(x)}{\pi \cdot R^2(x)} = \frac{\dot{V}_0}{\pi \cdot R^2(x)} \cdot \frac{x}{L} \tag{12.21}$$

and

$$\bar{v}_{slot}(x) = \frac{\dot{V}_{tot}}{B \cdot H} = \frac{\dot{V}_0}{L \cdot H} \tag{12.22}$$

the mean residence time in the pipe results in

$$\bar{t}_v(x = L) = \int_0^L \frac{1}{\bar{v}_{pipe}(x)} \cdot dx \tag{12.23}$$

Assuming that the residence time of the melt should be as short as possible on all flow paths, the minimum residence time for this equation is now sought, and the first derivative of the equation is determined and set to zero.

$$\frac{d\bar{t}_v}{dR} = \frac{d}{dR}\left[\int_0^L \frac{1}{\bar{v}_{pipe}(x)} dx\right] = \frac{d}{dR}\left[\int_0^L \frac{\pi \cdot R^2(x)}{\dot{V}_{pipe}(x)} dx\right] = 0 \tag{12.24}$$

This minimum search leads to the solution of the geometries for the radius of the distributor (pipe):

$$R(x) = R_0 \left(\frac{x}{L}\right)^{\frac{1}{3}} \tag{12.25}$$

and the x-dependent length of the island (slot):

$$y(x) = y_0 \left(\frac{x}{L}\right)^{\frac{2}{3}} \tag{12.26}$$

A further condition for designing the contour of the distributor pipe under the assumption of a uniform melt load can be formulated by assuming that the representative shear rate in the distributor pipe is constant.

$$\dot{\gamma}_{rep} = \frac{4\dot{V}_{pipe}(x)}{\pi R^3(x)} e_{pipe} = \text{constant} \tag{12.27}$$

That is, with

$$\dot{\gamma}_{rep}(x) = \dot{\gamma}_{rep}(x = L) = \text{constant} \tag{12.28}$$

it follows again:

$$R(x) = R_0 \left(\frac{x}{L}\right)^{\frac{1}{3}} \tag{12.29}$$

After adding the pressure losses in the pipe and in the island, and finding the first derivative, the following is obtained with $\frac{d\Delta p(x)}{dx} = 0$ for the curve of the island (slot):

$$y(x) = \frac{\overline{\eta}_{pipe}}{\overline{\eta}_{slot}} \frac{H^3 L^{\frac{4}{3}} x^{\frac{2}{3}}}{\pi R_0^4} \tag{12.30}$$

and

$$y_0 = y(x = L) = \frac{\overline{\eta}_{pipe}}{\overline{\eta}_{slot}} \frac{H^3 L^2}{\pi R_0^4} \tag{12.31}$$

The following applies to the maximum pipe radius R_0:

$$R_0 = \left(\frac{\overline{\eta}_{pipe}}{\overline{\eta}_{slot}} \frac{H^3 B^2}{4\pi y_0}\right)^{\frac{1}{4}} \tag{12.32}$$

This equation can only be solved iteratively, since the viscosity of the melt in the distributor pipe depends on the radius R, which is a variable quantity $\left(\overline{\eta}_{pipe} = f(R_x)\right)$. One way to initiate the calculation is to use the viscosities in the pipe and in the island as the starting value for the first calculation step, that is, $\overline{\eta}_{slot} = \overline{\eta}_{pipe}$. In the second step, the calculated radius R is used to calculate a viscosity, which is then used in the second step in the equation for R_0. These calculation steps are repeated until the results converge towards a limit value for R_0.

The method described is referred to as operating-point dependent. The term "operating-point dependence" has already been explained in the context of the design of hot runner systems. The design is also carried out here under certain boundary conditions—that is, for example, for a defined material, a volume flow rate, etc. As soon as these boundary conditions change, the system is no longer balanced.

Theoretically, the abovementioned equations can be extended so that they enable a design that is independent of the operating point. For this purpose, the condition of the same representative shear rate in the pipe and in the island can be introduced. This is

$$\dot{\gamma}_{\text{slot}} = \dot{\gamma}_{\text{pipe}} \tag{12.33}$$

From which it follows:

$$\overline{\eta}_{\text{slot}} = \overline{\eta}_{\text{pipe}} \tag{12.34}$$

Under this assumption, the viscosities are not included in the equation for calculating the pipe radius. The result for R_0 is:

$$R_0 = \left(\frac{2}{3\pi} \frac{e_{\text{pipe}}}{e_{\text{slot}}} LH^2 \right)^{\frac{1}{3}} \tag{12.35}$$

and for y_0:

$$y_0 = \left(\frac{3}{2} \frac{e_{\text{pipe}}}{e_{\text{slot}}} \right)^{\frac{4}{3}} \left(\pi HL^2 \right)^{\frac{1}{3}} \tag{12.36}$$

Since the prerequisite for balancing is that the pressure loss from each flow path must be the same, it is sufficient to consider one flow path to calculate the total pressure loss in the system. The simplest calculation of the pressure loss is then carried out by considering the center of the island ($x = L$). There, the island has the length y_0. If we calculate using the width L, it should be noted that the volume flow rate is $\dot{V}_0 = \frac{\dot{V}_{\text{tot}}}{2}$. The same result would be obtained if the complete island width B and the total volume flow rate $\dot{V}_{\text{tot}}$ were used in the calculation.

$$\Delta p = \frac{12 \cdot \overline{\eta}_{\text{slot}} \cdot \dot{V}_0 \cdot y_0}{L \cdot H^3} = \frac{12 \cdot \overline{\eta}_{\text{slot}} \cdot \dot{V}_{\text{tot}} \cdot y_0}{B \cdot H^3} \tag{12.37}$$

The procedure for designing extrusion molds is briefly explained below:

1. First of all, the maximum permissible pressure loss must be known or specified. Furthermore, the island gap height H must be known or a value must be assumed. After calculating a representative shear rate and determining a representative viscosity, the maximum island length y_0 in the center of the mold, $x = L$, can be calculated.

2. In the next step, the maximum distributor radius R_0 is determined. Since the viscosity in the pipe, which depends on R, is included in the equation, an iterative procedure is useful. As a starting value, the viscosity in the pipe is set equal to the viscosity in the island: $\overline{\eta}_{slot} = \overline{\eta}_{pipe}$. The actual viscosity in the pipe is then calculated and a new R_0 is determined. This step is repeated until a convergence criterion for R_0 is reached.

3. The operating-point independence of the mold can then be estimated. For this estimation, the variables that, as already mentioned, define the operating-point dependence of the system are changed. These include the material (viscosity) and the volume flow rate. With the previously determined R_0, y_0 is recalculated for a different material, for example, and compared with the value calculated in 1. The difference between the results can be used to make a statement about the operating point.

According to Michaeli [2], a major disadvantage of the design methods presented for coat-hanger distributors is that the calculation of the pressure loss in the distributor is not based on the exact distributor length, but on the distributor length projected onto the outlet gap. According to Michaeli, this can lead to insufficient extrudate thickness in the outer areas of the mold in the case of distributors with a great island length.

12.4.4 Numerical Balancing

In addition to the analytical design, in which the curve of the distributor is calculated as a function of the current coordinate x while specifying certain boundary conditions (such as a possible low dwell time), the design can also be carried out numerically. For this purpose, the mold geometry is broken down into individual segments for the island and for the pipe (Figure 12.27). The volume flow rate and the pressure are then balanced for each segment [2].

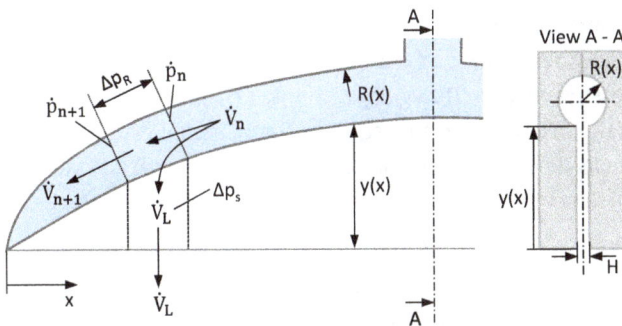

Figure 12.27 Coordinate system of a coat-hanger distributor [2]

In the island, for each segment the result is the same volume flow rate $\dot{V}_L$. This is calculated by dividing the volume flow rate for one half of the island by the number of segments. In the distributor, the volume flow rate decreases from segment to segment, starting from the melt inlet, by the volume flow rate $\dot{V}_L$ of the island segments. As a result, for each segment n, the relationship between the incoming and the outgoing volume flow rate in the distributor must be as follows:

$$\dot{V}_n = \dot{V}_{n+1} + \dot{V}_L \tag{12.38}$$

The volume flow rate $\dot{V}_L$ is given for each segment. According to Hagen–Poiseuille, it is directly related to the pressure in the center of the segment and the slot geometry.

$$\dot{V}_L = f\left[(\bar{p} - p_0), H(x), y(x)\right] \tag{12.39}$$

The pressure in the center of the segment is calculated as the mean value from the pressures on the segment nodes:

$$\bar{p} = \frac{p_n + p_{n+1}}{2} \tag{12.40}$$

In the distributor, the pressures add up from segment to segment. The following therefore applies:

$$p_n = p_{n+1} + \Delta p_R \tag{12.41}$$

The pressure in the distributor again can be described using the Hagen–Poiseuille equation:

$$\Delta p_R = f\left[\dot{V}_n, \dot{V}_{n+1}, R(x)\right] \tag{12.42}$$

The procedure for numerical balancing can be carried out in different ways. One possibility is to first determine the maximum island length y_0 for a given volume flow rate, permissible pressure loss, and material data. Half of the volume flow rate is then divided by the number of defined segments for one half of the mold. This gives the volume flow rate for each island segment and also for each pipe section. It is possible to calculate the representative shear rate and the representative viscosity for each segment of the mold. The pressure on each flow path can now be considered. Depending on the pressure losses in the pipe and the subsequent island, a pipe geometry and an island length result for each flow path. The use of a program is generally recommended at this point.

The numerical balancing of a distribution system has some significant advantages:

- It is theoretically possible to assume a different island height H for each island segment. This means that there is greater geometric freedom with regard to the outlet geometry (e.g., the thickness of the foil or plate).

- It is also possible to take into account the pressure losses that occur at cross-sectional jumps, the so-called inlet or extensional pressure losses. These can occur, for example, when overflowing from the distributor pipe into the island gap. With numerical balancing, it would be possible to implement Cogswell's approach (see Chapters 6 and 13) for calculating the extensional pressure losses at these points without any problems.

Figure 12.28 shows an example of a typical pressure distribution in a slot die.

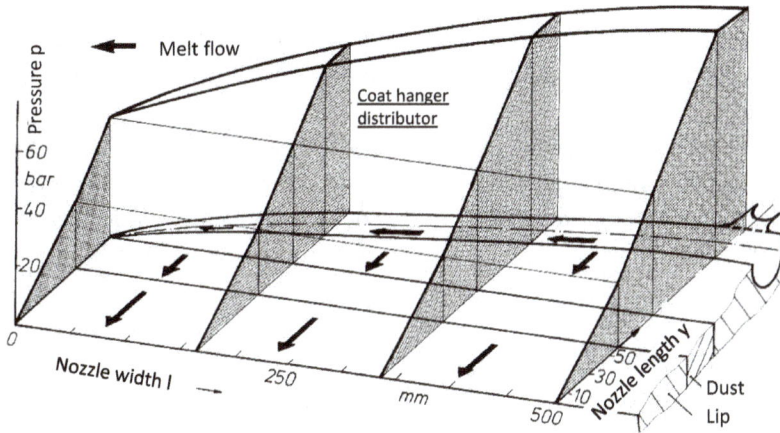

Figure 12.28 Pressure distribution in a slot die [2]

12.4.5 *Task*: Analytical Balancing of a Fish-Tail Distributor

Lay out a wide slot mold with a fish-tail distributor. The exit gap width B of the mold is 1000 mm and the gap height H is 3 mm. The material used is a PE-LD of the type Lupolen 1800 H. The material data is as follows:

Material Data (Carreau Coefficients)

- $A = 16{,}800$ Pa·s
- $B = 1.8$ s
- $C = 0.63$
- Reference temperature $T_{ref} = 170\,°C$
- Standard temperature $T_s = 10\,°C$
- Density (115 °C) $\rho = 801$ kg/m^3
- Coefficient of thermal extension $\alpha = 0.69 \cdot 10^{-3}$ 1/°C

The *wanted* quantities are the maximum distributor radius R_0 and the maximum island length y_0 for a mass temperature of 200 °C and a mass flow rate of 250 kg/h if the maximum pressure loss in the mold should be 100 bar.

12.4.6 *Task*: Analytical Balancing of a Coat-Hanger Distributor

A design independent of the operating point is to be obtained for a wide slot nozzle with a coat hanger. The calculation data can be taken from the previous task.

The *wanted* quantities are the (operating-point-independent) maximum distributor radius R_0, the maximum island length y_0, the average residence time in the island, and the pressure requirement.

Then use software such as Microsoft Excel® to create a design that is operating-point dependent for this coat-hanger distributor of the wide slot nozzle. Plot the curve $y(x)$ and $R(x)$ for the operating-point-dependent design of the coat hanger. Also show the result of the previous task (Section 12.4.5), namely $R(x)$ for the fish-tail distributor.

12.4.7 *Task*: Numerical Balancing of a Wide Slot Nozzle
with a Coat-Hanger Distributor with Segments

Use software such as Microsoft Excel to *create* a program for numerical balancing of a wide slot nozzle with a coat-hanger distributor (Figure 12.29). The input variables should be the geometric, material, and processing data. The output variable is the geometric progression of the distributor system.

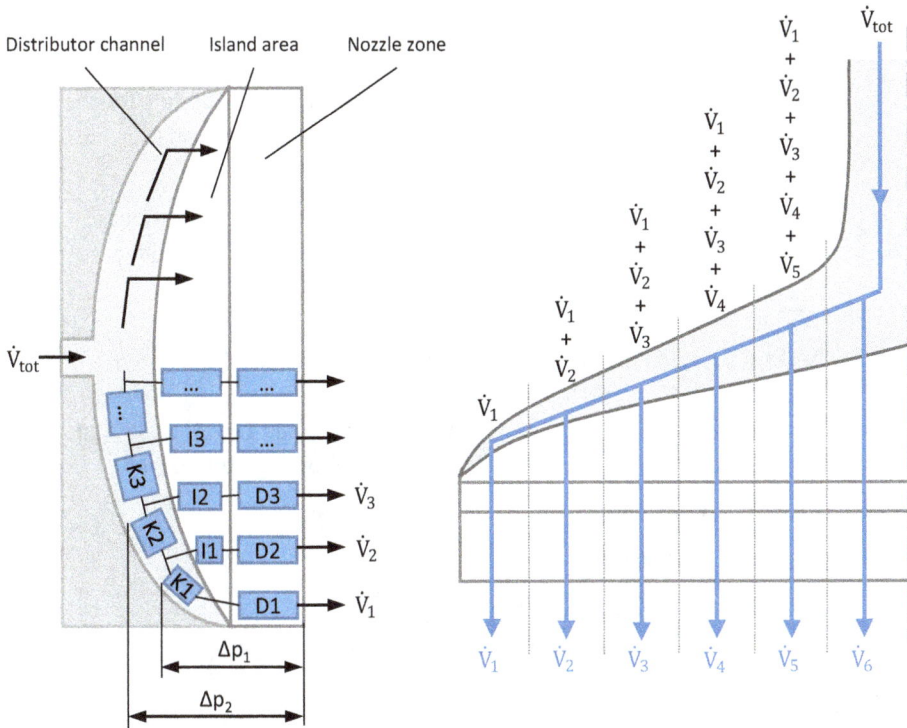

Figure 12.29 Volume flow rate curve in a coat-hanger distributor

This task is to be solved numerically using software.

Procedure:

- Divide the distributor into two symmetrical halves and then divide one half into segments. The number of segments should be variable in the software file.

- For each segment, the diameter of the distributor pipe must be considered constant.

- The pressure on all flow paths in the segments must be the same. Thus, there are parallel isobars in the island.

- The volume flow rate is divided in the island. In the distributor pipe, these volume flow rates add up from the edge to the center.

- The analytical equations should not be used for the design of $R(x)$ and $y(x)$. For numerical balancing, the equations from the "Important Rheological Formulas" collection (see chapter at the beginning of the book) should be used. This means that the pressure losses in the pipe and in the slot (island) are added up for each flow path for the design.

- The following input data should be variable, i.e., changeable in the software spreadsheet:
    - *Material data* for the Carreau or power-law approach, parameters for the WLF or the Arrhenius function for the temperature dependence and the density
    - *Processing parameters:* mass temperature, mass flow, etc.
    - *Geometric parameters:* slot width, film thickness, number of segments
    - *Permissible pressure loss*

12.4.8 *Task*: Calculation of the Output Rate of an Extruder

You are extruding a tubular film made of polyethylene (see Figure 12.30 and Table 12.4). The ring slot die for extruding the film has the following dimensions:

- Inner diameter $D_i = 497$ mm
- Outer diameter $D_a = 500$ mm
- Length of the annular gap $L = 400$ mm

The processing temperature is 220 °C.

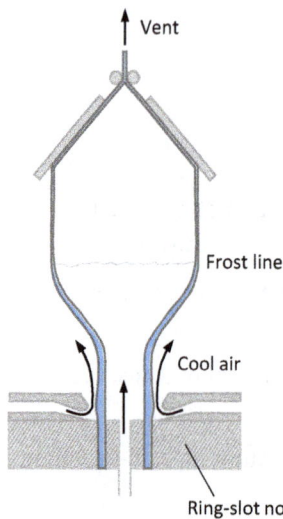

Figure 12.30
Principle for the production of tubular film [7]

Table 12.4 Material Data of the PE-HD

Density	$\rho = 817\ kg/m^3$
Reference temperature	$T_{ref} = 200\ °C$
Flow exponent	$m = 2.26$
Fluidity	$\phi = 7.5 \cdot 10^6\ (mm^{2m})/(N^m s)$
Flow activation energy	$E_0 = 41.77\ kJ/mol$
Universal gas constant	$R = 8.31 \cdot 10^{-3}\ kJ/(mol\ K)$

Question: What is the output of the extrusion system in kg/h if the available pressure is 40 bar?

12.4.9 *Task*: Design of a Slot Die

Plastic is extruded through a slot die of length L, width B, and height H and with the pressure at the outlet $p_{flow\ path\ end} = 0$ bar (see Figure 12.31). The volume flow rate is given.

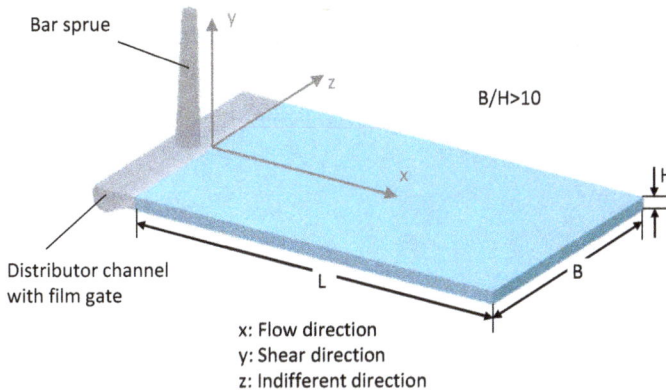

Bar sprue

y

z

B/H>10

H

Distributor channel with film gate

x

L

B

x: Flow direction
y: Shear direction
z: Indifferent direction

Figure 12.31 Geometry of a slot die

Calculate the forces exerted on the upper and the lower nozzle walls in the horizontal and the vertical directions.

Note: Consider the upper wall and the forces acting on it in the horizontal and the vertical directions (see Figure 12.31). Assume the following conditions:

- The force of gravity is negligible.

- The influence of the lateral nozzle walls is negligible.

- This is a fully developed, steady-state, laminar layered flow.

- Elastic influences are negligible.

- The flow behavior of the plastic is described by Ostwald and de Waele's power-law approach.

- The representative viscosity method can be used to determine the shear pressure losses.

Given data:

$L = 200\,\text{mm}$	$B = 20\,\text{mm}$	$H = 2\,\text{mm}$
$\dot{V} = 1\,\text{cm}^3/\text{s}$	$K = 30\,000\ \text{Pa·s}^{0.36}$	$n = 0.36$

References

[1] Unger, P.: *Heißkanal-Technik*. Hanser, Munich, 2004

[2] Michaeli, W.: *Extrusionswerkzeuge für Kunststoffe und Kautschuk: Bauarten, Gestaltung und Berechnung*. Hanser, Munich · Vienna, 1991

[3] Menges, G.; Michaeli, W.; Mohren, P.: *Spritzgießwerkzeuge: Auslegung, Bau, Anwendung*. Hanser, Munich, 2007

[4] Limper, A.: *Process Engineering in Thermoplastic Extrusion*. Hanser, Munich, 2012

[5] Johannaber, F.; Michaeli, W.: *Handbuch Spritzgießen*. Hanser, Munich, 2004

[6] Michaeli, W.; Dombrowski, U.; Hüsgen, U.; Kalwa, M.; Meier, M.; Schwenzer, C.: *Extrusionswerkzeuge für Kunststoffe und Kautschuk: Bauarten, Gestaltung und Berechnungsmöglichkeiten*. Hanser, Munich, 2009

[7] Greif, H.; Limper, A.; Fattmann, G.: *Technologie der Extrusion: Lern- und Arbeitsbuch für die Aus- und Weiterbildung*. Hanser, Munich, 2017

13 Shear and Extensional Pressure Losses at Cross-Sectional Transitions

When a viscoelastic melt flows through a nozzle, it strives to adopt the most energetically favorable flow shape. As a result, as soon as the cross section of a flow channel changes, the extensional flow components must be taken into account in addition to the shear flow effects. This can be the case, for example, during injection molding in the area of the machine nozzle, the sprue, the gate, and in the hot runner system. In these areas, a cylindrical geometry cannot always be assumed. Particularly in the case of the machine nozzle, it becomes clear that the cross section is reduced in the direction of the flow. As a consequence, at such flow points, in addition to the shear flow (shear pressure losses), the extensional flow effects (extensional pressure losses) must also be taken into account [1–3].

Considering Figure 13.1, Equation 13.1 is obtained for a cone to determine the pressure loss due to shear:

$$p_{sh} = \left(\frac{\dot{V}}{\pi}\right)^n \left(\frac{3n+1}{n}\right)^n \cdot \frac{2K}{3n} \cdot \frac{L}{(R_0 - R_1)} \cdot \frac{1}{R_1{}^{3n}} \cdot \left[1 - \left(\frac{R_1}{R_0}\right)^{3n}\right] \tag{13.1}$$

where the following parameters are used:

- Geometric parameters:
  - R_0: initial radius
  - R_1: end radius
  - L: nozzle length
- Material parameters for Ostwald and de Waele's power-law approach:
  - n: viscosity exponent
  - K: consistency factor of the material

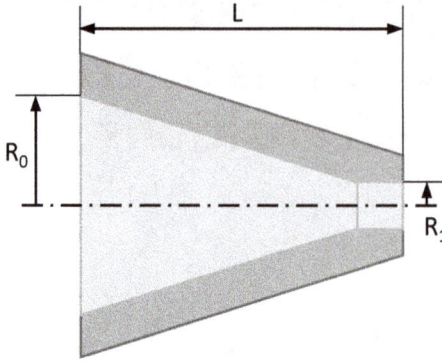

Figure 13.1

Approximation method for the calculation of conical nozzles

If R_0 and R_1 are the same value, there is no change in cross section. In this case, the equation of Hagen–Poiseuille for a pipe flow extended by the power-law approach of Ostwald and de Waele can be applied.

$$\dot{V} = \frac{\pi \cdot R^{m+3} \cdot \phi \cdot \Delta p^m}{2^m \cdot (m+3) \cdot L^m}$$
(13.2)

$$\Delta p_{sh} = \frac{\dot{V}^{\frac{1}{m}} \cdot L \cdot 2 \cdot (m+3)^{\frac{1}{m}}}{\pi^{\frac{1}{m}} \cdot \phi^{\frac{1}{m}} \cdot R^{\frac{m+3}{m}}}$$
(13.3)

with $m = \frac{1}{n}$ and $m \geq 1$

In addition to the pressure losses due to shear, the unidirectional stretching (extension) of the material in the direction of the flow and the associated extensional pressure losses must also be taken into account. The extensional viscosity μ can be used to determine the extensional pressure loss p_{ex} in a cone [4]:

$$p_{ex} = \frac{4 \cdot \mu \cdot \dot{V}}{3 \cdot \pi} \cdot \frac{R_0 - R_1}{L} \cdot \frac{1}{R_1^3} \cdot \left[1 - \left(\frac{R_1}{R_0}\right)^3\right]$$
(13.4)

For very low shear rates (Newtonian range of the flow curve), the extensional viscosity μ initially can be determined approximately using the zero viscosity η_0 (Trouton ratio). The following applies:

$$\mu = 3 \cdot \eta_0$$
(13.5)

The zero viscosity can be determined using the Carreau–WLF equation. In this approach, the zero viscosity is contained directly as a material quantity in the numerator as the parameter A or P_1 and is read off as a value on the y axis (Figure 13.2).

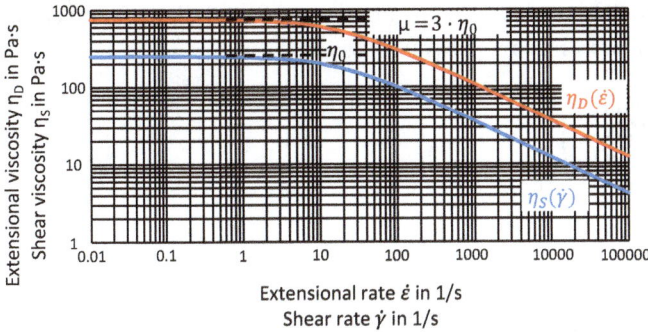

Figure 13.2 Flow curve for an PE-LD at 210 °C (Lupolen 1800 S)

If the plastic melt flows through a conical nozzle, the shear pressure p_{sh} and the extensional pressure p_{ex} can be additively superimposed as a first approximation to calculate the pressure p_{tot} loss:

$$p_{tot} = p_{sh} + p_{ex} \tag{13.6}$$

Figure 13.3 shows an example of a calculation of the shear and the extensional pressure losses for a polystyrene 164 N as a function of the inlet and the outlet radii. The greater the difference shown in the radius ratio, the more the inlet pressure losses (extensional pressure losses) outweigh the shear pressure losses.

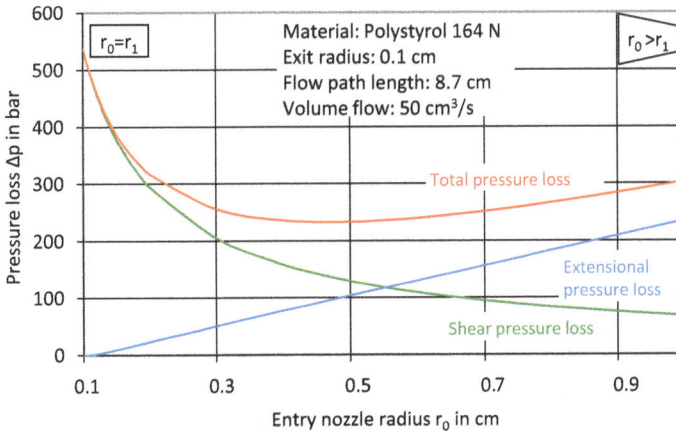

Figure 13.3 Calculated values for shear/extensional and total pressure losses

On the other hand, the optimum nozzle geometry can also be designed with a view to minimizing pressure loss. This requires an integral calculation of the pressure loss, that is, the pressure losses in the nozzle are added up segment by segment and an analytical solution is used to determine the minimum pressure.

Equation 13.7 makes it possible to obtain the contour-optimized design of a machine nozzle under the condition of minimum pressure loss:

$$x = \left(\frac{\mu}{\eta}\right)^{\frac{1}{2}} \cdot \frac{R_1 \cdot \sqrt{2}}{3n - 1} \cdot \left[\left(\frac{R(x)}{R_1}\right)^{\frac{3n-1}{2}} - 1\right] \tag{13.7}$$

or

$$R(x) = R_1 \cdot \left[\frac{x}{\left(\frac{\mu}{\eta}\right)^{\frac{1}{2}} \cdot \frac{R_1 \cdot \sqrt{2}}{3n-1}} + 1\right]^{\frac{2}{3n-1}} \tag{13.8}$$

where the following parameters are used:

- Geometric parameters:
  - x: coordinate in the flow direction
  - $R(x)$: variable radius in the direction of the flow
  - R_1: end radius
- Material parameters:
  - η: shear viscosity at the nozzle outlet
  - μ: extensional viscosity
  - n: viscosity exponent from the power-law approach according to Ostwald and de Waele

 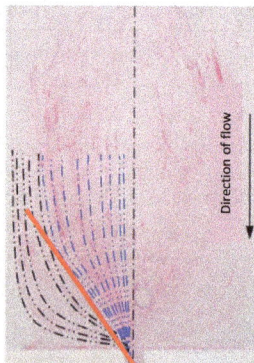

V1: T_{wz}=40 °C, v_{in}=5 mm²/s   V2: T_{wz}=40 °C, v_{in}=90 mm²/s

Direction of flow

Figure 13.4 Comparison of the nozzle inlet flow lines

If the cross section does not change linearly but abruptly, the plastic melt will not follow this jump, but will form secondary vortices in the corner areas and the most energetically favorable flow shape. This was already discussed in Chapter 6 when determining the extensional viscosity via the inlet pressure losses using F. N. Cogswell's approach [4–6].

13.1 *Task*: Extensional and Shear Pressure Losses

The molded parts shown in Figure 13.5 are three test specimens for impact tests and a so-called "quarter circle disk" (bottom right) to determine the volume shrinkage of the plastic. As part of the task, the injection pressure (before the injection screw) required to fill the molded part "quarter circle disk" is *to be calculated.* This pressure is also relevant for the choice of the holding pressure ($p_{hold} \sim 0.3$–$0.6 \cdot p_{injec}$) in reality!

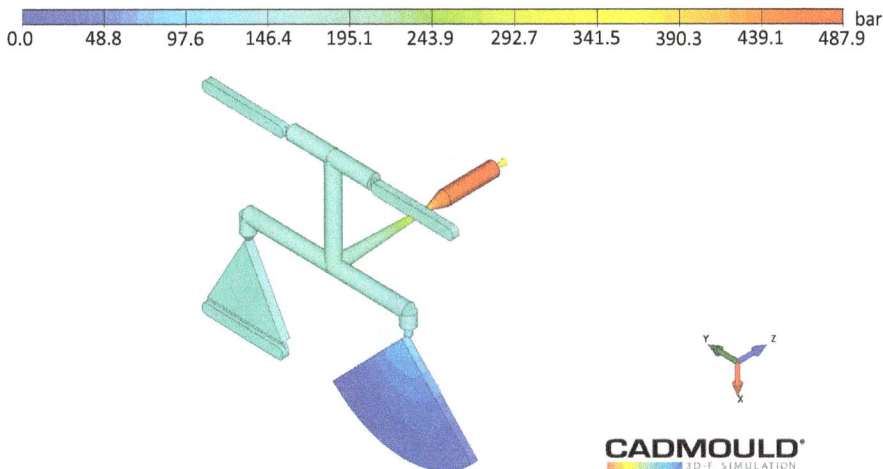

Figure 13.5 Example of a sprue system

Material data:

Total volume flow rate	30 cm³/s
Volume flow rate in the "quarter circle disk"	10 cm³/s
Diameter of the screw cylinder	30 mm
Nozzle diameter	3 mm
Nozzle length	40 mm
Average diameter of the central sprue bar	6 mm
Length of the sprue bar	50 mm
Diameter of the left distributor	8 mm
Length of the left distributor	50 mm
Gate diameter	2 mm
Gate length	5 mm
Molded part radius	60 mm
Molding wall thickness	4 mm
Material data	BASF polystyrene 168N
Ground temperature	230 °C

All cross-sectional transitions can be assumed to be conical.

References

[1] Menges, G.; Michaeli, W.; Mohren, P.: *Spritzgießwerkzeuge: Auslegung, Bau, Anwendung*. Hanser, Munich, 2007

[2] Lichius, U.; Schmidt, L.: *Rechnergestütztes Konstruieren von Spritzgießwerkzeugen: systematisches Entwickeln von Betriebsmitteln, Aufbau und Funktion von Spritzgießwerkzeugen*. Vogel, Würzburg, 1986

[3] Kleinecke, K.-D.: On the influence of fillers on the rheological behavior of high-molecular poly-ethylene melts: 2. investigations in the inlet flow. *Rheologica Acta*, 1988, 27

[4] Cogswell, F. N.: Converging flow of polymer melts in extrusion dies. *Polymer Engineering & Science*, 1972, Vol. 12, 1

[5] N. N.: Polymer Melt Rheology: A Guide for Industrial Practice. University Microfilms International, Ann Arbor, Michigan, 1992

[6] N. N.: Tensile deformations in molten polymers. *Rheologica Acta*, 1969, Vol. 8, 2

14 Rheological Mold Design for Injection Molding with the Filling Pattern Method

14.1 Reasons for a Graphical Process

The quality of injection-molded parts can be affected by air pockets or so-called weld lines, which occur when different melt streams flow together while filling the molded part. Air pockets lead to burn marks in the molded part if the cavity is insufficiently vented. Weld lines, for example, can be mechanical weak points that reduce part strength. Particularly with dark-color materials, weld lines and flow lines often appear as unsightly streaks on the surface of the molded part.

In the following, a method with which the designer can detect critical areas of the molded part even before the mold is built will be presented [1]. Air pockets and weld lines can be localized with sufficient accuracy and avoided by suitable design measures such as changing the gate position or type, arranging venting elements in the mold, or arranging and dimensioning ribs and flow aids [2].

14.2 Model of the Graphical Mold Filling Process

The model of the filling process is based on wave propagation theory, according to which every particle excited to oscillate is itself an exciter of oscillations and thus ensures the propagation of the wave. According to Huygens's principle (wave theory), each point on a wave surface can be regarded as the starting point of a new so-called elementary wave and each wave surface can, in turn, be regarded as the envelope of elementary waves. According to this wave propagation theory, refraction phenomena—that is, changes in the direction of wave propagation—can be explained very clearly by differences in the velocity of wave propagation in different media, and simple laws can be derived from this.

During the filling process, the flow front is now regarded as a wave front. From this wave front, each point is regarded as the starting point for a new wave. This means that the melt spreads around each point into the still-empty mold cavity in the form of a wave. The new flow front then results as the envelope of the individual waves. The driving force for the spread of the flow front is the injection pressure, which propagates from the gate to the flow front. Different propagation rates of elementary waves, which cause changes in direction, are described by different flow channel geometries during the filling process and therefore cause different flow resistances.

On the basis of this model and the rheological principles described below, a design method for the flow front of the melt was developed. Even if there is only a limited analogy between wave propagation and the physical principles of the filling process (this could only be considered physically and mathematically exact by means of momentum, energy, and continuity theorems as well as mathematical descriptions of the real material behavior), this model concept can still be regarded as a suitable aid in the design of the flow front courses.

14.3 Rheological Principles

The rheological basis for the method for determining the filling pattern is primarily the pressure requirement calculation for wall-adhering flows according to Hagen–Poiseuille. For laminar flow, the following applies to the plate:

$$\dot{V} = \frac{H^3 \cdot B \cdot \Delta p}{12 \cdot \eta \cdot L} \tag{14.1}$$

With the continuity equation $\dot{V} = \bar{v} \cdot A = \bar{v} \cdot B \cdot H$, it follows:

$$\frac{\Delta p}{L} = \frac{12 \cdot \bar{v} \cdot H \cdot B \cdot \eta}{H^3 \cdot B} = \frac{12 \cdot \bar{v} \cdot \eta}{H^2} \tag{14.2}$$

The pressure loss from the start of the flow path (gate) to the end of the flow path under consideration (flow front) must be the same for each flow path. Thus, it follows:

$$\Delta p = p_{\text{gate}} - p_{\text{flow front}} = \text{constant} \tag{14.3}$$

This means that, assuming 1 and 2 correspond to two flow paths,

$$\Delta p_1 = \Delta p_2 \tag{14.4}$$

It follows from this:

$$\frac{12 \cdot \overline{v_1} \cdot \eta_1 \cdot L_1}{H_1^2} = \frac{12 \cdot \overline{v_2} \cdot \eta_2 \cdot L_2}{H_2^2} \tag{14.5}$$

As a simplification, an isothermal flow is assumed. This means that cooling effects and dissipation are neglected. Furthermore, it is assumed that the viscosity is the same for all flow paths. If, for example, the power-law approach is applied to describe the viscosity, this assumption would be confirmed under the condition of constant mass temperature (isothermal conditions). This proof is to be provided in the following task.

The following therefore applies:

$$\eta_1 = \eta_2 \tag{14.6}$$

It follows:

$$\frac{\overline{v_1} \cdot L_1}{H_1^2} = \frac{\overline{v_2} \cdot L_2}{H_2^2} \tag{14.7}$$

Now, the average velocity is replaced by the flow path per time interval:

$$\overline{v} = \frac{L}{\Delta t} \tag{14.8}$$

This is how we obtain

$$\frac{L_1 \cdot L_1}{\Delta t_1 \cdot H_1^2} = \frac{L_2 \cdot L_2}{\Delta t_2 \cdot H_2^2} \tag{14.9}$$

Since the time intervals under consideration are of equal size, the following applies: $\Delta t_1 = \Delta t_2$. Thus, it follows:

$$\frac{L_1^2}{H_1^2} = \frac{L_2^2}{H_2^2} \tag{14.10}$$

or

$$\frac{L_1}{H_1} = \frac{L_2}{H_2} \tag{14.11}$$

Then, the "preflow" of the melt (the flow of the melt in the two directions L_1 and L_2) behaves like the corresponding wall thicknesses (H_1 and H_2). This simple relationship (L/H = constant) is valid for a wide range of injection molding conditions.

With the help of this equation, a graphical filling pattern method has been developed. It can be assumed that, in Figure 14.1, each point of the flow front (1) is the outlet for a concentric flow (2) with the radius ΔL. The envelope curve (3) around these circles is the new flow front.

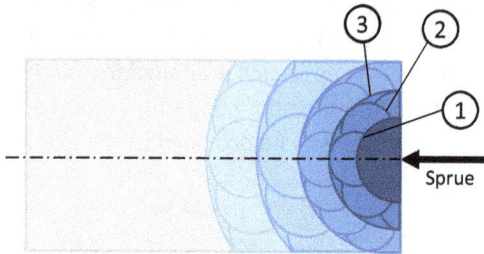

Figure 14.1 Swelling spread of the melt [2]

The filling pattern method is sufficiently accurate for practical requirements. It can be used for all thermoplastics. The limits are at very low flow-front rates and small cross sections of the molded part, because there is a risk of partial freezing of the melt. Due to the freezing of the melt, the actual velocity at the affected points is lower than the previously designed velocity. Values for limit cross sections cannot be specified due to the complicated dependences.

This simple approach was also the basic equation for the first computer-aided simulation programs to predict weld lines and air pockets.

First of all, the molded part must be unrolled (Figure 14.2). The overflow over an edge into an adjacent surface must be taken into account.

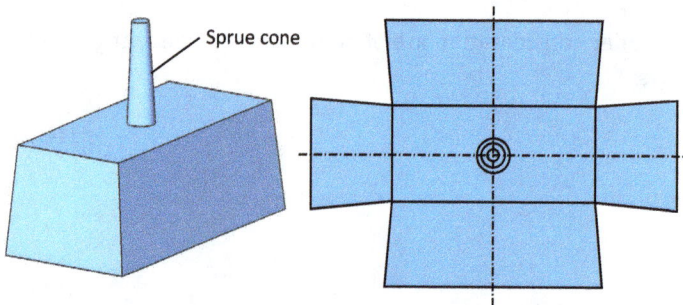

Figure 14.2 Box-shaped molded part with corresponding development

14.4 Example of the Filling Pattern Method

The flow front design is explained below as an example.

- The melt initially spreads in a swelling shape from the gate.

- The spread is indicated by concentric circles around the injection point.

- The radii of these circles increase in steps. The size of the steps must be selected by the designer so that a sufficiently precise design be possible.

- If the melt flow front reaches a side surface or a surface with a different wall thickness, a flow front circle is drawn in such a way that the melt touches the side surfaces.

The procedure is illustrated in Figure 14.3 using a plate with a local thin section on the right at the end of the flow path, as an example. The left side of this figure displays the filling of the molded part using the filling pattern method; the right side, using a FEM (finite element method) simulation program. The comparison of the simulation results with the filling pattern method results shows a good agreement. Starting from the gate, the melt flows in concentric circles until the melt flow front reaches the thin section. In the left-hand area, the melt flow front corresponding to the wall thickness ratio H_L = 6 mm to H_R = 3 mm will spread out over a distance L_L that is twice the one in the right-hand area of the thin section. The distance between the lines for the melt flow fronts on the left and right must be selected accordingly. As a result, the spread of the melt flow front in the left and the right molded part areas can be drawn in a circle. What still needs to be taken into account is the transition of the melt from the left to the right area, as the melt leads in the left area. For this purpose, starting from the point where the flow front arrives at the thin section, a circle is drawn with the distance corresponding to the spread in the thin section. Now a tangent to the circle is drawn in the thin section, with a connection to the circle of the thicker section. This is repeated until the end of the flow path is reached.

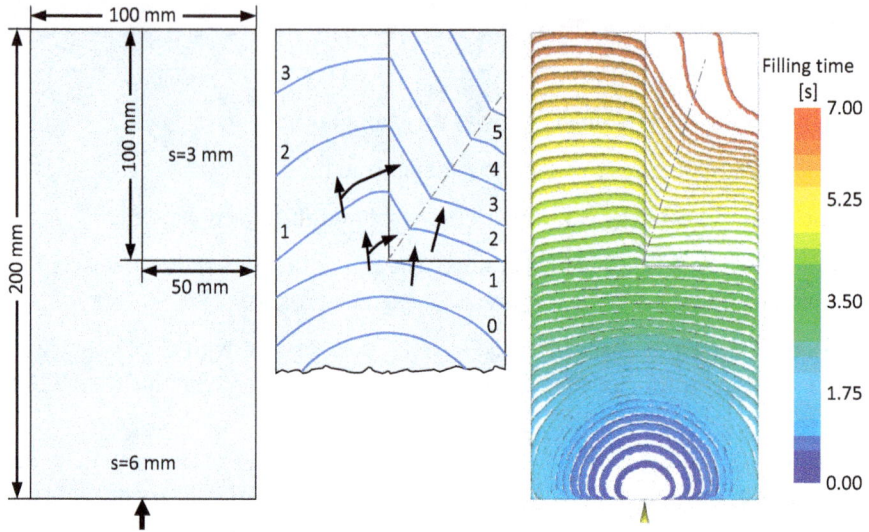

Figure 14.3 Filling pattern for a plate-shaped component created using the filling pattern method and a simulation program

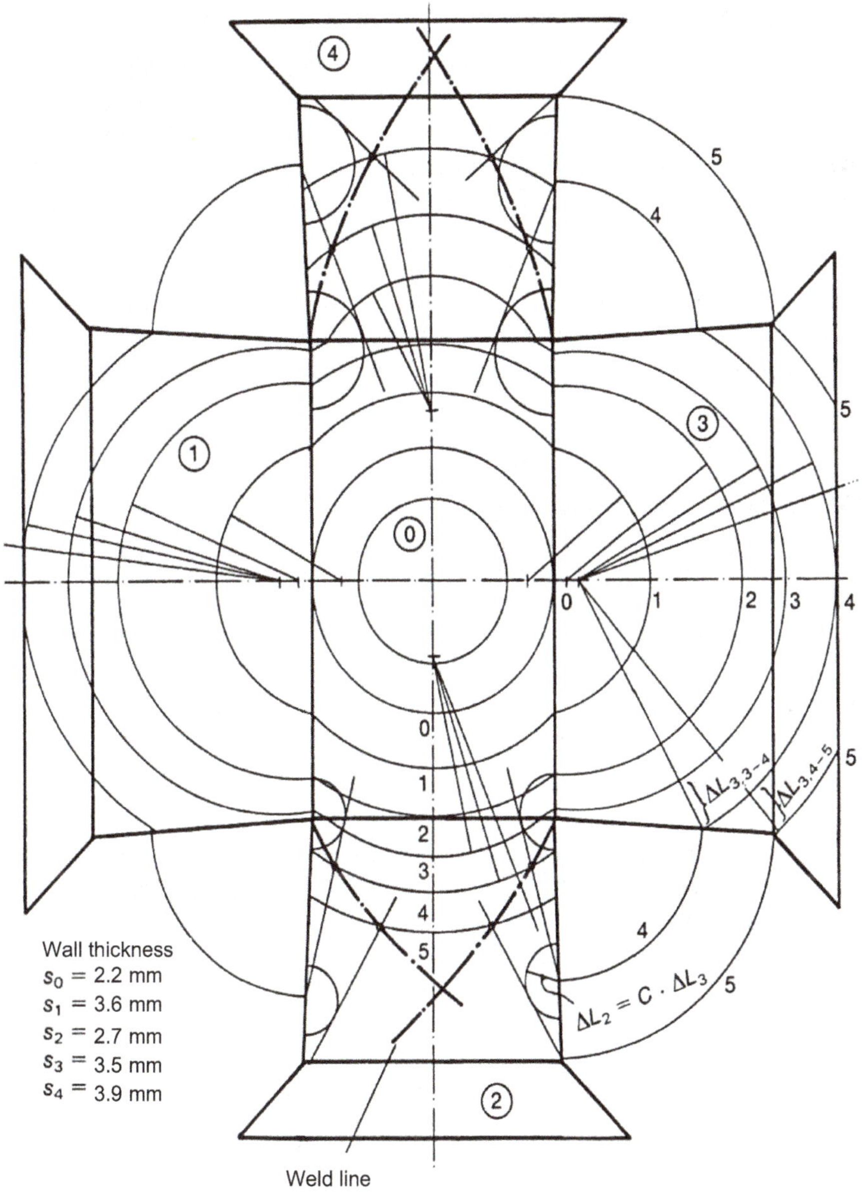

4
5
4
5
1
3
0
0
1
2
3
4
0
1
2
3
4
5
5
$\Delta L_{3,3-4}$
$\Delta L_{3,4-5}$
4
$\Delta L_2 = C \cdot \Delta L_3$
5
2
Wall thickness
s_0 = 2.2 mm
s_1 = 3.6 mm
s_2 = 2.7 mm
s_3 = 3.5 mm
s_4 = 3.9 mm
Weld line

Figure 14.4 Filling pattern method for a box-shaped component [2]

14.5 *Task*: Proof of the Independence of the Filling Pattern Method from the Shear-Thinning Viscosity

You want to injection-mold two parts with different wall thicknesses ($H_2 = H_1/2$) but the same flow path lengths L_1 and L_2 and the same width in one mold ((1+1)-cavity; see Figure 14.5). The sprue/gate system is the same for both molded parts.

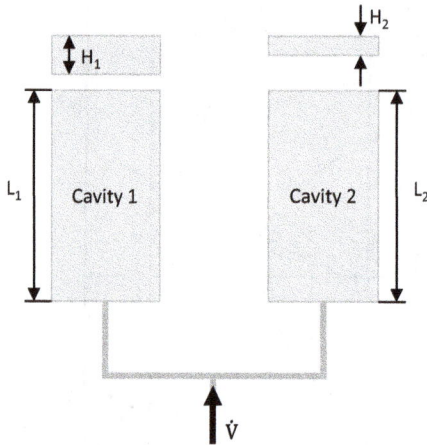

Figure 14.5 Cavity geometry

Question: How far has the melt progressed in the second cavity when the first cavity is filled (same time interval!)?

Initially assume a material with Newtonian flow behavior, and then, a polymer melt with shear-thinning viscous flow behavior ($m = 4$). First, set up the basic equation as in the previous section, that is, the following applies:

$$\Delta p_1 = \Delta p_2 \tag{14.12}$$

Insert the Hagen–Poiseuille equation into this equation. The shear-thinning viscosity of the plastic melt should then be taken into account. This is done here by substituting the power-law approach of Ostwald and de Waele into the Hagen–Poiseuille equation.

References

[1] Lichius, U.; Schmidt, L.: *Rechnergestütztes Konstruieren von Spritzgießwerkzeugen: systematisches Entwickeln von Betriebsmitteln, Aufbau und Funktion von Spritzgießwerkzeugen*. Vogel, Würzburg, 1986

[2] Menges, G.; Michaeli, W.; Mohren, P.: *Spritzgießwerkzeuge: Auslegung, Bau, Anwendung*. Hanser, Munich, 2007

15 Screw Flows

15.1 Introduction and Models

For manufacturers of films, pipes, or even sheets, the plasticizing capacity (kilograms per hour) of an extruder is of great importance; the same applies to the injection molder. They must know the plasticizing flow (grams per second) of the injection molding machine in order to be able to estimate whether the machine is capable of processing the required shot weight in the time available (dosing time). In the following, the flow types of a screw flow (Figure 15.1) will be considered and analyzed. The aim is to mathematically describe and illustrate the types of flow that occur.

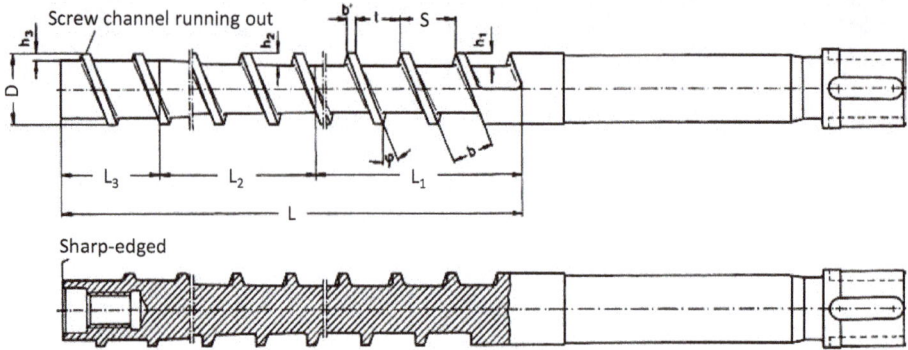

D	Screw diameter		**S**	Screw increase	0.8 to 1 D
L	Profiled screw length	16 to 20 D	**h₁**	Channel depth in the feeding zone	ca. 0.1 D
L₁	Feeding zone (Core diameter constant)	0.6 L	**h₂**	Channel depth in the compression zone	
L₂	Compression zone (Core diameter conical)	0.2 L	**h₃**	Channel depth in the metering zone (Core diameter conical)	ca. 0.05 D
L₃	Metering zone (Core diameter constant)	0.2 L	**b'**	Bar width	ca. 0.1 D

Figure 15.1 Geometry of a three-zone screw [1]

15.1.1 Melting Model According to Maddock

A melting model that describes the flow processes in screws is the Maddock model (Figure 15.2). Due to heat conduction and friction, a melt film forms on the channel wall. This melt film increases in thickness until it is scraped off the active flank of the screw, where a melt vortex forms. This melt vortex becomes wider in the direction of the screw channel and reduces the width of the solids present. The melting process ends when the solids width x becomes zero [1].

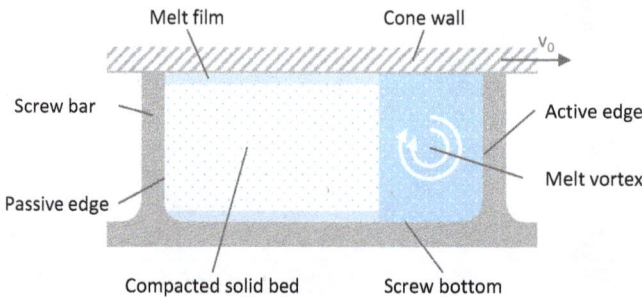

Figure 15.2 Melting model according to Maddock [1]

The volume flow that is conveyed by a screw is made up of three volume flows. It results from the drag flow, the pressure flow, and the leakage flow.

- As a rule, the *leakage flow* can be neglected, at least in the case of new (i.e., unworn) units.

- The *drag flow* results from the screw rate and the geometric data of the screw and is always linear.

- The *pressure flow* has a parabolic velocity profile and is in the direction opposite to the drag flow during injection molding.

15.1.2 The Two-Plate Model of the Drag Flow

In the two-plate model, the screw channel of the real geometry is unwound in the plane and the reference system is introduced at rest on the screw wall. Imagine that the cylinder wall, represented as a plate, moves over the screw flights in relation to the unwound channel at the circumferential velocity v_0, at the angle ϕ. The circumferential velocity v_0 is broken down into its components (Figure 15.3).

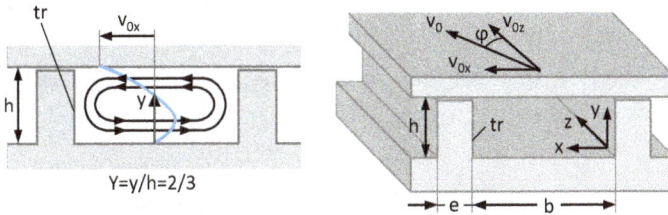

Figure 15.3 Flow types and two-plate model

Boundary Conditions and Assumptions
- The flow is incompressible.
- The flow is laminar and in a steady-state.
- The flow profile is fully developed hydrodynamically.
- The leakage flow over the screw flights, the velocity profile in the y direction, and the inertial forces are neglected.
- No consideration of normal stress effects, i.e., assumption of constant pressure gradients across the channel cross section.
- The channel curvature and the screw flights influence are taken into account later via correction factors.
- In simple approaches, isothermal flow is also assumed.

With the boundary conditions met, the momentum equation is reduced to the equilibrium of forces on the volume element. Assuming that the channel height is considerably smaller than the channel width, the influence of the screw flights is initially neglected, and the following result is obtained:

$$\frac{\partial \tau_{yx}}{\partial y} = \frac{\partial p}{\partial x} \tag{15.1}$$

and

$$\frac{\partial \tau_{yz}}{\partial y} = \frac{\partial p}{\partial z} \tag{15.2}$$

While the pressure gradient $\partial p/\partial x$ (Equation 15.1) perpendicular to the screw flights is caused by the velocity component v_{0x}, the pressure gradient $\partial p/\partial z$ (Equation 15.2) is responsible for the pressure buildup in the direction of the channel. These equations can be solved independently of each other for Newtonian media. For shear-thinning viscous fluids, the material laws (Carreau–WLF, Ostwald and de Waele, etc.) must be taken into account.

The mathematical description of the melt flow is then based on the simplified (one-dimensional) momentum balance of the two-plate flow. By inserting a flow law and integrating it twice, the velocity distribution is obtained as a function of the channel coordinate y. A further integration provides the volume flow rate.

As a first approximation, the flow rate can thus be calculated as follows, neglecting the leakage flow via the screw flights.

Procedure:

1. Assume an incompressible flow. This means that the pressure is linear. Then, the following applies: $\frac{\partial p}{\partial z} = \frac{\Delta p}{L}$

2. This assumption is used in Equation 15.2, which is then integrated once (indefinite integration).

3. The first constant of integration is determined using the limits of integration. This gives an equation that describes the relationship between the shear stress and the pressure.

4. In this equation, the shear stress is replaced by a yield law, e.g., by the power-law approach of Ostwald and de Waele (see Chapter 8, Equation 8.3).

5. The following applies to the shear rate: $\dot{\gamma} = \frac{\partial v}{\partial y}$

6. This equation is then integrated (indefinite integration) and the velocity curve for shear-thinning viscous materials is obtained.

7. A final, definite integration of the velocity curve results in the volume flow rate for the pressure flow.

8. The drag flow is obtained from the screw velocity and the geometric data of the screw.

15.2 *Task*: Calculation of the Velocity Profile of a Screw Flow

The result of the mathematical solution should be a superposition of the velocity curves for a drag flow and a pressure flow (see left-hand diagram in Figure 15.4). The leakage flow can be neglected. The pressure flow should be a function of the shear-thinning viscosity.

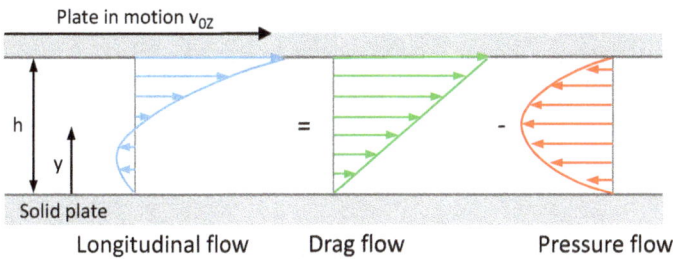

Besides the drag and pressure flow, leakage flow theoretically also has to be taken into consideration.

Figure 15.4 Velocity curves for a screw flow

Create a program (using software such as Microsoft Excel ®) with the following boundary conditions:

Input data (variable):

- Geometric data of the screw
- Material data (flow laws)
- Screw turnaround velocity
- Back pressure

Output data:

- Pressure and drag flow as well as the superposition as a graphic
- See Figure 15.5.

Velocity curve of a screw flow				
Machine data			Material data	
Screw diameter	mm	30	Consistency factor	K in kg/ms^(2·n)	15890
Screw pitch	°	17			
Screw speed	1/s	0.5	Flow exponent	m	2.1
Pressure decline	bar	100			
Length	mm	100	Temperature shift factor	a$_T$	0.8
Screw depth	mm	1.5			

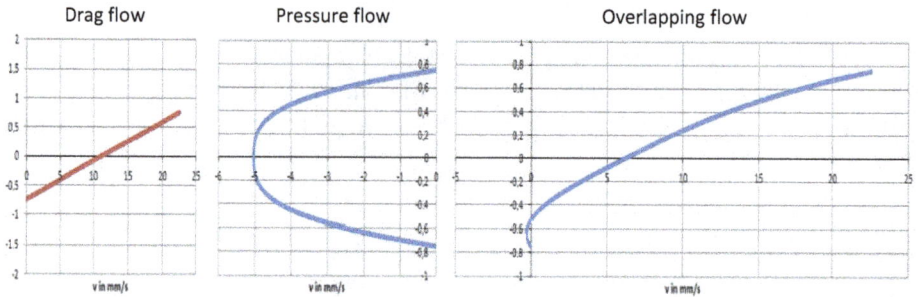

Figure 15.5 Flow types and two-plate model

References

[1] Johannaber, F.; Michaeli, W.: *Handbuch Spritzgießen*. Hanser, Munich, 2004

16 Problems with the Flow of Plastic Melts and Solutions

16.1 Flow Problems in Multilayer Flows

16.1.1 Rearrangement of the Melts

The problem of rearrangement in the coextrusion of fluids with different viscoelastic properties was already investigated in the early 1970s [1–3]. The content of this work was both the consideration of the occurrence of rearrangement as a function of the flow channel geometry, the flow properties of the melt, and the operating point, and the development of models for the processes taking place as well as the computational description of the flow processes. In the following, the manifestations of rearrangements will be discussed in more detail.

16.1.2 Phenomenology of Rearrangement

When two fluids with different viscosities flow in parallel, laminar flow, the fluid with the lower viscosity tends to displace the higher-viscosity fluid from the wall. This phenomenon is known as rearrangement. The rearrangement leads to a shift in the position of the boundary layer at the contact point during parallel flow. Figure 16.1 shows the occurrence of the rearrangement for a two-layer flow.

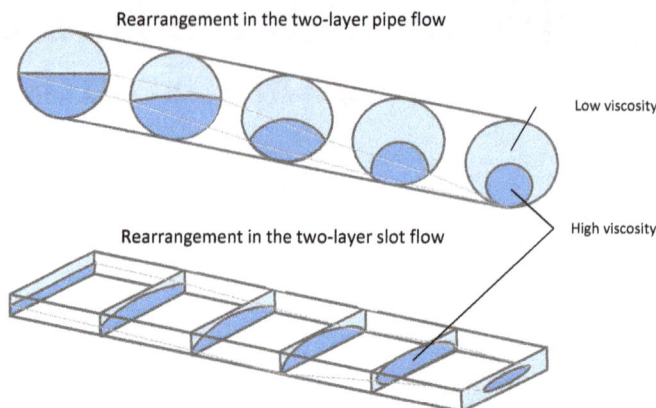

Figure 16.1 Rearrangements in two-layer flows

The rearrangement in the two-layer pipe flow begins with a bulge of the high-viscosity layer in the low-viscosity layer. The low-viscosity layer flows increasingly around the high-viscosity layer. In the next step, the rearrangement is so pronounced that the high-viscosity layer has lost all contact with the wall. The final state is a two-layer flow with concentric boundary layers. In the two-layer slot flow, the boundary layer initially only changes in the corners of the flow channel. As the rearrangement becomes more pronounced, the highly viscous layer is pushed further and further away from the wall.

The final state is almost a symmetrical three-layer flow, where the highly viscous middle layer has no wall contact even in the area of the side walls of the flow channel. The degree of rearrangement is influenced by the ratio of the viscosities of the fluids and the flow path. Rearrangement is therefore a particular problem in adapter coextrusion, as adapter coextrusion dies have long multilayer flow channels due to their design [4].

16.1.3 Models for the Development of Rearrangement

One reason for the occurrence of the rearrangement is the tendency to minimize the pressure loss and thus the energy consumption in the multilayer flow. The melt layers try to arrange themselves in such a way that the pressure loss is as small as possible and, thus, as little energy as possible is dissipated [4].

In the case of two-layer flow with lateral wall contact of both layers, the reason for the rearrangement can be the discontinuity of the wall shear stresses in the area of the boundary layer of fluids with different viscosities. This can be seen from a consideration of the shear stresses acting on the lateral surfaces in a two-layer flow (grid area in Figure 16.2) [5].

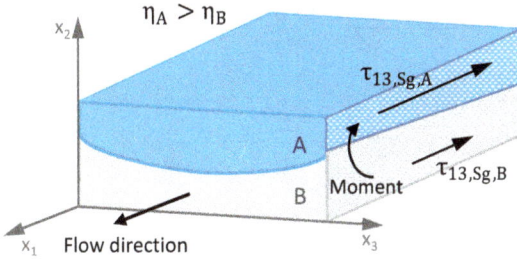

Figure 16.2 Wall layer stress of a two-layer flow [5]

The most energetically favorable arrangement of the two-layer flow with lateral wall contact of both layers is achieved when only the low-viscosity layer has wall contact and thus forms a kind of lubricating film for the high-viscosity layer. Due to the adhesion of the melts at the side wall and at the boundary layer, the shear rate is equally high for both melts in the side wall area:

$$\dot{\gamma}_{13,Sg,A} = \dot{\gamma}_{13,Sg,B} \tag{16.1}$$

However, this means that the wall shear stresses must be unequal if the viscosity of the fluids is unequal:

$$\tau_{13,Sg,A} = \eta_A \cdot \dot{\gamma}_{13,Sg,A} > \tau_{13,Sg,B} = \eta_B \cdot \dot{\gamma}_{13,Sg,B} \tag{16.2}$$

For

$$\eta_A > \eta_B \tag{16.3}$$

the jump in the wall shear stress results in Figure 16.2: it pulls the low-viscosity layer in the wall area in the direction of the high-viscosity layer.

The opposing force to the driving moment is represented by the extensional stresses, which result from a displacement of the boundary layer in the course of the flow and the associated velocity components perpendicular to the main flow (only in a parallel layer flow are the extensional stresses equal to zero). Since the rearrangements are largely related to the stress ratios on the side walls, the width-to-height ratio of the flow channel also has an influence on the characteristics of the rearrangement. The smaller the B/H ratio of the flow channel, the greater the influence of the side walls on the stress field of the flow and thus also the tendency to rearrangement. Conversely, rearrangements in flow channels with a large B/H ratio are very weak [4].

According to Michaeli [4], if a volume flow rate is specified, the layer thickness ratio $\frac{d_2}{d_1}$ for a two-layer flow with known viscosity η can be calculated using Equation 16.4 (Figure 16.3). The procedure is the same as for the derivation of the Hagen–Poiseuille equation, with the only difference being that two melt flows of different viscosities are considered in one channel. The prerequisite for the derivation is the adhesion condition at the two boundary layers.

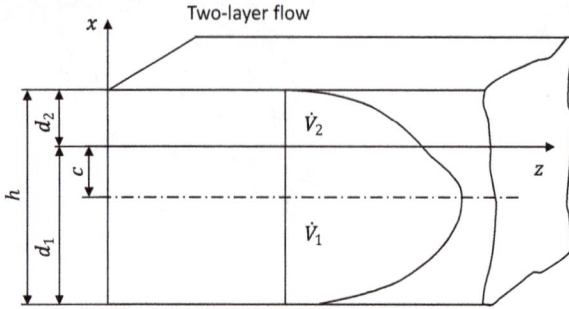

Figure 16.3 Two-layer flow according to Michaeli [4]

The derivation of the volume flow rate ratios $\frac{\dot{V}_2}{\dot{V}_1}$ and the corresponding pressure gradient $\frac{\partial p}{\partial z}$ of a two-layer flow is as follows, according to [4]:

$$\frac{\dot{V}_2}{\dot{V}_1} = \frac{\left(\frac{d_2}{d_1}\right)^4 + 4 \cdot \frac{\eta_2}{\eta_1} \cdot \left(\frac{d_2}{d_1}\right)^3 + 3 \cdot \frac{\eta_2}{\eta_1} \cdot \left(\frac{d_2}{d_1}\right)^2}{3 \cdot \frac{\eta_2}{\eta_1} \cdot \left(\frac{d_2}{d_1}\right) + 4 \cdot \frac{\eta_2}{\eta_1} + \frac{d_2}{d_1} + \left(\frac{\eta_2}{\eta_1}\right)^2} \tag{16.4}$$

$$\frac{\partial p}{\partial z} = \frac{12 \cdot \dot{V} \cdot \eta_1 \cdot \left(-\frac{\eta_2}{\eta_1} - \frac{d_2}{d_1}\right)}{\left(3 \cdot \left(\frac{d_2}{d_1}\right)^2 + 4 \cdot \frac{d_2}{d_1} + \frac{\eta_2}{\eta_1}\right) \cdot d_1^3} \tag{16.5}$$

Figure 16.4, Figure 16.5, Figure 16.6, and Figure 16.7 show the layer thicknesses in a two-layer flow as a function of the volume flow rate and the viscosity ratio of the two components, as calculated with a CFD (computational fluid dynamics) program.

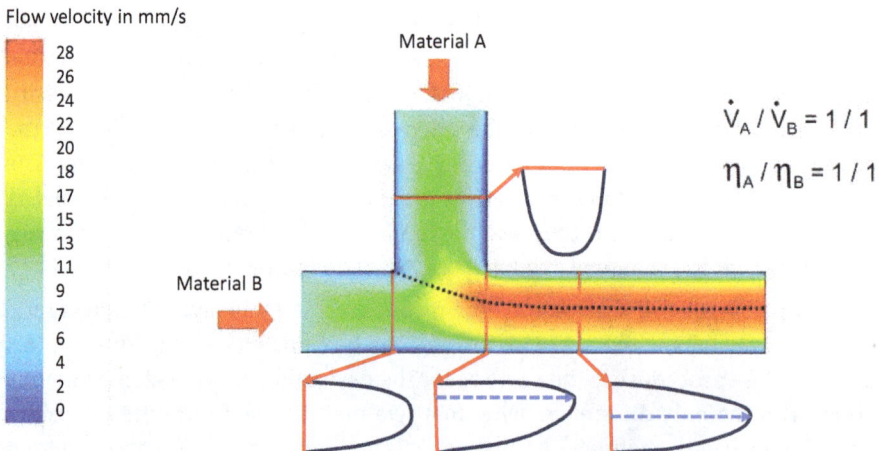

Figure 16.4 Layer thicknesses and flow rates of a two-layer flow

Flow velocity in mm/s

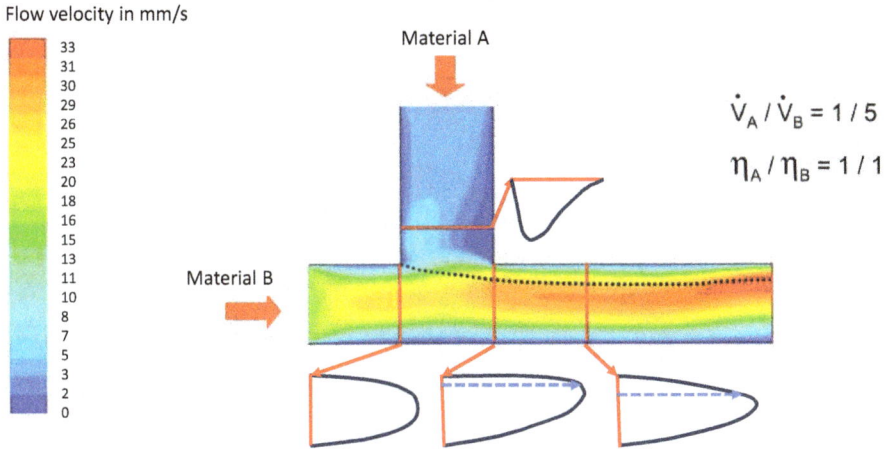

Figure 16.5 Layer thicknesses and flow rates of a two-layer flow

Flow velocity in mm/s

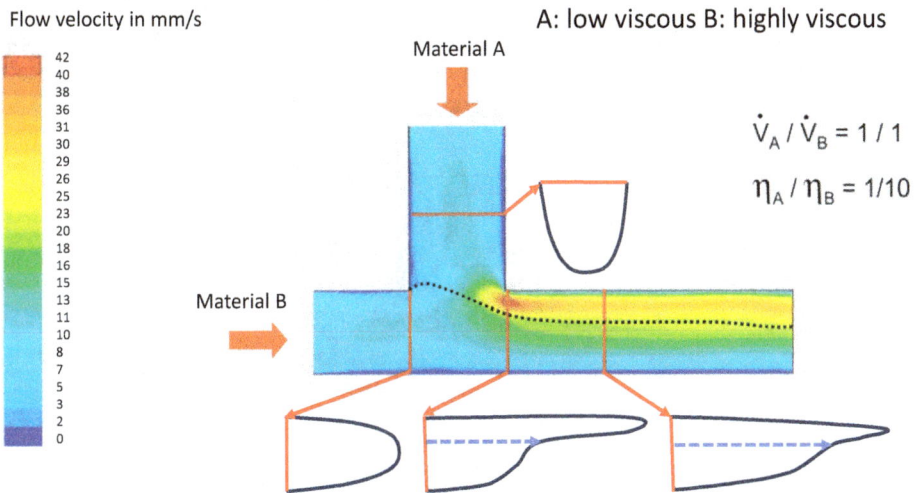

Figure 16.6 Layer thicknesses and flow rates of a two-layer flow

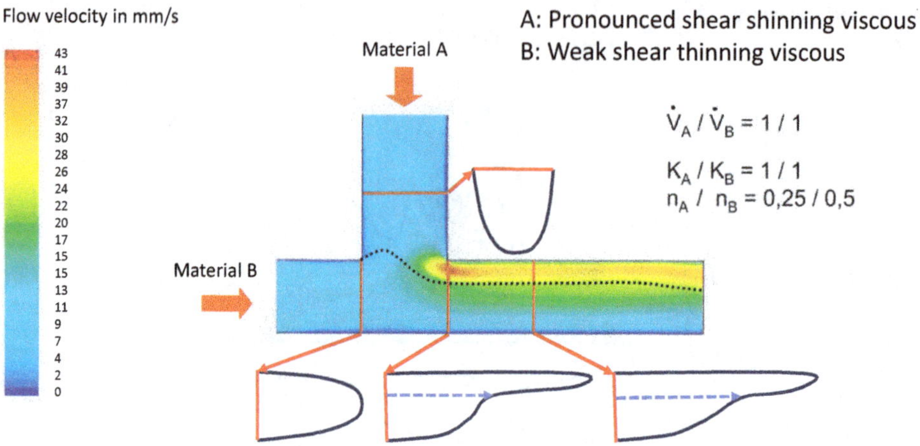
Flow velocity in mm/s
43
41
39
37
32
30
28
26
24
22
20
17
15
15
13
11
9
7
4
2
0
Material A
Material B
A: Pronounced shear shinning viscous
B: Weak shear thinning viscous
V̇A / V̇B = 1 / 1
KA / KB = 1 / 1
nA / nB = 0,25 / 0,5

Figure 16.7 Layer thicknesses and flow rates of a two-layer flow

In coextrusion blow molding, up to six layers are used, for example, for the production of automotive tanks (Figure 16.8). In addition to the cover layers (in this case, PE-HD), a barrier material (EVOH), a bonding agent (PE-LLD), and a layer of recycled material are also used. The recycled material consists of all materials.

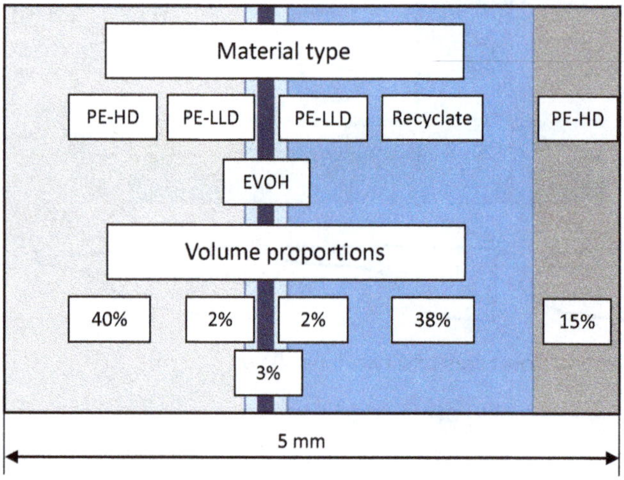
Material type
PE-HD
PE-LLD
PE-LLD
Recyclate
PE-HD
EVOH
Volume proportions
40%
2%
2%
38%
15%
3%
5 mm

Figure 16.8 Coextrusion container made of six layers with layer structure for a fuel tank

In contrast to the model considered above, in which the layers have lateral contact, the layers have no lateral wall contact in coextrusion blow molding. In this case, the theory of lateral torque cannot be held responsible for the rearrangement. Nevertheless, even in coextrusion blow molding, the layers tend to flow instabilities and to rearrangement (Figure 16.9). These effects and their causes are described and discussed below.

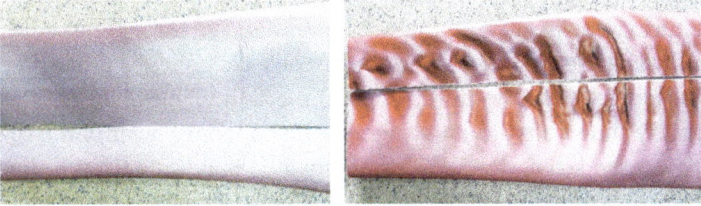

Figure 16.9 Stable and unstable flows during coextrusion

Figure 16.10 displays the viscosity curve for plastics used in the coextrusion process. The measured flow curves show that the viscosities of the PE-HD cover layers and the recycled material in the typical processing ranges of coextrusion blow molding ($1\,s^{-1}$ to $10\,s^{-1}$) are significantly higher than the viscosities of the EVOH barrier material and the bonding agent.

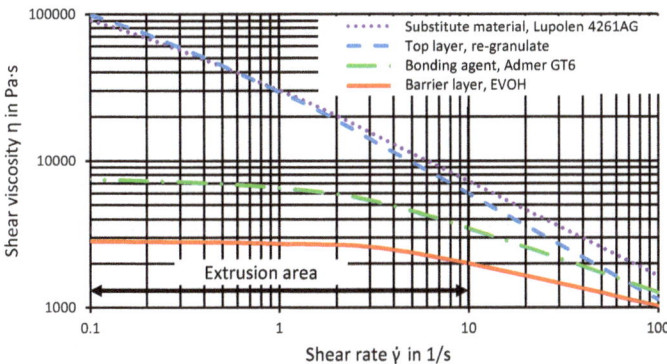

Figure 16.10 Viscosity as a function of the shear rate of plastics in a multilayer flow

Studies of coextrusion with varying material combinations have shown that stable and unstable process areas can be set depending on the throughput. The instabilities can also be generated with two identical materials with the same properties [4]. This suggests that the generation of the flow problem does not depend exclusively on the viscosity or the viscosity ratio. In addition, the location of the instability flow problem could not be clearly determined. The origin is located either in the island area of the distributor or in the area of confluence. Furthermore, it was found that the adhesion of two different adjacent polymers is not necessarily guaranteed. For this reason, the model of failure of the so-called boundary layer is used here.

This flow phenomenon is an instability that occurs at the interface of two polymer melts flowing side by side. The boundary layer instability and rearrangement effects are not related. They can be recognized as a wave-shaped or parabolic structure between the layers. The formation of such an instability depends on many factors that make up the manufacturing process of a multilayer composite. The throughput of individual layers, the flow channel geometry, and the viscoelastic polymer properties are some of the most important factors influencing the origin of instabilities. The boundary layer instability shows a typical waviness in the boundary layer of the co-extrudates. Depending on the strength of the instability, this manifests itself in the form of a changing frequency and amplitude. This results in parabolic structures that are oriented in the direction of the flow. Figure 16.11 shows a schematic representation of the boundary layer instability that is generated depending on its characteristics.

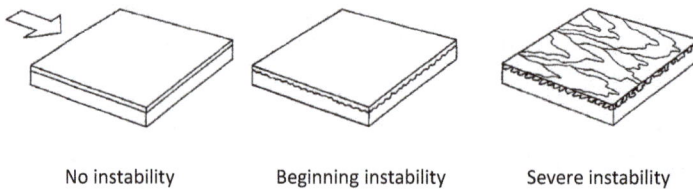

No instability   Beginning instability   Severe instability

Figure 16.11 Schematic representation of the resulting boundary layer instabilities [1]

Two general theories have been developed for the formation of boundary layer instabilities. The first deals with the pulsation of the flow and the viscoelastic properties of the melt. The second theory is concerned with the failure of the boundary layer.

Pulsation of the Flow and Viscoelastic Properties of the Melt

This assumption involves an oscillation of the velocity profile of a two-layer flow in the boundary layer. The unsteady flow is caused by the interaction between small pressure fluctuations and the viscoelastic properties of the melt. Due to a fluctuating mass flow or the material inhomogeneity, the instability is stimulated and the elastic components of the melt lead to propagation.

When investigating these assumptions, it was found that the triggered pressure fluctuations in a two-layer flow are not sufficient to create instability. The effect of unsteady mass transport is almost completely damped out.

Failure of the Boundary Layer

In contrast to the previous theory, it is the stress on the boundary layer caused by the resulting flow forces that can lead to a failure of adhesion between the layers. Some studies on multilayer flows have shown that, under certain conditions, flow instabilities occur when a shear stress limit is exceeded. However, a general definition for a shear stress limit has not been found. The limit value for the failure of the boundary layer varies depending on the material.

The results obtained were used to develop a model for the loss of adhesion at the boundary layer. The basis of this model is the polymer chain behavior in a multilayer flow, which is explained using Figure 16.12.

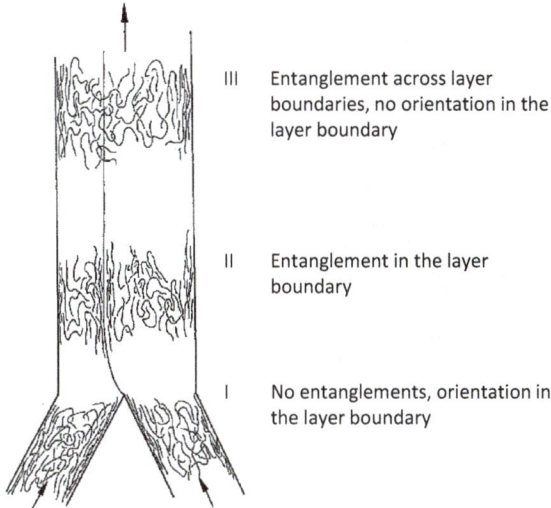

III   Entanglement across layer boundaries, no orientation in the layer boundary

II   Entanglement in the layer boundary

I   No entanglements, orientation in the layer boundary

Figure 16.12
Model for adhesion at the boundary layer [4]

The first area (I) represents a separate melt flow with wall contact. In the area of the single-layer flow, the entanglements of the polymer chains near the wall are loosened and the chains are oriented in the direction of the flow. As the shear rate increases over the height of the flow channel, the orientation towards the center of the flow channel decreases. In the second area (II), the two melt streams come together and a displaced boundary layer is formed. In the boundary layer, the polymer chains begin to relax due to the change in velocity and the loss of wall contact. This means that the orientation decreases and convolutions form in the individual streams. Due to this process, no adhesion can be established between the layers. After a certain (relaxation) time, the orientation of the chains in the boundary layer has completely decreased. New entanglements have formed across the layers, which enable adhesion.

Based on this assumption, it is obvious that the boundary layer forms a weak point due to the lack of interlacing and the pronounced orientation. If there are large differences in velocity when they meet, the layers can slip and boundary layer instabilities occur.

For this reason, it is not sufficient to consider only the viscous properties of the plastics. Investigations carried out in the context of coextrusion with a test mold have shown that the elastic properties of the plastics must also be taken into account. Figure 16.13 shows the measurement of the viscoelastic properties for the aforementioned materials—PE-HD (Lupolen 4261 AG), recycled pellets, a PE-LLD bonding agent (Admer), and a barrier material (EVAL)—using oscillation rheometry. The storage modulus and the loss modulus are registered as a function of the angular frequency. It is clear that the bonding agent and the barrier material have predominant viscous properties in the shear rate ranges relevant for extrusion blow molding (1 s^{-1} to 10 s^{-1}). In contrast, the PE-HD and the recycled material show pronounced elastic properties in the processing range.

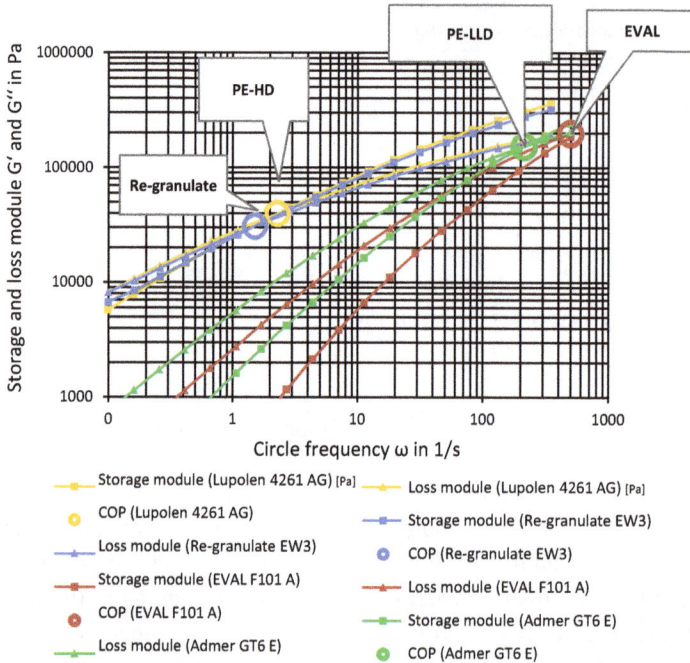

Figure 16.13 Crossover point (COP) of storage and loss moduli for plastics in a multilayer flow

Figure 16.14 illustrates the flow processes when the plastics involved in a three-layer flow, with different properties, meet. The bonding-agent component located between the surface layer and the recycled pellets has almost exclusively viscous properties and is responsible for the rearrangement effects at the boundary layer depending on the volume flow rate.

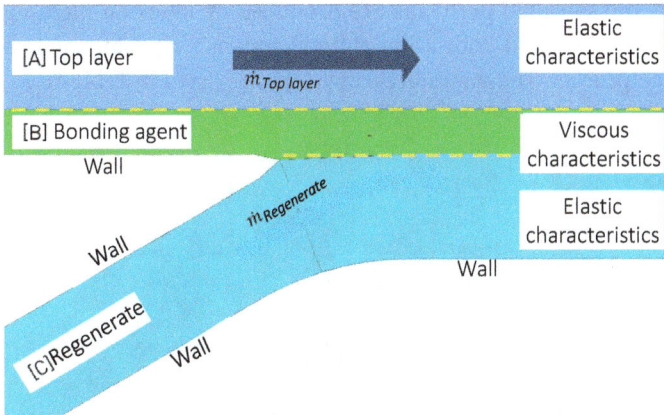

Figure 16.14 Encounter of plastics with different viscoelastic properties in the test carrier

16.2 Formation of Layer Thicknesses during Sandwich Injection Molding

The product quality of the molded parts produced by sandwich injection molding depends, among other things, on the distribution of the core component in the skin component (Figure 16.15). The surface distribution of the core component in the skin component plays a key role here. Product quality can only be guaranteed if the core component reaches all areas of the molded part that are relevant for product quality.

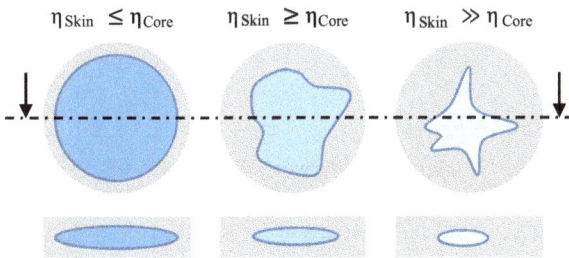

Figure 16.15 Two-dimensional distribution of the core component in the skin component as a function of the viscosity ratio

One example of this is the barrier quality feature. The required barrier properties, for example, of a packaging, can only be sufficiently fulfilled if the core component (e.g., EVOH, PA, etc.) actually reaches all relevant molded-part areas. Packaging produced using the stretch blow molding process should serve as an illustration of this. A first example was the filling of beer in polyethylene terephthalate (PET) bottles.

In the first process step, the so-called preforms are produced by injection molding. Figure 16.16 shows an injection mold with a hot runner system for producing preforms using the sandwich process. The manufacturing steps are listed in Table 16.1.

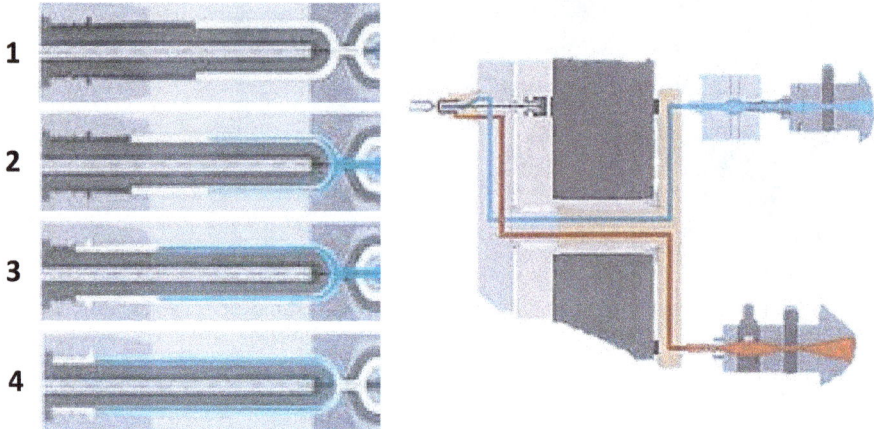

Phase	Function	Time	Filling level
1	1st injection phase component A	Start of 1st parallel phase for components A and B	57 %
2	Injection phase component B	20 % component B injected	79 %
3	Injection phase component B	Start of 2nd parallel phase for components A and B	94 %
4	2nd injection phase component A and holding pressure	End of holding pressure	100 %

Figure 16.16 Production of sandwich components with hot runner technology (see Table 16.1)

Table 16.1 Example of Filling for Figure 16.16

Phase	Function	Time	Filling level
1	1st injection phase component A	Start of 1st parallel phase for components A and B	57 %
2	Injection phase component B	20 % component B injected	79 %
3	Injection phase component B	Start of 2nd parallel phase for components A and B	94 %
4	2nd injection phase component A and holding pressure	End of holding pressure	100 %

PET is generally used as the skin component here. In order to achieve a sufficient shelf life (with protection concerning oxygen, CO_2, H_2O, etc.) of the product, a further barrier layer must be added. This can be achieved by subsequent coating (physical vapor deposition (PVD), SiO_2) or by the so-called sandwich process. In the sandwich process, a second component with very good barrier properties (e.g., EVOH or PA, in this case) is injected into the first component (PET, in this case). The distribution of the core component in the skin component is of great importance (Figure 16.17).

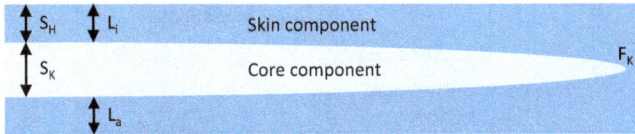

F_K: Location of flow front

$\frac{L_i}{L_a}$: Location of core component ($\frac{L_i}{L_a} = 1 \rightarrow$ core component is centered)

$\frac{S_K}{S_H}$: Layer thickness ratio ($\frac{S_K}{S_H} = 0 \rightarrow$ no barrier layer)

Figure 16.17 Layer thickness distribution of the core component in the skin component

On the one hand, the required barrier property can only be fulfilled if all relevant areas of the molded part are reached by the core component and the layer thickness of the core component is sufficiently large. On the other hand, the core component must not break through the skin component at any point, as this could jeopardize food approval (e.g., by the FDA, U.S. Food and Drug Administration). In addition, barrier materials such as EVOH are priced in a higher segment. Given the high quantities and a material cost share of over 80 %, it is essential to use as little core material as possible, with optimum distribution. The number of factors influencing the layer thickness distribution is manifold, as Figure 16.18 illustrates.

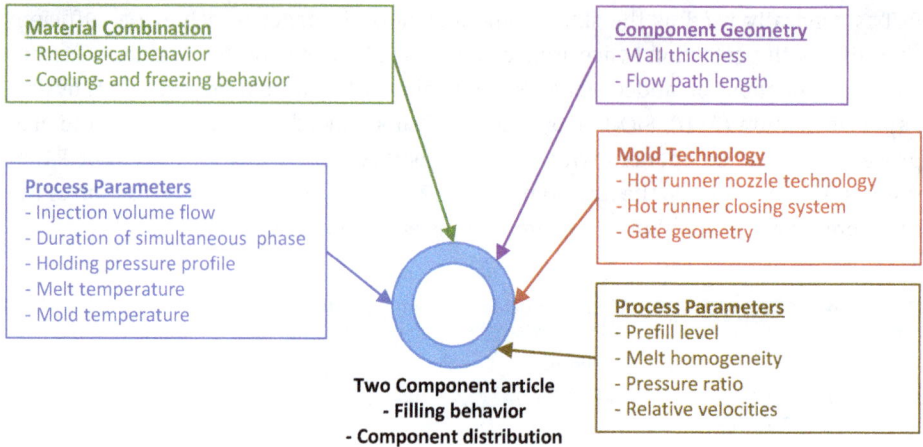

Figure showing factors influencing layer thickness distribution:

Material Combination
- Rheological behavior
- Cooling- and freezing behavior

Component Geometry
- Wall thickness
- Flow path length

Process Parameters
- Injection volume flow
- Duration of simultaneous phase
- Holding pressure profile
- Melt temperature
- Mold temperature

Mold Technology
- Hot runner nozzle technology
- Hot runner closing system
- Gate geometry

Process Parameters
- Prefill level
- Melt homogeneity
- Pressure ratio
- Relative velocities

Two Component article
- Filling behavior
- Component distribution

Figure 16.18 Factors influencing the layer thickness distribution during sandwich injection molding

In this context, the rheological effects on the formation of the core component in the skin component will be examined in more detail in this section. Figure 16.19 shows a preform that was produced using the sandwich process. PET was employed here as the skin component. Polyamide (PA) was used for the core component.

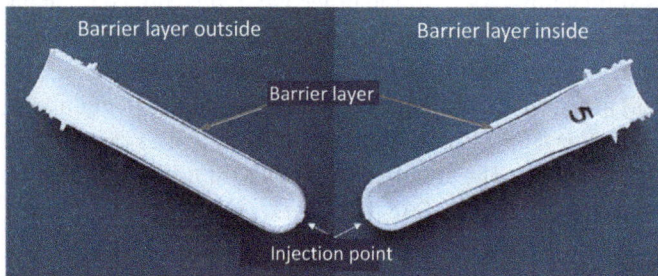

Barrier layer outside   Barrier layer inside

Barrier layer

Injection point

Figure 16.19 Position of the barrier layer in the preform; skin component: PET (white); core component: PA (black)

As the two pictures show, the thickness of the barrier layer is very small. This is also illustrated by Figure 16.20 and Figure 16.21. Here, the process was only varied in such a way that the core component was not only injected into the skin component in a second step, but this process was repeated. The process sequence is shown in Figure 16.20.

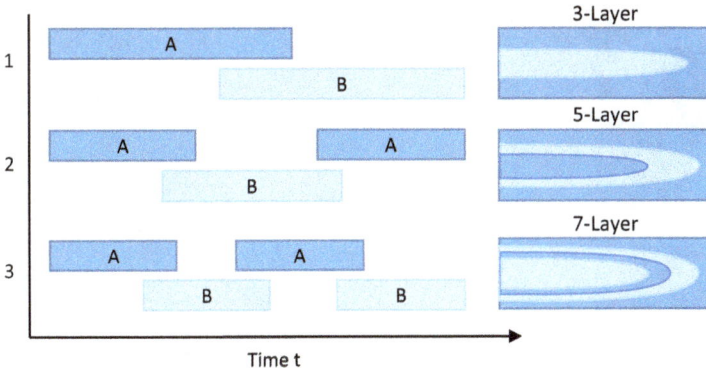

Figure 16.20 Options for manufacturing sandwich components

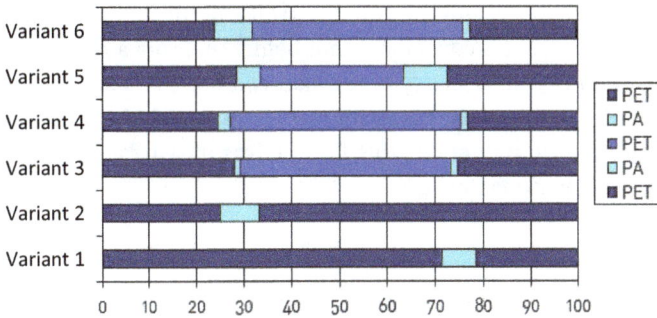

Figure 16.21 Distribution of layers (skin/core) during sandwich injection molding of barrier molded parts

Figure 16.21 shows the distribution of the core component (PA) in the radial direction measured in the preform. It can be clearly seen that the layer thickness of PA is significantly smaller than that of PET. It is also clear that the position of the barrier layer can be influenced. However, this is not done via the rheological properties, but via the control of the valve gate systems in the hot runner system.

As already mentioned, the areal spread of the core component in the skin component also plays an important role. In addition to the pre-filling of the cavity in the first process step, the viscosity ratio of the core component to the skin component has a decisive effect on the distribution. Figure 16.22 shows a preform made from PET (clear/transparent) and PA (black) using the sandwich injection molding process. It is evident that the melt distribution of the core component at the end of the flow path is anything but uniform.

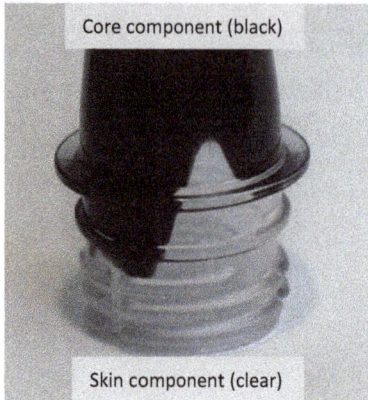

Figure 16.22
Melt distribution at the end of the flow path of a preform

Figure 16.23 shows the qualitative velocity profiles in sandwich injection molding as a function of viscosity. In this case, it is not the velocity profile at the boundary layer where the two components meet that is considered, but behind it. From a rheological point of view, the flow is stable if the viscosity of the skin component is less than or equal to that of the core component. If the viscosity ratio is reversed, that is, a low-viscosity core component flows in the middle, this flow strives to rearrange itself. It seeks the energetically more favorable flow state.

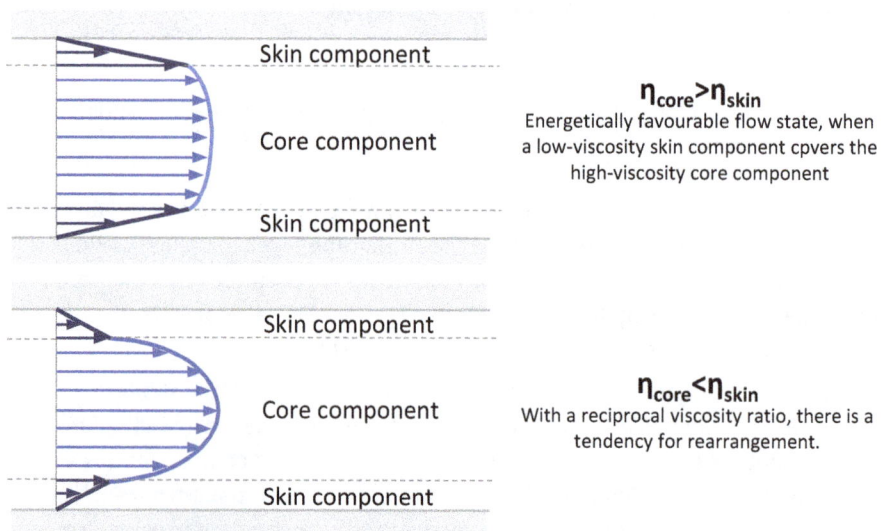

Figure 16.23 Qualitative velocity curves of the core and skin components as a function of the viscosity ratio

The effect that the different viscosity ratios of the core component to the skin component have on the layer thickness distribution of the two components in the molded part will be examined in more detail below.

In order to investigate the influence of viscosities on the layer thickness distribution, molded sheet parts were produced using the sandwich injection molding process. Two types of polypropylene with different viscosities were used. In order to better visualize the result, the two types were colored (core component: black; skin component: white). The molded parts were produced without holding pressure in order to rule out any possible influence of the holding pressure on the layer thickness formation.

As the pictures in Figure 16.24 illustrate, the profile at the flow front and also the residual wall thickness of the skin component behind the flow front depend on the viscosity ratio of the two polymer types. In the left-hand diagram, the viscosity of the core component is lower than that of the skin component. The flow front profile of the low-viscosity core component is "blunt" (flat) and the residual wall thickness of the skin component is small. If the viscosity ratio is reversed, this changes. A high-viscosity core component has a "pointed" (sharp) profile at the flow front and tends to have a larger residual wall thickness. This phenomenon can be linked to the swelling flow behavior of the plastic melt at the interface where the two components meet. A swelling flow is not only found at the flow front of the skin component, but also at the flow front of the core component. The processes that take place at this flow front are shown in Figure 16.25.

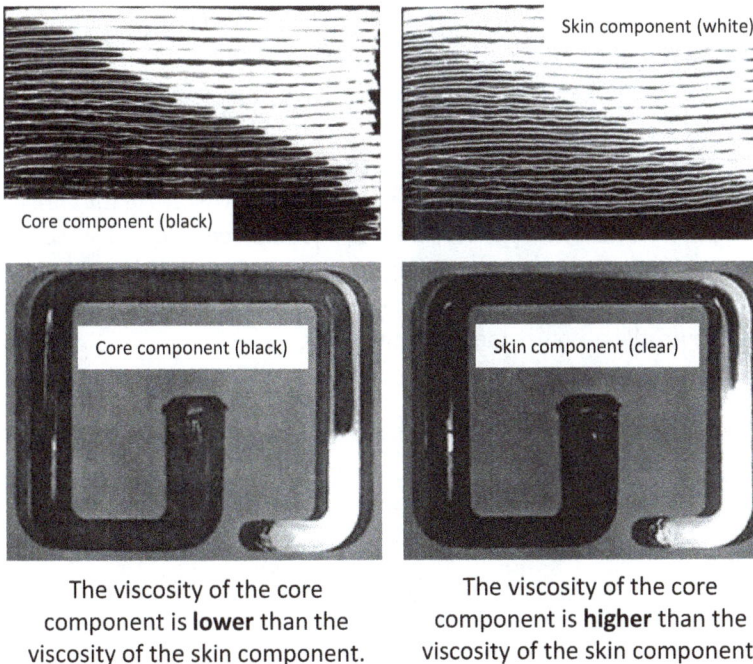

The viscosity of the core component is **lower** than the viscosity of the skin component.

The viscosity of the core component is **higher** than the viscosity of the skin component.

Figure 16.24 Formation of the core component in the skin component as a function of the viscosity ratio

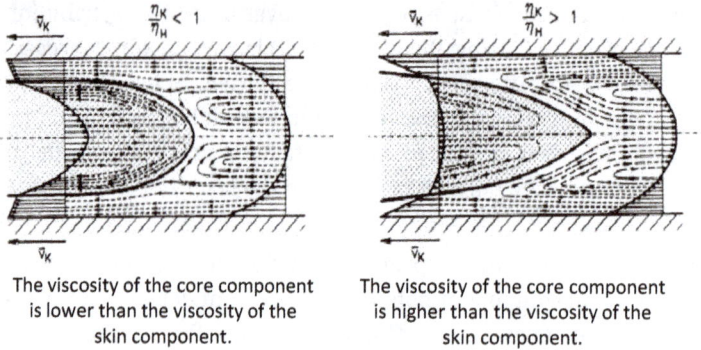

The viscosity of the core component is lower than the viscosity of the skin component.

The viscosity of the core component is higher than the viscosity of the skin component.

Figure 16.25 Flow processes at the flow fronts of the skin and core components as a function of the viscosity ratio

If a low-viscosity core component meets a high-viscosity skin component, the melt is deflected outwards at the flow front, to a greater or lesser extent depending on the viscosity of the core component due to the swelling flow process. This leads to a "blunt" flow profile at the flow front and to a higher displacement process of the outer component in the direction of the flow. If the viscosity of the core component is higher than that of the skin component, this deflection process takes place to a lesser extent. The result is a more "pointed" flow profile and consequently a larger residual wall thickness, as less melt of the skin component is displaced by the core component in the direction of the flow.

If a sandwich-molded part is cut along the flow direction, the cross section of the part always has the parabolic shape of the core component. The profile of the parabola depends on the rheological properties of the materials used. Figure 16.26 shows the dependence of the boundary layer profiles on the ratio of the viscosities of the core to the skin materials for a circular component. This ratio is referred to as the viscosity ratio $k_\eta = \eta_{core}/\eta_{skin}$. For $k_\eta < 1$, the core material has a lower viscosity than the skin material, and vice versa. The core material flow front has a "blunt" profile for a ratio $k_\eta < 1$. Conversely, the profile becomes more "pointed" if $k_\eta > 1$.

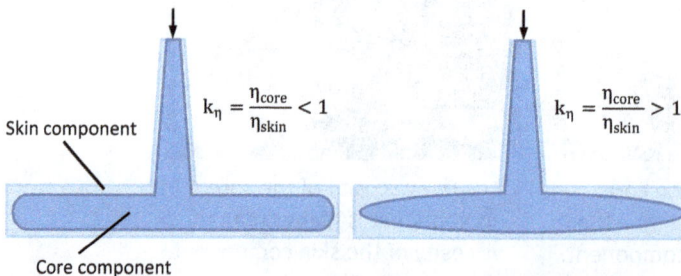

$$k_\eta = \frac{\eta_{core}}{\eta_{skin}} < 1$$

$$k_\eta = \frac{\eta_{core}}{\eta_{skin}} > 1$$

Skin component

Core component

Figure 16.26 Boundary layer profile as a function of the viscosity ratio [6]

The viscosity ratio of extensional viscosity to shear viscosity is decisive for this process. With a high extensional viscosity, the deflection processes (swelling flow) are met with high resistance. This means that the deflection processes only take place to a limited extent. This applies to the core and the skin components. If we take a low-viscosity core component and a high-viscosity skin component as an example, this leads to the following statement: At the flow front of the core component, it is deflected outwards due to the low extensional viscosity. In contrast, the skin component cannot be easily deflected outwards in front of the flow front of the core component due to its high viscosity. The result is a thin surface layer of the skin component and a high proportion of the core component.

Figure 16.27 shows the results of the investigation on the influence of the extensional and the shear viscosities on the formation of the residual wall thickness with the gas injection technique (GIT). The core component here is a gas (nitrogen), which has a very low viscosity and "no" shear stresses can be transferred to the surface layer by the flow. Nevertheless, the results can be applied to the two-component sandwich technology as a very good approximation.

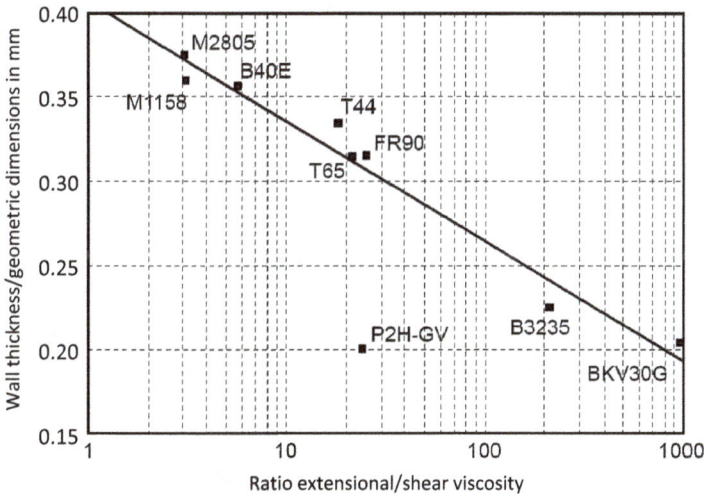

Figure 16.27 Correlation between the residual wall thickness of the GIT molded part and the ratio of extensional and shear viscosities [7]

The flow processes at the boundary between the core component and the skin component are also decisive for this effect (Figure 16.28). Since the gas injection technique involves a gas as the core component, the shear and the extensional flow processes upstream of the gas bubble, that is, those of the skin component, are primarily decisive for the formation of the residual wall thickness. As Figure 16.27 illustrates, the residual wall thickness decreases with increasing extensional viscosity.

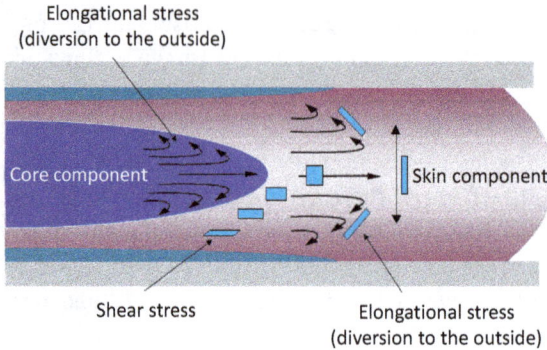

Figure 16.28 Flow processes in a multicomponent sandwich flow

The reason for this is that the deflection of the skin component to the outside of the mold wall is reduced by the high extensional viscosity. As a result, the residual wall thickness decreases with increasing extensional viscosity. Once this extensional flow process has been completed, no further flow processes take place behind the gas bubble flow front, as no shear stresses (very low viscosity) are transferred from the gas bubble to the melt. In consequence, the residual wall thickness remains constant to a good approximation, except for shrinkage effects [8, 9].

16.3 Normal Stress Effects, Pressure Losses, and Flow Instabilities

Figure 16.29 shows the first normal stress coefficient ψ_1 of a polyethylene melt plotted against the shear rate. The coefficient was measured over almost six decades of the shear rate. As the figure illustrates, the normal stress coefficient depends very strongly on the shear rate. It is very high for low shear rates and decreases as the shear rate increases.

Figure 16.29 shows that for shear-thinning viscous plastic melts, the coefficient of normal stress decreases even more than the viscosity. At high shear rates, the first normal stress coefficient is very small, but this does not immediately mean that the normal stress difference is also very small at high shear rates.

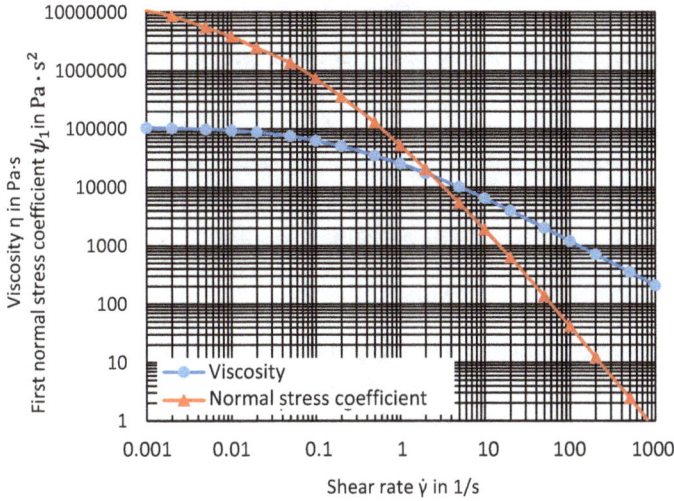

Figure 16.29 Normal stress coefficients over shear rate for a polyethylene melt

In Figure 16.30, the shear stress τ and the first normal stress difference N_1 are plotted as a function of the shear rate. This diagram shows the typical situation for nearly all plastic melts. Very low shear rates result in

- Small τ
- Very small N_1
- τ greater than N_1

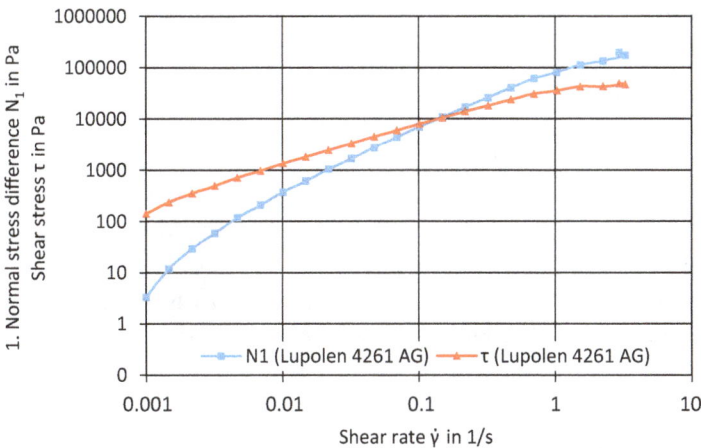

Figure 16.30 Shear stress and first normal stress difference for a Lupolen 4261 (PE-HD)

However, even at low shear rates, N_1 and τ are equal. From a shear rate of approximately $0.1\,\text{s}^{-1}$, the first normal stress difference N_1 is much greater than the shear stress τ.

As plastic melts are predominantly processed at high shear rates, this means that the normal stresses predominate during processing. The normal stresses are greater than the shear stresses. This indicates that the stress state is determined more by the normal stress coefficient than by the viscosity. As a result, viscoelastic flow phenomena often influence the processing of plastic melts in technical processes. In consequence, it is useful to know the first normal stress coefficient to describe the flow anomalies.

It can be shown that the viscosity function $\eta = f(\dot{\gamma})$ and the first normal stress coefficient ψ_1 are linked by a simple integral relationship. With Equation 16.6, the first normal stress coefficient can be calculated directly from the viscosity function without having to know the relaxation time spectrum. For practical purposes, the integral equation is formulated as the sum of finite differences (difference between two measured values), according to which the first normal stress coefficient can then be calculated directly from measured viscosities [10].

$$\Psi_1(\dot{\gamma}) = 2 \int\limits_{\eta\left(\frac{\dot{\gamma}}{k}\right)}^{\eta_\infty} \frac{d\eta}{\dot{\gamma}} \tag{16.6}$$

$$\Psi_1(k\dot{\gamma}) = 2 \sum \frac{\Delta\eta_i}{\dot{\gamma}_i} \tag{16.7}$$

The typical flow phenomena (melt fracture, for example) are observed in fluids with pronounced elastic properties. Since melt fracture usually only occurs at very high shear rates, the first normal stress difference responsible for this cannot be measured or can only be measured with difficulty. With the approach described above, it is also possible to investigate and describe these relationships.

Table 16.2 lists the critical flow data for a selection of very different plastics for which melt fracture was measured when used in a circular capillary. The values show that melt fracture can occur under very different flow conditions. As can be seen here, the quotient obtained from the first normal stress difference divided by the shear stress is a good criterion for describing the melt fracture that occurs. For all flow conditions in which the first normal stress difference N_1 is more than 4.5 times greater than the shear stress τ, the onset of flow instabilities in the capillary must be expected. This stability criterion, also known as the Weissenberg number (Equation 16.8) stability criterion, means that the first normal stress difference is essentially responsible for the melt fracture.

$$\text{Weissenberg number} = \frac{N_1}{\tau} \tag{16.8}$$

Table 16.2 Critical Flow Data of Different Plastic Melts at Melt Fracture [10]

Material	Temperature [°C]	Shear rate [s⁻¹]	Viscosity η [Pa·s]	Shear stress τ [Pa]	Normal stress coefficient ψ [Pa·s²]	Normal stress N_1 [Pa]	Weissenberg number N_1/τ [-]
LDPE	150	$2.45 \cdot 10^3$	$6.94 \cdot 10^1$	$1.70 \cdot 10^5$	$1.25 \cdot 10^{-1}$	$7.50 \cdot 10^5$	4.41
PMMA	200	$1.24 \cdot 10^3$	$3.50 \cdot 10^2$	$4.33 \cdot 10^5$	$1.30 \cdot 10^0$	$2.00 \cdot 10^6$	4.62
PE	200	$2.46 \cdot 10^3$	$1.46 \cdot 10^2$	$3.58 \cdot 10^5$	$2.94 \cdot 10^{-1}$	$1.78 \cdot 10^6$	4.97
SB	200	$4.92 \cdot 10^2$	$3.27 \cdot 10^2$	$1.61 \cdot 10^5$	$3.01 \cdot 10^0$	$7.29 \cdot 10^5$	4.53
ABS	200	$1.10 \cdot 10^3$	$1.92 \cdot 10^2$	$2.11 \cdot 10^5$	$8.26 \cdot 10^{-1}$	$1.00 \cdot 10^6$	4.74
PS	200	$7.50 \cdot 10^3$	$3.00 \cdot 10^1$	$2.25 \cdot 10^5$	$1.75 \cdot 10^{-2}$	$9.84 \cdot 10^6$	4.38
HDPE	190	$1.50 \cdot 10^2$	$2.07 \cdot 10^3$	$3.10 \cdot 10^5$	$6.44 \cdot 10^1$	$1.45 \cdot 10^6$	4.68
PA	220	$2.64 \cdot 10^4$	$1.75 \cdot 10^1$	$4.62 \cdot 10^5$	$1.95 \cdot 10^{-2}$	$2.02 \cdot 10^6$	4.37
						(mean) $\overline{R} = 4.63$ SD = ±0.26	

Furthermore, it can be shown that there is a direct coupling between N_1 and the inlet pressure drop p_E. The pressure curve when flowing through a nozzle can be described with a simple equation. In this equation, the viscous shear pressure losses and the elastic inlet pressure losses are added:

$$p(L) = p_s(L) + p_E \qquad (16.9)$$

The shear pressure loss $p_s(L)$ increases linearly with the flow path length; p_E is a constant value. The inlet pressure loss p_E increases very strongly with increasing shear rate.

In addition to the shear stress and the first normal stress difference N_1, Figure 16.31 shows the inlet pressure loss p_E for a melt. The inlet pressure losses can be described well by the extrapolated curve of the first normal stress difference N_1. Here, P is the proportionality factor between the first normal stress difference and the inlet pressure loss [10].

$$p_E(\dot{\gamma}) = P \cdot N_1(\dot{\gamma}) \qquad (16.10)$$

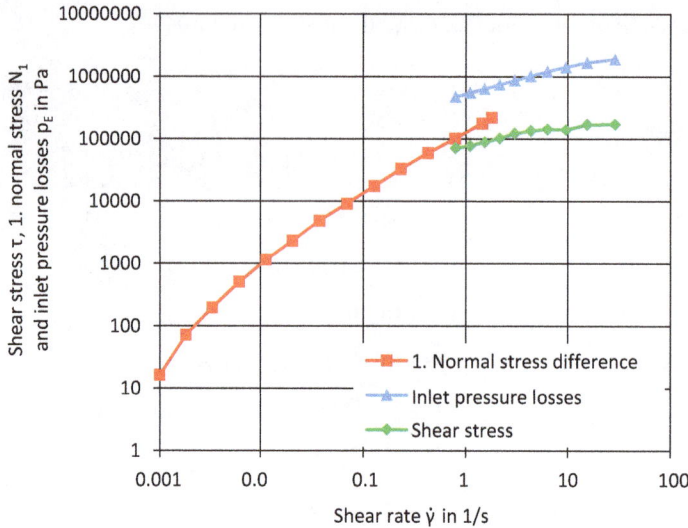

Figure 16.31 Shear stress, first normal stress difference, and inlet pressure loss of Lupolen 4261 (PE-HD)

In our example, the inlet pressure loss p_E for Lupolen 4261 (PE-HD) is approximately three times greater than the extrapolated curve of N_1. Accordingly, P would assume the value three. The following applies:

$$p(L) = \frac{2 \cdot L}{R} \cdot \tau + P \cdot N_1 = \frac{2 \cdot L}{R} \eta \cdot \dot{\gamma} + P \cdot \psi_1 \cdot \dot{\gamma}^2 \qquad (16.11)$$

This simple correlation has been tested and confirmed on many polymers.

16.3.1 *Task*: Total Pressure Drop in an Extrusion Die

You want to produce polystyrene granulate on a pelletizing system using an extruder. To do this, the melt flows at 170 °C through a perforated plate with $N = 200$ holes. The holes have a length of $L = 10$ mm and a diameter of $D = 3$ mm. The mass throughput $\dot{m}$ is 217.44 kg/h. The density at 170 °C is $\rho = 0.975$ g/cm^3.

Questions:

1. What is the total pressure drop Δp_{total} across the perforated plate?

    The calculation must be based on the true wall shear stress. The exact values for the viscosity and other quantities can be determined by interpolation using Table 16.3. The proportionality between the first normal stress difference N_1 and the inlet pressure loss can be assumed to be $P = 3.5$.

2. How large does Δp_{total} become if the bore length is halved?

3. Is melt fracture to be expected?

See Figure 16.32.

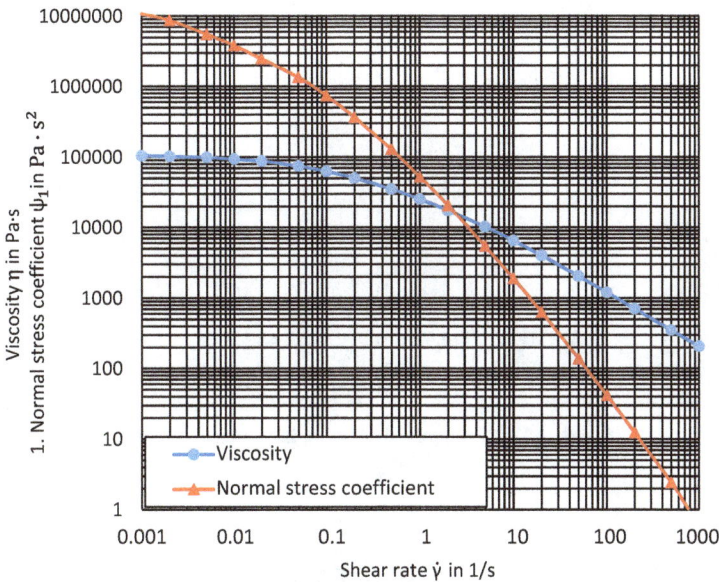

Figure 16.32 Curves of the normal stress coefficient and the viscosity as a function of the shear rate for polystyrene, $T = 170\,°C$

Table 16.3 Material Characteristics of Polystyrene ($T = 170\,°C$)

i	$\dot{\gamma}\,[s^{-1}]$	$\eta\,[Pa \cdot s]$	$\tau\,[Pa]$	$\psi_1\,[Pa \cdot s^2]$	$N_1\,[Pa]$
1	$1 \cdot 10^{-3}$	$1.04 \cdot 10^5$	$1.04 \cdot 10^2$	$1.07 \cdot 10^7$	$1.07 \cdot 10^1$
2	$2 \cdot 10^{-3}$	$1.02 \cdot 10^5$	$2.04 \cdot 10^2$	$8.67 \cdot 10^6$	$3.47 \cdot 10^1$
3	$5 \cdot 10^{-3}$	$9.84 \cdot 10^4$	$4.92 \cdot 10^2$	$5.52 \cdot 10^6$	$1.38 \cdot 10^2$
4	$1 \cdot 10^{-2}$	$9.38 \cdot 10^4$	$9.38 \cdot 10^3$	$3.84 \cdot 10^6$	$3.84 \cdot 10^2$
5	$2 \cdot 10^{-2}$	$8.78 \cdot 10^4$	$1.76 \cdot 10^3$	$2.48 \cdot 10^6$	$9.93 \cdot 10^3$
6	$5 \cdot 10^{-2}$	$7.49 \cdot 10^4$	$3.74 \cdot 10^3$	$1.36 \cdot 10^6$	$3.39 \cdot 10^3$
7	$1 \cdot 10^{-1}$	$6.29 \cdot 10^4$	$6.29 \cdot 10^3$	$7.39 \cdot 10^5$	$7.39 \cdot 10^3$
8	$2 \cdot 10^{-1}$	$5.05 \cdot 10^4$	$1.01 \cdot 10^4$	$3.62 \cdot 10^5$	$1.45 \cdot 10^4$
9	$5 \cdot 10^{-1}$	$3.50 \cdot 10^4$	$1.75 \cdot 10^4$	$1.28 \cdot 10^5$	$3.18 \cdot 10^4$

Table 16.3 Material Characteristics of Polystyrene (T = 170 °C) (continued)

i	$\dot{\gamma}\,[\mathrm{s}^{-1}]$	$\eta\,[\mathrm{Pa \cdot s}]$	$\tau\,[\mathrm{Pa}]$	$\psi_1\,[\mathrm{Pa \cdot s^2}]$	$N_1\,[\mathrm{Pa}]$
10	$1 \cdot 10^0$	$2.54 \cdot 10^4$	$2.54 \cdot 10^4$	$5.22 \cdot 10^4$	$5.22 \cdot 10^4$
11	$2 \cdot 10^0$	$1.76 \cdot 10^4$	$3.51 \cdot 10^4$	$2.06 \cdot 10^4$	$8.26 \cdot 10^4$
12	$5 \cdot 10^0$	$1.02 \cdot 10^4$	$5.10 \cdot 10^4$	$5.45 \cdot 10^3$	$1.36 \cdot 10^5$
13	$1 \cdot 10^1$	$6.51 \cdot 10^3$	$6.51 \cdot 10^4$	$1.89 \cdot 10^3$	$1.89 \cdot 10^5$
14	$2 \cdot 10^1$	$4.01 \cdot 10^3$	$8.01 \cdot 10^4$	$6.30 \cdot 10^2$	$2.52 \cdot 10^5$
15	$5 \cdot 10^1$	$2.04 \cdot 10^3$	$1.02 \cdot 10^5$	$1.37 \cdot 10^2$	$3.43 \cdot 10^5$
16	$1 \cdot 10^2$	$1.20 \cdot 10^3$	$1.20 \cdot 10^5$	$4.16 \cdot 10^1$	$4.16 \cdot 10^5$
17	$2 \cdot 10^2$	$7.06 \cdot 10^2$	$1.41 \cdot 10^5$	$1.23 \cdot 10^1$	$4.93 \cdot 10^5$
18	$5 \cdot 10^2$	$3.52 \cdot 10^2$	$1.76 \cdot 10^5$	$2.44 \cdot 10^0$	$6.10 \cdot 10^5$
19	$1 \cdot 10^3$	$2.08 \cdot 10^2$	$2.08 \cdot 10^5$	$7.23 \cdot 10^{-1}$	$7.23 \cdot 10^5$
20	$2 \cdot 10^3$	$1.20 \cdot 10^2$	$2.40 \cdot 10^5$	$2.21 \cdot 10^{-1}$	$8.84 \cdot 10^5$

16.3.2 Effects during Extrusion due to Exceeding the Critical Shear Stress Limit

The processing of plastics is accompanied by flow phenomena that cannot be explained using classical hydromechanics. These flow phenomena are typical of plastic solutions and melts and are attributable to their internal structure.

Plastic melts can absorb a maximum amount of elastic deformation. If the shear stress (normal stress) is increased further, melt fracture occurs. The melt strand emerging from a capillary is no longer smooth but more or less wavy. With increasing shear stress τ_w on the nozzle wall, irregularities occur from a critical wall shear stress. With plastic melts, these cannot be explained by the normal transition from a laminar flow to a turbulent flow. In the case of viscoelastic media, a stability criterion is decisive for this phenomenon.

In Equation 16.8, a critical key figure was defined as a stability criterion. This figure, known as the Weissenberg number, describes the ratio of normal stress to shear stress. The aim is to define a ratio at which, for example, critical surface effects or melt fracture occur (Figure 16.33).

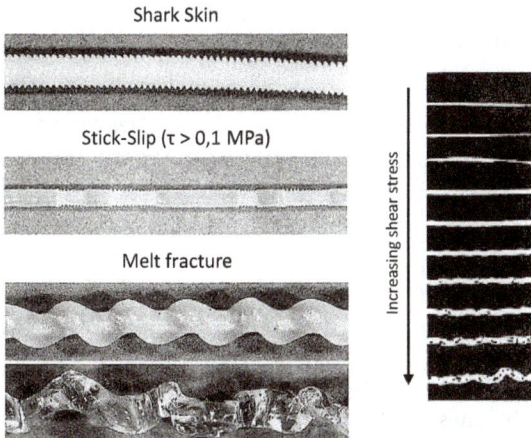

Shark Skin

Stick-Slip (τ > 0,1 MPa)

Melt fracture

Increasing shear stress

Figure 16.33
Various surface effects of extrudates [11]

16.3.3 Effects during Injection Molding due to Exceeding the Critical Shear Stress Limit

High shear rates can occur at points of small flow cross sections with a large melt volume flow rate. Furthermore, the shear stress assumes high values. If a critical shear stress limit is exceeded, this will have a negative effect on the properties of the molded part. The polymer is damaged by the high mechanical and thermal stresses. Moreover, excessive shear stresses in the vicinity of the flow channel wall can cause the already frozen surface layers to peel off. This leads to a so-called delamination of the outer layers, as shown in Figure 16.34.

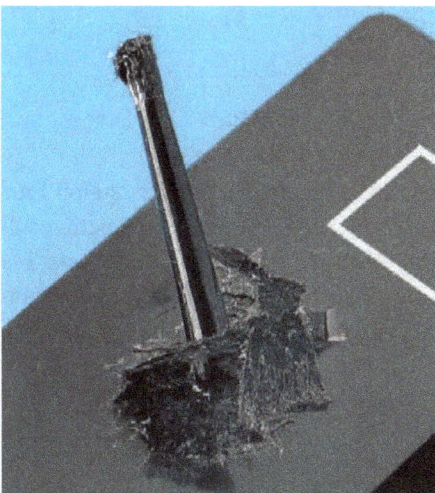

Figure 16.34
Delaminated material layers due to exceeding the limit shear stress

If the pressure in a flow channel is known, the shear stress can be calculated quickly with known geometric data. This has already been described in detail in Chapter 9 in the derivation of the Hagen–Poiseuille equation.

The shear stress is always maximum at the channel wall, while it assumes the value zero in the middle of the channel. The equations for the wall shear stress τ_W for the pipe flow and the flow in the rectangular channel are as follows:

$$\tau_W = \frac{\Delta p \cdot H}{2 \cdot L} \qquad\qquad\qquad (16.12)$$

$$\tau_W = \frac{\Delta p \cdot R}{2 \cdot L} \qquad\qquad\qquad (16.13)$$

If, for example, a pressure of 1000 bar is used in Equation 16.13, this results in a flow path length L of 300 mm and a wall thickness of 3 mm. There is a wall shear stress in the rectangular channel of $\tau_W = 0.5 \frac{N}{mm^2}$. Values for a critical wall shear stress, which should not be exceeded in order not to damage the plastic, are contained in the material databases of the simulation programs, for instance. As an example, an ABS–PC blend (T88 2N) can be listed here, for which a maximum permissible wall shear stress of $\tau_W = 0.4 \frac{N}{mm^2}$ is specified. As a result, this material would be damaged under the processing conditions assumed above. Table 16.4 gives some guide values for limit shear stresses.

Table 16.4 Guide Values for the Limit Shear Stress

Material type	Name	Limit shear stress
PMMA	Plexiglas 7N	0.37 N/mm²
ABS	Terluran 877T	0.32 N/mm²
PP	Hostalen 5200	0.17 N/mm²
PS	Polystyrene 168N	0.09 N/mm²
PS	Polystyrene 454H	0.06 N/mm²

Since the specification of the critical wall shear stress is usually not very helpful for the injection molder, material manufacturers often create so-called flow-path–wall-thickness diagrams. In these diagrams, a filling pressure limit is set, and it ensures that the critical shear stress limit is not exceeded. Figure 16.35 shows such a diagram for a polycarbonate. A pressure of 650 bar is specified here as the filling pressure limit.

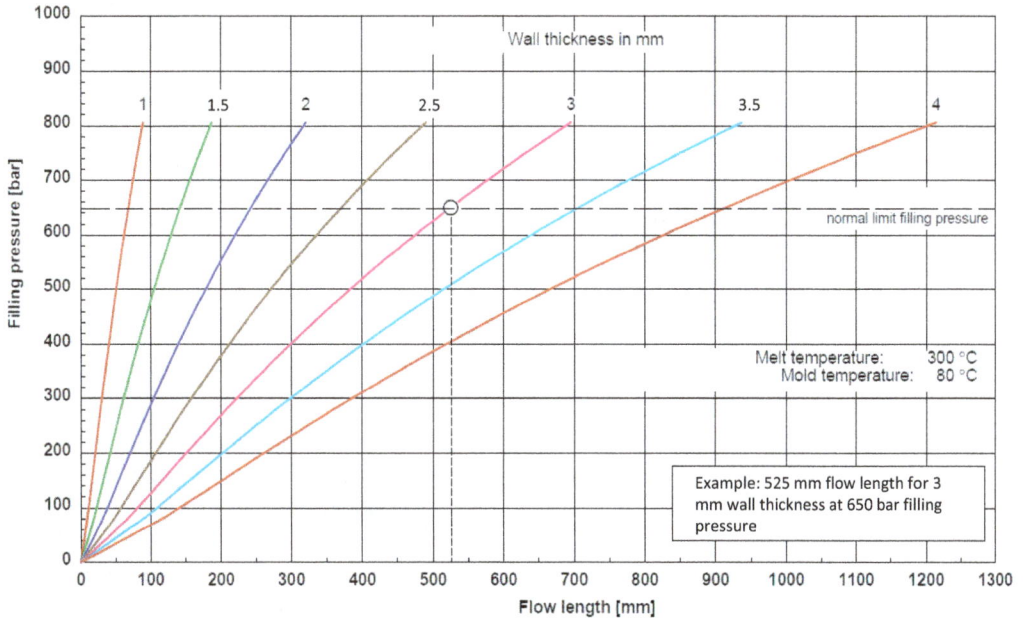

Figure 16.35 Flow-path–wall-thickness diagram for a polycarbonate

16.3.4 Wall Slip (Stick-Slip Effect)

Until now, it was assumed that the flowing fluids adhere to the wall (Figure 16.36 (a)). However, this condition cannot always be assumed for certain plastic melts. Figure 16.36 (c) shows the flow profile of a fluid with wall slip w_g. This wall slip effect can have different causes. Lubricants (kerosene, wax) and other additives can create a zone of the wall layer in the edge area with significantly lower viscosity ($\eta_1 \ll \eta_2$) (Figure 16.36 (b)). The slipping film causes a profile shift. The same phenomenon can be observed under certain circumstances using thermosets. With thermosets, a relatively cold plastic is injected into a heated mold [12–17].

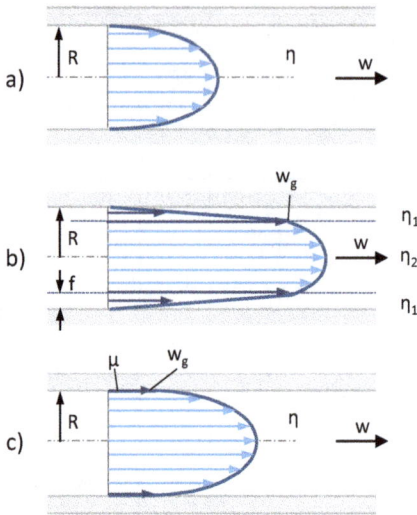

Figure 16.36
Velocity profile of a capillary flow under different boundary conditions

The time- and temperature-dependent viscosity curve of a thermoset is shown qualitatively in Figure 16.37. First, the viscosity (curve A) decreases due to the increasing temperature (contact of the thermoset with the hot mold). At the same time, the cross-linking reaction (curve B) starts. Depending on the time and the cross-linking kinetics, the viscosity of the thermoset increases again. As a result, a parabolic curve C is obtained for the viscosity.

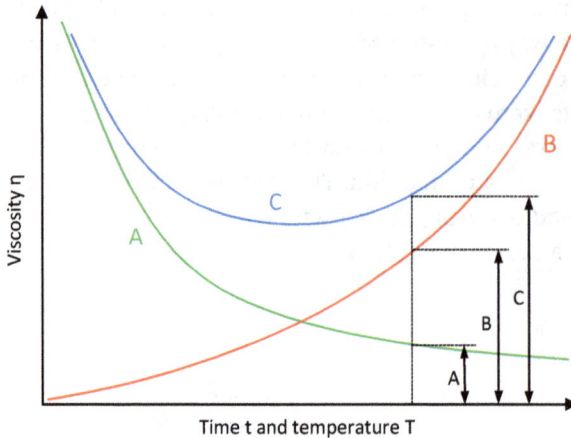

Figure 16.37 Viscosity curve of a thermoset as a function of time and temperature

Depending on the point where we are on the parabola (e.g., before the minimum) and the velocity of the cross-linking reaction, a low-viscosity lubricating film forms in the surface layer, that is, on the mold wall. The high-viscosity core slides on this low-viscosity slipping film (Figure 16.38, right).

Thermoplast	Duroplast

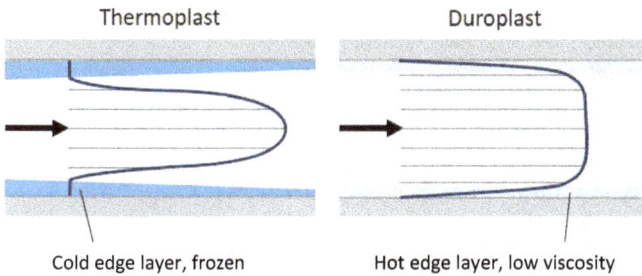

Cold edge layer, frozen   Hot edge layer, low viscosity

Figure 16.38 Velocity profile of a thermoplastic and a thermoset with edge layers

Especially with organically filled thermosets (wood, for instance), it can be observed that a so-called uncompacted area forms behind the flow front (Figure 16.39). This may be related to the moisture absorption of the filler. Furthermore, the filler wood is compressible. During the injection process, it is pressurized and thus compressed. The filler expands at the flow front and is therefore also responsible for the uncompacted area. As soon as the pressure behind the flow front increases and cross-linking begins, the so-called compaction line is created, and wall adhesion is restored.

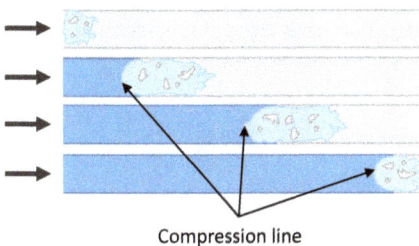

Compression line

Figure 16.39 Filling behavior of an organically filled thermoset with uncompacted area in front of the compaction line [18]

A wall slip or stick-slip effect cannot be calculated with computer-aided simulation programs. These generally assume wall adhesion. The derivation of the Hagen–Poiseuille equation is also based on the assumption of wall adhesion.

References

[1] Schrenk, W. J.; Bradley, N. L.; Alfrey, T.: Interfacial flow instabilities in multilayer coextrusion. *Polymer Engineering and Science*, 1978, pp. 620–623

[2] Han, C. D.: *Rheology in Polymer Processing*. Academic Press Inc, New York, 1976

[3] Cogswell, F. N.: *Polymer Melt Rheology – A Guide to Industrial Practice*. George Godwin Limited, London, 1981

[4] Michaeli, W.: *Extrusionswerkzeuge für Kunststoffe und Kautschuk: Bauarten, Gestaltung und Berechnung*. Hanser, Munich · Vienna, 1991

[5] Meier, M.: Fließprobleme in Mehrschichtströmungen, Ein Beitrag zur rheologischen Auslegung von Coextrusionswerkzeugen. Dissertation RWTH Aachen, Aachen, 1990

[6] Johannaber, F.; Michaeli, W.: *Handbuch Spritzgießen*. Hanser, Munich, 2004

[7] Johannaber, F.; Konejung, K.; Platschke, R.: Formation of the wall thickness in the gas injection technique GIT. ATI 970 Bayer, Leverkusen, 1995

[8] Schröder, T.: Neue Aspekte bei der Herstellung von Kunststoffformteilen mit der Gasinjektionstechnik (GIT), Dissertation at Rheinisch-Westfälischen Technischen Hochschule Aachen, Aachen, 1996

[9] Findeisen, H.: Ausbildung der Restwanddicke und Prozeßsimulation bei der Gasinjektionstechnik, Dissertation at Rheinisch-Westfälischen Technischen Hochschule Aachen, Aachen, 1997

[10] Pahl, M.; Gleißle, W.; Laun, H.-M.: *Praktische Rheologie der Kunststoffe und Elastomere*. VDI-Verlag GmbH, Düsseldorf, 1995

[11] Osswald, T.; Rudolph, N.: *Polymer Rheology: Fundamentals and Applications*. Hanser, Munich, 2014

[12] Schröder, T.: Duroplast-Spritzgießen mit Gasinnendruck. *Kunststoffe*, 1993, 83, 2, pp. 91–94

[13] Schröder, T.: A promising start has been made. *Kunststoffe Synthetics*, 1993, 2, pp. 12–16

[14] Schröder, T.: Die Gasinnendrucktechnik für härtbare Formmassen. *Österr. Kunststoff-Zeitschrift*, 1993, 24, 3/4, pp. 61–64

[15] Schröder, T.: Airmould injection molding now also for thermosets. *Plastverarbeiter*, 1993, 44, 2, pp. 18–24

[16] Schröder, T.: Influence of process parameters on gas bubble formation. *Plastverarbeiter*, 1993, 44, 11, pp. 102–112

[17] Schröder, T.: The gas injection technique for curable molding compounds – influence of process parameters on gas bubble formation. *Plastverarbeiter*, 1993, 44, 11, pp. 102–112

[18] Thienel, P.; Hoster, B.: Ermittlung der Füllbildkonstruktion härtbarer Formmassen mit einem Sichtwerkzeug, in: Niemann, K.; Schröder, K. (Eds.): *Spritzgießen von Duroplasten*. Hüthig, Heidelberg, 1994

17 Material Parameters

17.1 Power-Law (Ostwald and de Waele) Approach

Polymer type	Trade name	Temperature T [°C]	Approximation range: $\dot{\gamma}_{min}$ to $\dot{\gamma}_{max}$ [s^{-1}]	Flow exponent m [–]	Fluidity ϕ [Pa^m·s^{-1}]
ABS	Novodur PH-AT/217	220	10 to 20,000	2.6	$1.34 \cdot 10^{-11}$
ABS	Novodur PK/8233	220	10 to 20,000	2.68	$2.19 \cdot 10^{-12}$
ABS	Novodur PKT/7112	220	20 to 5000	2.95	$1.49 \cdot 10^{-14}$
ABS	Novodur PL-AT/253TL	220	20 to 20,000	2.43	$1.58 \cdot 10^{-10}$
ABS	Novodur PM/199	220	20 to 20,000	2.74	$9.40 \cdot 10^{-13}$
ABS	Novodur PMT	220	20 to 10,000	2.9	$8.13 \cdot 10^{-14}$
ABS	Novodur P2T/716	220	20 to 10,000	3.17	$1.09 \cdot 10^{-15}$
ABS/PC	Bayblend T45MN	240	10 to 2000	2.01	$4.24 \cdot 10^{-9}$
ABS/PC	Bayblend T65MN	280	10 to 2000	1.47	$1.62 \cdot 10^{-5}$
ABS/PC	Bayblend T85MN	260	10 to 1000	1.5	$3.32 \cdot 10^{-6}$
CA	Cellidor S 100-33	190	50 to 2000	2.11	$1.26 \cdot 10^{-8}$
CA	Cellidor S 200-22	210	10 to 2000	2.81	$1.13 \cdot 10^{-12}$
CA	Cellidor S 200-27	190	20 to 100	2.01	$1.18 \cdot 10^{-8}$
CAB	Cellidor B 500-5	230	5 to 2000	1.61	$4.33 \cdot 10^{-6}$
CAB	Cellidor B 500-15	210	5 to 2000	1.51	$2.87 \cdot 10^{-5}$

Polymer type	Trade name	Temperature T [°C]	Approximation range: $\dot{\gamma}_{min}$ to $\dot{\gamma}_{max}$ [s^{-1}]	Flow exponent m [−]	Fluidity ϕ [Pa^m·s^{-1}]
CAB	Cellidor B 932-05	200	5 to 500	2.48	$1.25 \cdot 10^{-11}$
CAB	Cellidor B 932-10	200	5 to 500	2.19	$4.17 \cdot 10^{-10}$
CP	Cellidor CP 330-08	210	20 to 1000	3.23	$3.9 \cdot 10^{-15}$
CP	Cellidor CP 400-10	190	5 to 2000	2.71	$2.08 \cdot 10^{-12}$
CP	Cellidor CPVP KL1-3406-10	210	100 to 10,000	3.18	$2.73 \cdot 10^{-14}$
CP	Cellidor CPVP KL1-3800-05	240	20 to 5000	2	$5.65 \cdot 10^{-8}$
CP	Cellidor CPVP KL1-3800-10	240	20 to 5000	2.3	$1.19 \cdot 10^{-9}$
HDPE	Hostalen GF 4760	200	1 to 260	1.84	$2.4 \cdot 10^{-8}$
HDPE	Hostalen GM 5010 T2	200	1 to 1600	2.26	$1.79 \cdot 10^{-10}$
LDPE	Lupolen 1840 D	190	2 to 6000	2.76	$2.0 \cdot 10^{-12}$
PA	Durethan B 31 SK	230	20 to 2000	1.17	$4.79 \cdot 10^{-4}$
PA	Durethan BKV 15	250	50 to 5000	1.46	$2.54 \cdot 10^{-5}$
PA	Durethan BKV 50	250	10 to 500	1.66	$4.02 \cdot 10^{-7}$
PA	Durethan BVK 30	250	20 to 2000	1.52	$5.34 \cdot 10^{-6}$
PA	Vestamid D 14	230	76 to 10,100	1.71	$1.13 \cdot 10^{-6}$
PA	Vestamid D 22	250	76 to 4040	2.31	$3.47 \cdot 10^{-11}$
PA	Vestamid E 40 M-93	170	76 to 4040	1.72	$2.12 \cdot 10^{-6}$
PA	Vestamid E 62 M-83	190	76 to 6070	2.37	$1.2 \cdot 10^{-10}$
PA	Vestamid L 1600	200	25 to 10,100	1.4	$5.36 \cdot 10^{-5}$
PA	Vestamid L 1700	200	25 to 10,100	1.54	$6.08 \cdot 10^{-6}$
PA	Vestamid L 2101	200	25 to 3030	1.89	$6.55 \cdot 10^{-9}$
PA	Vestamid L 2124	200	76 to 6070	2.44	$1.62 \cdot 10^{-11}$
PA	Vestamid X 3715	200	76 to 4040	2.24	$6.38 \cdot 10^{-11}$
PA	Minlon 11C40NC10	290	10 to 10,000	1.38	$8.07 \cdot 10^{-5}$
PA	Zytel 101FNC10	290	10 to 10,000	1.16	$3.98 \cdot 10^{-3}$

Polymer type	Trade name	Temperature T [°C]	Approximation range: $\dot{\gamma}_{min}$ to $\dot{\gamma}_{max}$ [s^{-1}]	Flow exponent m [–]	Fluidity ϕ [Pa^m·s^{-1}]
PA 12	Vestamid L 2102	200	20 to 3000	1.89	$6.55 \cdot 10^{-9}$
PBT	Pocan B 1305	240	10 to 4000	1.08	$2.35 \cdot 10^{-3}$
PBT	Pocan B 1505	250	8 to 2000	1.12	$6.21 \cdot 10^{-4}$
PBT	Pocan B 3225	250	8 to 2000	1.42	$2.57 \cdot 10^{-5}$
PBT	Pocan B 3235	250	8 to 1000	1.38	$3.93 \cdot 10^{-5}$
PBT	Pocan B 4225	250	8 to 2000	1.38	$4.9 \cdot 10^{-5}$
PBT	Pocan B 4235	250	8 to 1000	1.42	$1.78 \cdot 10^{-5}$
PBT	Pocan B 7375	250	8 to 2000	1.14	$5.01 \cdot 10^{-4}$
PBT	Pocan B 7503	250	8 to 2000	1.1	$1.16 \cdot 10^{-3}$
PBT	Pocan S 1506	250	4 to 1000	1.51	$3.53 \cdot 10^{-6}$
PBT	Vestodur HI15	240	253 to 8090	2.13	$1.5 \cdot 10^{-9}$
PBT	Vestodur HI17	240	253 to 6070	1.92	$1.9 \cdot 10^{-8}$
PBT	Vestodur HI19	240	253 to 10,100	2.54	$7.91 \cdot 10^{-12}$
PBT	Vestodur 1000	250	759 to 10,100	2.08	$8.43 \cdot 10^{-9}$
PBT	Vestodur 2000	250	759 to 10,100	2.19	$4.72 \cdot 10^{-10}$
PBT	Vestodur 3000	250	759 to 10,100	2.2	$1.16 \cdot 10^{-10}$
PC	Makrolon 2400	300	10 to 2000	1.06	$2.17 \cdot 10^{-3}$
PC	Makrolon 2600	300	20 to 2000	1.12	$6.52 \cdot 10^{-4}$
PC	Makrolon 2800	300	10 to 1000	1.07	$9.1 \cdot 10^{-4}$
PC	Makrolon 3100	300	40 to 2000	1.18	$1.74 \cdot 10^{-4}$
PC	Makrolon 8320	300	10 to 2000	1.38	$1.02 \cdot 10^{-5}$
PC	Makrolon 9415	300	10 to 4000	1.33	$3.57 \cdot 10^{-5}$
PE	Lupolen 2410 T	160	1 to 10,000	1.62	$1.17 \cdot 10^{-5}$
PE	Lupolen 5021 D	180	2 to 1000	2.35	$7.71 \cdot 10^{-11}$
PE	Lupolen 6031 MX	180	10 to 10,000	1.86	$2.56 \cdot 10^{-7}$
PE	Baylon V 18 E 464/564	160	10 to 1000	2.68	$1.96 \cdot 10^{-12}$
PE	Baylon V 18 H 460/564	160	10 to 1000	2.36	$1.4 \cdot 10^{-10}$

Polymer type	Trade name	Temperature T [°C]	Approximation range: $\dot{\gamma}_{min}$ to $\dot{\gamma}_{max}$ [s^{-1}]	Flow exponent m [–]	Fluidity ϕ [Pa$^m \cdot$s^{-1}]
PE	Baylon V 18 H 464/564	160	10 to 2000	2.40	$8.79 \cdot 10^{-11}$
PE	Baylon V 19 D 760 S	160	20 to 10,000	3.02	$2.32 \cdot 10^{-14}$
PE	Baylon V 22 D 464	160	10 to 400	2.58	$2.64 \cdot 10^{-12}$
PE	Baylon V 22 E 464	160	10 to 800	2.61	$2.48 \cdot 10^{-12}$
PE	Baylon V 22 F 564/464	160	10 to 2000	2.57	$9.86 \cdot 10^{-12}$
PE	Baylon 21 H 460	120	20 to 10,000	3.11	$6.84 \cdot 10^{-15}$
PE	Baylon 23 L 430	260	20 to 4000	1.51	$3.13 \cdot 10^{-5}$
PE	Baylon 28 D 780	180	20 to 8000	2.85	$2.37 \cdot 10^{-13}$
PE	Vestolen A 3512 R	180	3 to 95	1.56	$2.76 \cdot 10^{-7}$
PE	Vestolen A 5041 R	180	3 to 225	2.48	$7.93 \cdot 10^{-12}$
PE	Vestolen A 5561 P	210	3 to 84	2.32	$4.46 \cdot 10^{-11}$
PE	Vestolen A 6016	180	10 to 1010	1.56	$4.01 \cdot 10^{-6}$
PE	Vestolen A 6017	170	6 to 1360	1.48	$1.35 \cdot 10^{-5}$
PE	Vestolen A 6042 F	180	3 to 800	2.33	$1.36 \cdot 10^{-10}$
PE	Hostalen GC 7260	200	3 to 1770	1.36	$5.23 \cdot 10^{-5}$
PE	Hostalen GF 4760	190	1 to 180	1.82	$2.7 \cdot 10^{-8}$
PE	Hostalen GM 5010 T2	200	1 to 1600	2.26	$1.79 \cdot 10^{-10}$
PE	Hostalen GM 7255	200	1 to 40	1.96	$9.29 \cdot 10^{-10}$
PE	Hostalen GM 9240 HT	200	1 to 100	2.32	$4.26 \cdot 10^{-11}$
PET	Rynite 935 NC 10	290	10 to 10,000	2.19	$3.71 \cdot 10^{-9}$
PMMA	Resarit 840 ZK	210	1 to 2150	3.17	$1.59 \cdot 10^{-15}$
PMMA	Resarit 844	210	2 to 5180	2.29	$4.69 \cdot 10^{-10}$
PMMA	Plexiglas HW 55	240	3 to 6220	2.66	$6.42 \cdot 10^{-12}$
PMMA	Plexiglas ZK 37	200	0.7 to 135	2.63	$2.87 \cdot 10^{-13}$
PMMA	Plexiglas ZK 38	200	0.7 to 132	2.72	$1.07 \cdot 10^{-13}$
PMMA	Plexiglas 7 H	180	0.02 to 2.5	2.04	$1.75 \cdot 10^{-11}$
POM	Ultraform N 2200	190	50 to 3320	1.94	$4.74 \cdot 10^{-8}$

Polymer type	Trade name	Temperature T [°C]	Approximation range: $\dot{\gamma}_{min}$ to $\dot{\gamma}_{max}$ [s^{-1}]	Flow exponent m [–]	Fluidity ϕ [Pa^m·s^{-1}]
POM	Delrin 500 NC-10	215	10 to 10,000	1.44	$2.98 \cdot 10^{-5}$
POM	Hostaform C 2521	180	1 to 225	1.39	$4.0 \cdot 10^{-6}$
POM	Hostaform C 9021	180	14 to 1190	1.50	$4.85 \cdot 10^{-6}$
POM	Hostaform C 9021 GV1/30	180	60 to 3480	2.05	$6.88 \cdot 10^{-9}$
POM	Hostaform C 13021	180	14 to 1870	1.43	$1.37 \cdot 10^{-5}$
POM	Hostaform C 27021	180	30 to 7100	1.60	$5.19 \cdot 10^{-6}$
POM	Hostaform C 32021	180	30 to 8240	1.52	$1.74 \cdot 10^{-5}$
POM	Hostaform T 1020	180	1 to 424	1.70	$1.56 \cdot 10^{-7}$
PP	Novolen 1125 NX	200	2 to 5000	2.25	$9.05 \cdot 10^{-9}$
PP	Novolen 1325 MX	200	1 to 2000	1.75	$7.97 \cdot 10^{-7}$
PP	Vestolen P 2300	250	3 to 16,800	2.27	$1.79 \cdot 10^{-8}$
PP	Vestolen P 2330	250	3 to 17,900	2.10	$1.57 \cdot 10^{-7}$
PP	Vestolen P 3200	250	3 to 15,100	2.21	$2.0 \cdot 10^{-8}$
PP	Vestolen P 5200	200	20 to 9050	3.60	$6.09 \cdot 10^{-16}$
PP	Hostalen PPH 1022	200	1 to 500	2.69	$1.96 \cdot 10^{-12}$
PP	Hostalen PPH 1050	200	1 to 430	2.62	$5.35 \cdot 10^{-12}$
PP	Hostalen PPH 1050	230	1 to 1000	2.61	$1.27 \cdot 10^{-11}$
PP	Hostalen PPN VP 7180 TV 20	220	1 to 18,600	2.52	$1.92 \cdot 10^{-10}$
PP	Hostalen PPN VP 7190 TV 40	220	1 to 13,300	2.60	$4.87 \cdot 10^{-11}$
PP	Hostalen PPN VP 7790 GV2/30	210	1 to 10,000	3.46	$2.14 \cdot 10^{-15}$
PP	Hostalen PPN 1034	200	4 to 16,100	3.39	$4.76 \cdot 10^{-15}$
PP	Hostalen PPN 1060	230	1 to 13,700	2.41	$6.55 \cdot 10^{-10}$
PP	Hostalen PPN 1752	230	2 to 12,000	2.40	$7.72 \cdot 10^{-10}$
PP	Hostalen PPN 7118 TV 20-10	220	2 to 10,900	2.51	$2.2 \cdot 10^{-10}$
PP	Hostalen PPN 8008/10	200	1 to 10,700	2.54	$1.07 \cdot 10^{-10}$

Polymer type	Trade name	Temperature T [°C]	Approximation range: $\dot{\gamma}_{min}$ to $\dot{\gamma}_{max}$ [s⁻¹]	Flow exponent m [–]	Fluidity ϕ [Pa^m·s⁻¹]
PP	Hostalen PPR 1042	200	1 to 2440	2.21	$4.86 \cdot 10^{-9}$
PP	Hostalen PPR 7342 FL	210	2 to 10,000	2.68	$4.08 \cdot 10^{-11}$
PP	Hostalen PPT 1070	200	1 to 16,600	2.29	$3.51 \cdot 10^{-9}$
PP	Hostalen PPT 8027	200	20 to 10,100	2.93	$3.63 \cdot 10^{-12}$
PP	Hostalen PPU VP 7180 TV 10	220	2 to 9430	2.09	$6.9 \cdot 10^{-8}$
PP	Hostalen PPU VP 7190 TV 40	220	1 to 9680	2.06	$1.01 \cdot 10^{-7}$
PP	Hostalen PPU 1080 F	200	3 to 2120	1.84	$6.72 \cdot 10^{-7}$
PP	Hostalen PPU 1734	200	3 to 10,500	2.00	$1.49 \cdot 10^{-7}$
PP	Hostalen PPW 1780 S 1	200	14 to 9810	1.98	$3.54 \cdot 10^{-7}$
PS	Polystyrene 144	170	0.2 to 1000	2.27	$6.0 \cdot 10^{-10}$
PS	Polystyrene 168 N	230	20 to 16,000	4.22	$7.64 \cdot 10^{-19}$
PS	Polystyrene 427 D	190	0.2 to 20,000	3.00	$2.02 \cdot 10^{-13}$
PS	Polystyrene 432 B Natural	190	0.5 to 10,000	2.16	$8.06 \cdot 10^{-9}$
PS	Polystyrene 454 H Natural	190	5 to 10,000	4.22	$2.26 \cdot 10^{-19}$
PS	Polystyrene 456 M Q 13	190	3 to 4830	3.65	$7.16 \cdot 10^{-17}$
PS	Polystyrene 473 D Natural	190	0.2 to 5000	2.03	$2.14 \cdot 10^{-8}$
PS	Polystyrene 476 L	190	5 to 5000	3.55	$4.8 \cdot 10^{-16}$
PS	Polystyrene 2710 Nature	190	5 to 20,000	3.49	$5.05 \cdot 10^{-16}$
PS	Vestyron 214-31	200	12 to 17,600	3.92	$2.77 \cdot 10^{-17}$
PS	Vestyron 314-31	180	14 to 8390	3.63	$1.88 \cdot 10^{-16}$
PS	Vestyron 512-31	180	13 to 10,900	3.90	$1.27 \cdot 10^{-17}$
PS	Vestyron 620-31	200	3 to 23,100	3.07	$3.78 \cdot 10^{-13}$
PVC	Vestolit EK 2857	180	12 to 11,900	3.00	$9.8 \cdot 10^{-13}$

Polymer type	Trade name	Temperature T [°C]	Approximation range: $\dot{\gamma}_{min}$ to $\dot{\gamma}_{max}$ [s^{-1}]	Flow exponent m [–]	Fluidity ϕ [Pa^m·s^{-1}]
PVC	Vestolit HI-EF 5174	180	11 to 19,800	7.41	$1.92 \cdot 10^{-40}$
PVC	Vestolit MF 1838	180	12 to 68,600	5.99	$1.67 \cdot 10^{-31}$
PVC	Vestolit MSPLZ	180	11 to 106,000	4.85	$8.04 \cdot 10^{-25}$
PVC	Vestolit V 1745/D	180	11 to 86,900	6.08	$2.95 \cdot 10^{-32}$
PVC	Vestolit V 5541/A	180	50 to 1390	8.75	$1.12 \cdot 10^{-49}$
PVC	Vestolit W 617	180	12 to 11,600	2.56	$8.62 \cdot 10^{-10}$

17.2 Carreau–WLF Approach

Material	Trade name	A [Pa·s]	B [s]	C [–]	T_B [°C]	T_S [°C]
ABS	Novodur P2T/716	13,020	0.0913	0.73	220	
ABS	Novodur PH-AT/217	5640	0.154	0.635	220	
ABS	Novodur PK/8233	7700	0.0967	0.705	220	
ABS	Novodur PKT/7112	16,800	0.175	0.674	220	
ABS	Novodur PM/199	8530	0.141	0.672	220	
ABS	Novodur PMT/7516	10,390	0.132	0.686	220	
ABS	Novodur P2T	35,230	1.283	0.5337	210	123
ABS	Novodur PMT	9557	0.1493	0.6891	220	135
ABS	Terluran 877T	26,140	0.428	0.7725	220	80
ABS/PC	Bayblend T45MN	2913	0.03116	0.6129	250	151
ABS/PC	Bayblend T65MN	2332	0.04768	0.0511	260	154
ABS/PC	Bayblend T85MN	1785	0.05447	0.4001	265	154
ASA	ASA 757R	205,500	2.8	0.7692	180	140
CAB	Cellidor B 932-05	25,170	0.622	0.667	200	
CAB	Cellidor B 932-10	27,520	1.2	0.596	200	
CP	Cellidor CP 400-10	10,400	0.197	0.703	190	

Material	Trade name	A [Pa·s]	B [s]	C [–]	T_B [°C]	T_S [°C]
EMPP	Celtic P0505	3022	0.08926	0.709	220	29
HDPE	HDPE 6011L	1440	0.07904	0.571	190	−115
HDPE	HDPE GB6450	304	$7.01 \cdot 10^{-3}$	0.5653	200	−63
HDPE	HDPE GF4760	14,110	0.9084	0.5827	200	0
HDPE	PE A6017	467	0.01389	0.5367	220	−83
HDPE	PE V18E464	6140	0.3223	0.6103	200	−3
LDPE	Lotrene FB 3003	21,040	0.685	0.658	170	
LDPE	LDPE 11	5643	0.4519	0.6108	190	30
LDPE	LDPE 1800M	2584	0.2203	0.6173	170	2
LDPE	LDPE 1800S	430	0.1017	0.4732	190	24
LDPE	LDPE 1810D	14,100	1.825	0.5901	170	−15
LDPE	PE 1800H	5218	0.3206	0.6387	190	49
PA	Durethan BC40	1816	0.04752	0.4879	260	102
PA	Durethan B 31 SK	413	$1.79 \cdot 10^{-3}$	0.527	230	
PA	Durethan BKV 15	528	0.0205	0.39	250	
PA	Durethan BKV 30	2060	0.195	0.391	250	
PA	Durethan BKV 50	5630	0.313	0.443	250	
PA	Minlon 11 C 40 NC 10	369	$5.2 \cdot 10^{-3}$	0.5216	290	203
PA	Vestamid D14	455	$1.82 \cdot 10^{-3}$	0.741	230	
PA	Vestamid D22	11,910	0.101	0.635	250	
PA	Vestamid E40M-93	429	0.0142	0.495	170	
PA	Vestamid E62M-83	2810	0.0383	0.627	190	
PA	Vestamid L 1700	585	$1.25 \cdot 10^{-3}$	0.868	200	
PA	Vestamid L 1600	335	$7.06 \cdot 10^{-4}$	0.924	200	
PA	Vestamid L 2124	4000	0.0182	0.721	200	
PA	Vestamid X 3715	524,270	74.5	0.588	200	
PA	Zytel 101 F NC10	69	$4.82 \cdot 10^{-4}$	0.4848	290	183
PA	PA 6B3	514	$5.31 \cdot 10^{-5}$	0.4611	255	51

Material	Trade name	A [Pa·s]	B [s]	C [–]	T_B [°C]	T_S [°C]
PA	PA 6B5	13,400	0.08399	0.9028	255	239
PBT	Pocan B 1305	218	$1.47 \cdot 10^{-3}$	0.2186	240	
PBT	Pocan B 1505	427	$3.94 \cdot 10^{-3}$	0.348	250	
PBT	Pocan B 3225	959	0.0419	0.424	250	
PBT	Pocan B 3235	1060	0.153	0.301	250	
PBT	Pocan B 4225	805	0.115	0.292	250	
PBT	Pocan B 4235	1390	0.102	0.355	250	
PBT	Pocan B 7375	544	$4.92 \cdot 10^{-3}$	0.3131	250	
PBT	Pocan B 7503	378	0.0276	0.143	250	
PBT	Pocan S 1506	3170	0.41	0.353	250	
PBT	Vestodur 1000	331	$8.49 \cdot 10^{-4}$	0.764	250	
PBT	Vestodur HI15	1020	$2.45 \cdot 10^{-3}$	0.792	240	
PBT	Vestodur HI17	1410	$5.25 \cdot 10^{-3}$	0.671	240	
PBT	Vestodur HI19	1090	0.0027	0.783	240	
PC	Makrolon 2400	270	$1.2 \cdot 10^{-3}$	0.225	300	
PC	Makrolon 2600	488	$3.22 \cdot 10^{-3}$	0.292	300	
PC	Makrolon 2800	584	$2.59 \cdot 10^{-3}$	0.241	300	
PC	Makrolon 3100	820	$5.36 \cdot 10^{-4}$	0.979	300	
PC	Makrolon 8320	2700	0.15	0.295	300	
PC	Makrolon 9415	1220	0.0325	0.31	300	
PE	Lupolen 1800 H	20,310	1.71	0.628	150	
PE	Lupolen 1810 H	29,500	3.2	0.612	150	
PE	Lupolen 1812 E	41,430	2.86	0.645	150	
PE	Lupolen 1840 D	46,920	3.45	0.642	170	
PE	Lupolen 2000 H	24,160	1.74	0.63	150	
PE	Lupolen 2410 T	673	0.0317	0.607	160	
PE	Lupolen 2440 D	160,210	43.9	0.612	190	

Material	Trade name	A [Pa·s]	B [s]	C [–]	T_B [°C]	T_S [°C]
PE	Lupolen 5021 D	34,550	2.2	0.593	180	
PE	Lupolen 6031 MX	986	0.0183	0.624	180	
PE	Vestolen A 3512 R	11,880	0.0717	0.707	180	
PE	Vestolen A 5041 R	31,370	0.806	0.633	180	
PE	Vestolen A 6042 F	12,270	0.315	0.659	180	
PE	Vestolen A 6016	990	$8.76 \cdot 10^{-3}$	0.745	180	
PE	Vestolen A 6017	886	0.011	0.68	170	
PE	Vestolen P 4500	4770	0.202	0.722	210	
PE	Vestolen P 6500	19,120	0.464	0.794	210	
PE	Baylon 23L 430	663	0.269	0.352	260	
PE	Baylon 28D 780	9280	0.159	0.671	180	
PE	Baylon V18E 464	9530	0.629	0.601	190	
PE	Baylon V18E 464/564	16,570	0.48	0.639	160	
PE	Baylon V18H 460/564	8990	0.35	0.592	160	
PE	Baylon V18H 464/564	9750	0.405	0.585	160	
PE	Baylon V19D 760S	22,250	0.477	0.685	160	
PE	Baylon V22D 464	24,250	0.554	0.638	160	
PE	Baylon V22E 464	22,360	0.589	0.64	160	
PE	Baylon V22F 564/464	9270	0.256	0.632	160	
PE	Hostalen GD 6250	10,380	1.4	0.559	190	
PE	Hostalen LD D 1018	40,290	2.81	0.64	190	
PE	Hostalen GC 7260	906	0.0299	0.44	200	
PE	Hostalen GF 4760	24,370	1.95	0.497	190	
PE	Hostalen GM 7255	86,360	2.97	0.557	200	
PE	Hostalen GM 9240 HAT	94,120	6.38	0.59	200	
PES	PES 200P	839	$6.874 \cdot 10^{-6}$	1.613	350	224
PES	PES 420P	1266	0.0163	0.436	350	222

Material	Trade name	A [Pa·s]	B [s]	C [–]	T_B [°C]	T_S [°C]
PES	PES 600P	5722	0.04081	0.5532	350	236
PES	Victrex 300P	1986	$9.2 \cdot 10^{-3}$	0.6145	360	195
PES	Victrex 520P	1013	$4.04 \cdot 10^{-3}$	0.6734	350	223
PES	Victrex 530P	1406	$7.25 \cdot 10^{-3}$	0.7247	350	215
PET	Rynite 935 NC10	9154	1.534	0.5461	290	180
PMMA	Plexiglas 7N	14,740	0.111	0.817	210	
PMMA	Plexiglas 5N	1600	0.0161	0.759	210	
PMMA	Plexiglas 7H	2,230,000	31.3	0.748	180	
PMMA	Plexiglas ZK37	97,870	1.35	0.704	200	
PMMA	Plexiglas ZK38	99,590	1.41	0.712	200	
PMMA	Resarit 840 ZK	46,520	0.42	0.816	210	
PMMA	PMMA 7H	3445	0.03318	0.8404	260	210
PMMA	PMMA LG	1169	0.0181	0.7	225	159
POM	Delrin 500 NC-10	540	$3.85 \cdot 10^{-3}$	0.6512	215	−60
POM	Hostaform C 13021	1000	$6.91 \cdot 10^{-3}$	0.609	180	
POM	Hostaform C 2521	6300	0.0733	0.503	180	
POM	Hostaform C 27021	459	$3.27 \cdot 10^{-3}$	0.667	180	
POM	Hostaform C 32021	343	$2.39 \cdot 10^{-3}$	0.652	180	
POM	Hostaform C 9021	1370	0.0106	0.583	180	
POM	Hostaform C 9021 GV1	1480	$9.42 \cdot 10^{-3}$	0.694	180	
POM	Hostaform T 1020	13,290	1.35	0.443	180	
POM	Ultraform N2200	1130	0.0159	0.579	190	
POM	POM C9021	967	$5.55 \cdot 10^{-3}$	0.6484	200	−65
POM	POM C9021 GV1/30	2532	0.08479	0.4613	200	−57
PP	Hostalen PPN 1060	2070	$9.07 \cdot 10^{-2}$	0.702	230	
PP	Hostalen PPH 1022	35,970	0.762	0.821	190	
PP	Hostalen PPH 1050	23,720	0.381	0.83	200	
PP	Hostalen PPN 1060	2070	0.0907	0.702	230	

Material	Trade name	A [Pa·s]	B [s]	C [–]	T_B [°C]	T_S [°C]
PP	Hostalen PPN 1752	6530	0.192	0.743	200	
PP	Hostalen PPN 7118 TV	4210	0.16	0.71	220	
PP	Hostalen PPN 7190 TV	6630	0.212	0.733	220	
PP	Hostalen PPN 8008/10	7850	0.374	0.698	200	
PP	Hostalen PPN VP 1034	8590	0.246	0.757	200	
PP	Hostalen PPR 1042	5690	0.319	0.688	200	
PP	Hostalen PPR 7342 FL	3640	0.128	0.74	210	
PP	Hostalen PPT 1070	3230	0.118	0717	200	
PP	Hostalen PPT 8027	1900	0.0715	0.705	200	
PP	Hostalen PPU 1030F	1420	0.0863	0.609	200	
PP	Hostalen PPU 1734	1080	0.0313	0.713	200	
PP	Hostalen PPU 7180 TV	1360	0.0398	0.63	220	
PP	Hostalen PPW 1780 S1	422	0.0124	0.692	200	
PP	Novolen 1320 H	10,970	0.288	0.758	190	
PP	Novolen 1125 NX	2160	0.143	0.679	200	
PP	Novolen 1325 MX	2050	0.0522	0.728	200	
PP	Vestolen P 2300	1070	0.0866	0.652	250	
PP	Vestolen P 2330	738	0.0676	0.635	250	
PP	Vestolen P 3200	1330	0.0833	0.657	250	
PP	Vestolen P 5200	3770	0.0897	0.769	200	
PP	PP 1050	9696	0.1318	0.8104	230	−69
PP	PP 1060	45,300	4.828	0.682	220	0
PP	PP 1120H	3692	0.1324	0.7352	220	−63
PP	PP 1120HX	8289	0.7421	0.6665	220	33
PP	PP 1120L	2049	0.1064	0.6993	220	−16
PP	PP 1320H	4947	0.1389	0.7472	230	88
PP	PP 1320L	1995	0.05315	0.736	220	4

Material	Trade name	A [Pa·s]	B [s]	C [–]	T_B [°C]	T_S [°C]
PP	PP GSE16	5067	0.1411	0.739	200	−1
PP	PP GSE16	8880	0.4085	0.7076	225	56
PP	PP GWM101	1120	0.02839	0.7082	240	1
PP	PP GWM203	1806	0.04304	0.741	240	−23
PP	PP GWM213	1772	0.03866	0.7213	240	−48
PP	PP GWM22	1358	0.03931	0.7272	240	−30
PP	PP GYM202	462	0.027	0.639	240	−44
PP	PP LYM123	394	0.02077	0.6452	240	−60
PP	PP NVP1034	6999	0.3827	0.7273	230	5
PP	PP PXC31403	205	0.01535	0.6012	240	36
PP	PP W1762S1	271	0.01117	0.6289	230	43
PPO	Noryl PX1112	7662	0.06214	0.7849	260	181
PPO	Noryl PX1180	2,839,000	17.43	0.9643	260	283
PS	Polystyrene 165 H	60,110	1.43	0.811	170	105
PS	Polystyrene 168 N	9280	0.198	0.815	210	95
PS	Polystyrene 144	19,640	0.709	0.727	170	
PS	Polystyrene 2710 NAT	19,500	0.618	0.734	190	
PS	Polystyrene 427 D NAT	11,450	0.286	0.741	190	
PS	Polystyrene 432 B NAT	4610	0.136	0.732	190	
PS	Polystyrene 454 H NAT	21,610	0.589	0.796	190	123
PS	Polystyrene 473 D NAT	7440	0.257	0.716	190	
PS	Polystyrene 476 L	15,000	0.493	0.749	190	
PS	Vestyron 214-31	2974	0.0886	0.7773	210	110.3
PS	Vestyron 314-31	13,780	0.43	0.753	180	
PS	Vestyron 512-31	10,770	0.363	0.753	180	
PS	Vestyron 620-31	3173	0.09013	0.7647	210	101.7
PS	PS 143E	1077	0.03542	0.7525	220	101
PS	PS 158K	10,810	0.2939	0.7718	210	116

Material	Trade name	A [Pa·s]	B [s]	C [–]	T_B [°C]	T_S [°C]
PS	PS 165H	4571	0.1148	0.758	220	105
PS	PS 168N	3571	0.07441	0.8162	230	95
PS	PS 4000N	5266	0.2118	0.7328	220	150
PS	PS 454H	27,850	2.146	0.7041	200	123
PS	PS 475K	4674	0.234	0.7225	210	54
PVC	PVC 1745	49,710	0.238	0.8521	175	128
SAN	Luran 368 R	33,910	0.428	0.797	190	
SAN	SAN 360R	6356	8.895	0.7813	220	117
SB	SB 456M	49,040	0.5313	0.9074	210	195
SB	DFD 6600	83,240	7.99	0.674	170	
SB	Hostyren N 3000	6180	0.154	0.771	190	
SB	Hostyren S 3200	4960	0.121	0.766	190	

17.3 Cross–WLF Approach

Material	Trade name	Viscosity exponent n [–]	τ [Pa]	D_1 [Pa·s]	D_2 [K]	A_1 [–]	A_2 [K]	T_G [K]
ABS	Novodur P2H-AT	0.2461	102,380	$4.26 \cdot 10^{+12}$	373.15	30.585	51.6	95
ABS	Novodur P2M	0.2549	94,783.4	$1.24 \cdot 10^{+14}$	373.15	33.863	51.6	100
ASA	Luran S 797 S	0.1774	82,967.1	$8.49 \cdot 10^{+12}$	378.15	28.709	51.6	110
ASA	Luran S 757 R	0.2579	40,707.2	$5.67 \cdot 10^{+12}$	378.15	28.077	51.6	110
ABS/PC	Bayblend T45	0.3436	57,379.5	$1.35 \cdot 10^{+10}$	417.15	22.442	51.6	115
ABS/PC	Bayblend T65	0.333	86,936.8	$4.41 \cdot 10^{+10}$	417.15	24.388	51.6	125
ABS/PC	Bayblend T85	0.3529	88,326.6	$4.31 \cdot 10^{+10}$	417.15	23.793	51.6	135
HDPE	Vestolen A 6016	0.4843	324.393	$6.31 \cdot 10^{+28}$	153.15	63.211	51.6	140
HDPE	Vestolen A 6017	0.5747	46.1697	$8.43 \cdot 10^{+27}$	153.15	61.519	51.6	140
HDPE	Hostalen GB 7250	0.3433	142,000	$3.27 \cdot 10^{+17}$	153.15	39.064	51.6	129
LDPE	Lupolen 1800 H	0.2907	23,500	$2.64 \cdot 10^{+19}$	233.15	44.335	51.6	90

Material	Trade name	Viscosity exponent n [–]	τ [Pa]	D_1 [Pa·s]	D_2 [K]	A_1 [–]	A_2 [K]	T_G [K]
LDPE	Lupolen 1800 S	0.3773	23,500	$1.09 \cdot 10^{+16}$	233.15	37.252	51.6	90
LDPE	Lupolen 1810 D	0.289	23,700	$1.19 \cdot 10^{+19}$	233.15	41.381	51.6	99
LDPE	Lupolen 2410 T	0.2989	42,800	$1.05 \cdot 10^{+17}$	233.15	40.944	51.6	99
PA	Durethan BC 40	0.3218	125,206	$1.27 \cdot 10^{+16}$	323.15	36.479	51.6	170
PA	Vestamid L 2124 NF	0.15	345,000	$1.05 \cdot 10^{+9}$	310.15	17.232	51.6	140
PA	Zytel 101 F NC010	0.2538	614,000	$1.27 \cdot 10^{+13}$	323.15	31.178	51.6	229
PA	Ultramid B3S	0.2647	309,343	$1.68 \cdot 10^{+13}$	323.15	31.206	51.6	195
PA	Ultramid C3U	0.2231	282,548	$4.19 \cdot 10^{+25}$	353.15	68.843	51.6	220
PBT	Pocan B 1305	0.264	403,045	$1.18 \cdot 10^{+15}$	323.15	36.47	51.6	200
PBT	Pocan B 1505	0.2862	443,500	$1.24 \cdot 10^{+16}$	323.15	38.26	51.6	200
PBT	Pocan S 1506	0.3512	188,143	$1.8 \cdot 10^{+20}$	323.15	49.46	51.6	190
PBT	Vestodur 1000 NF	0.2532	316,000	$5.74 \cdot 10^{+17}$	323.15	44.646	51.6	160
PES	Ultrason E 2010	0.15	724,658	$2.29 \cdot 10^{+11}$	492.15	26.532	51.6	225
PES	Ultrason E 2010 G6	0.2032	554,603	$5.85 \cdot 10^{+12}$	492.15	29.746	51.6	225
PES	Victrex 3601 GL20	0.2035	308,380	$2.17 \cdot 10^{+12}$	492.15	29.804	51.6	210
PES	Victrex 4800 G	0.2475	340,800	$4.14 \cdot 10^{+10}$	492.15	22.942	51.6	210
PMMA	Plexiglas 7H	0.3439	5994.99	$1.35 \cdot 10^{+15}$	377.15	30.955	51.6	115
PMMA	Plexiglas 7N	0.2813	67,160.3	$3.26 \cdot 10^{+18}$	377.15	46.032	51.6	105
PMMA	Plexiglas 6N	0.2576	102,856	$1.62 \cdot 10^{+16}$	377.15	41.961	51.6	105
POM	Delrin 500 P NC010	0.1958	378,000	$7.29 \cdot 10^{+11}$	223.15	25.44	51.6	154
POM	Hostaform C 27021 GV3/30	0.444	182,420	$2.9 \cdot 10^{+16}$	223.15	37.943	51.6	145
POM	Hostaform C 9021 GV1/30	0.2931	187,790	$2.9 \cdot 10^{+14}$	223.15	31.436	51.6	146
POM	Hostaform C 13021 RM	0.198	247,532	$6.02 \cdot 10^{+11}$	223.15	25.056	51.6	144
POM	Ultraform N2310P	0.4383	37,205.8	$4.65 \cdot 10^{+23}$	223.15	57.802	51.6	145
PP	Hostalen PPR 1042 M	0.275	20,000	$2.4 \cdot 10^{+14}$	263.15	29.011	51.6	135
PP	Hostalen PPU 1752	0.3237	20,000	$4.78 \cdot 10^{+11}$	263.15	24.087	51.6	135

Material	Trade name	Viscosity exponent n [–]	τ [Pa]	D_1 [Pa·s]	D_2 [K]	A_1 [–]	A_2 [K]	T_G [K]
PP	Hostalen PPT 1052	0.3147	20,000	$1.48 \cdot 10^{+12}$	263.15	24.639	51.6	135
PP	Moplen HP500U	0.3131	28,654.8	$2.38 \cdot 10^{+14}$	263.15	33.915	51.6	130
PS	Styrolution PS 168N	0.1831	39,846.4	$5.99 \cdot 10^{+12}$	373.15	28.331	51.6	106
PS	Styrolution PS 158N	0.2056	36,244.5	$2.96 \cdot 10^{+12}$	373.15	28.314	51.6	106
PS	Sytrolution PS 495N	0.2437	42,499.6	$1.81 \cdot 10^{+11}$	373.15	26.298	51.6	98
SAN	Luran 368 R	0.2011	85,732.7	$3.77 \cdot 10^{+13}$	373.15	32.13	51.6	109

References

Characteristics for the Processing of Thermoplastics, VDMA Verband Deutscher Maschinen- und Anlagenbau e.V., Fachgemeinschaft Gummi- und Kunststoffmaschinen, Part 4 Rheology. Hanser, Munich, 1986, ISBN 3-446-14638-5

18 Solutions

18.1 Section 7.2.2.2, *Task*: Determination of the Viscosity for a Given Shear Rate Using a Master Curve

Solution

- Using Figure 7.10, the temperature shift factor is $a_T = 0.25$.
- In the next step, move the master curve in Figure 7.11.
  - Use the equations for this:

$$\eta_{\text{wanted}} = a_T \cdot \eta_{\text{reference}}$$

and

$$\dot{\gamma}_{\text{wanted}} = \frac{1}{a_T} \cdot \dot{\gamma}_{\text{reference}}$$

  - The curve in Figure 7.11 must shift downwards and to the right!
- The value for the temperature (240 °C) and for the required shear rate (400 s^{-1}) can now be read off the new curve. This is approximately 330 Pa·s.

18.2 Section 7.2.2.3, *Task*: Exercise on Temperature Shift Using Zero Viscosity

Solution: The value determined in this way should be approximately 100 Pa·s. Check the value using Figure 7.12.

Procedure in steps:

1. The zero viscosity for a temperature of 240 °C is determined using Figure 7.12, η_0 (240 °C) = 2000 Pa·s.

2. This value is multiplied by the given (wanted) shear rate of the wanted viscosity

    $$\eta_0 \, (240°C) \cdot \dot{\gamma}_{wanted} = 2000 \text{ Pa} \cdot \text{s} \cdot 1000 \text{ s}^{-1} = 2,000,000 \text{ Pa}$$

3. This value is applied to the x axis in the normalized representation to then go from it vertically upwards to the normalized flow curve (Figure 18.1).

4. Now a value is read on the y axis for the point on the normalized flow curve (Figure 18.1). Here, this is approximately 0.05.

5. This value is now multiplied by the zero viscosity for 240 °C.

    $$\frac{\eta}{\eta_0} \cdot \eta_0 \, (240°C) = 0.05 \cdot 2000 \text{ Pa} \cdot \text{s} = 100 \text{ Pa} \cdot \text{s}$$

The *result* is: $\eta_{tot} = \left(240°C, 1000 \text{ s}^{-1}\right) = 100 \text{ Pa} \cdot \text{s}.$

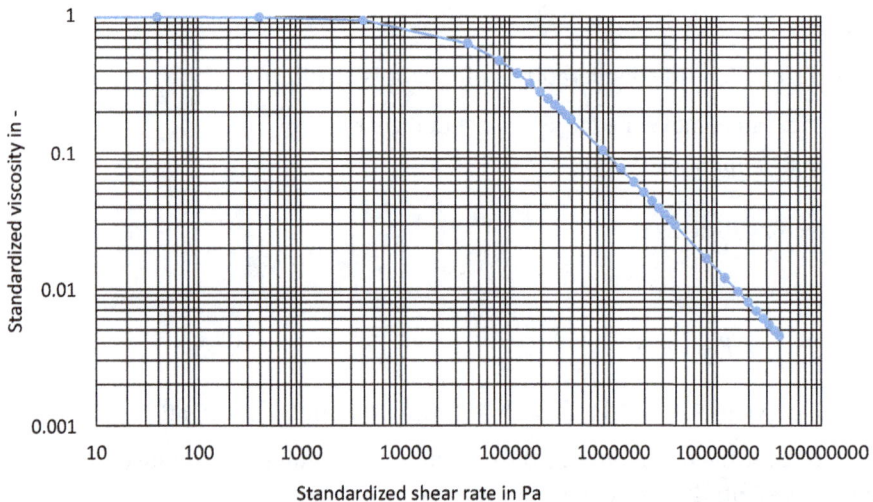

Figure 18.1 Normalized viscosity curve for 220 °C as a function of the normalized shear rate (master curve)

18.3 Section 7.2.4, *Task*: Calculation of the Temperature Shift Factor

Table 18.1 *Solution*: Task on Temperature Shift Factor

Results	190 °C	235 °C	280 °C
Reciprocal absolute temperature	0.00216	0.00198	0.00181
Arrhenius approach	1.65	0.63	0.28
WLF approach	1.72	0.63	0.29

18.4 Section 8.1, *Task*: Graphical Determination of the Constants of the Power Approach

Solution: The procedure is as follows:

1. First, a tangent is drawn to the selected point ($\dot{\gamma} = 10^2\,\mathrm{s}^{-1}$) of the flow curve. To increase the accuracy of the values to be read off, the auxiliary circle method should be used.

2. The gradient m is determined at this point using a gradient triangle. Two pairs of values are read off for the shear stress and the shear rate. The fact that m must always be greater than or equal to 1 ($1 < m < 5$) can be used to check the values.

3. The following applies: $m = \frac{\Delta \log \dot{\gamma}}{\Delta \log \tau}$, where m is the reciprocal gradient for $\tau = f(\dot{\gamma})$ and represents the measure of the shear-thinning viscosity of the material.

4. Then, n is calculated. The following is valid: $n = 1/m$ and $0 < n < 1$.

5. To determine the fluidity, the power-law approach is solved for ϕ. The pair of values (τ, $\dot{\gamma}$) for the previously determined point of the gradient m is inserted into this equation.

6. In the next step, the consistency factor K can be calculated. The units of the fluidity ϕ and the consistency factor K should be noted!

18.5 Section 8.2.1, *Task:* Graphical Determination of the Constants of the Carreau Approach

Table 18.2 *Solution*: Carreau Parameters Determined from the Graph

Parameter	A	B	C
Value	4800	0.56	0.53
Unit	Pa·s	s	-

18.6 Section 8.5, *Task*: Determination of the Consistency Factor and the Viscosity Exponent

Solution

- Viscosity exponent n of the Ostwald and de Waele power-law approach: $n = 0.25$.

- Consistency factor K: The value for K is obtained either as a single viscosity at a shear rate $\dot{\gamma} = 1\ \text{s}^{-1}$ or from the power approach: $K = 20{,}000\ \text{Pa} \cdot \text{s}^{0.25}$.

- Parameters A, B, and C in the Carreau approach:

  - A horizontal line for zero viscosity at 1000 Pa·s is drawn for the solution. This line intersects the viscosity curve. The point of intersection corresponds to the reciprocal transition shear rate B. The gradient C can be determined using Equation 8.11, $-|C| = n - 1$, or with the aid of Figure 8.8.

  - This results in the following values for the three parameters: $A = 1000\ \text{Pa} \cdot \text{s}$, $B = 0.02\ \text{s}$, $C = 0.75$.

18.7 Section 8.6, *Task*: Comparison of the Material Laws (Power Approach and Carreau Approach)

Solution: Figure 18.2 shows that the power approach is a linear equation that only describes the flow curve in certain areas. For low shear rates, the viscosity tends towards very large values. In contrast, the Carreau approach describes the flow curve quite accurately for all shear rates. This also applies to very low shear rates. The validity range (approximation range) of the power approach is between $40\ \text{s}^{-1}$ and $10{,}000\ \text{s}^{-1}$. A closer look at the flow curve in the shear-thinning viscous range reveals that the two curves do not coincide absolutely. The reason for this lies in the material coefficients that describe the gradient of the flow curve. In the power approach this is the parameter m; in the Carreau approach, the parameter P_3. The following applies: $n = 1/m$ and $n - 1 = -|C|$. These parameters do not match here.

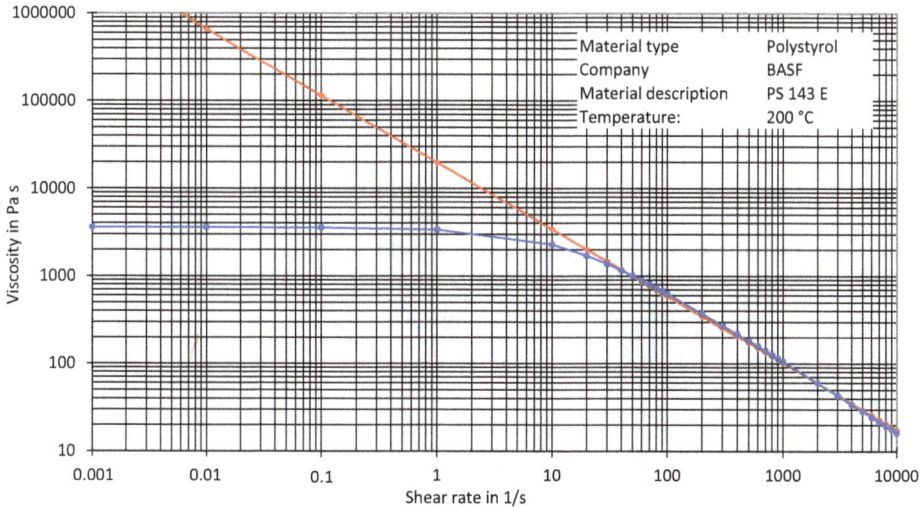

Figure 18.2 *Solution*

18.8 Section 9.4, *Task*: Effect of the Flow Channel on the Melt Volume Flow Rate

Solution

a) $\dot{V}_1 = 245\,\text{mm}^3/\text{s}$; $\dot{V}_2 = 270\,\text{mm}^3/\text{s}$

b) $\dot{V}_1 = 307\,\text{mm}^3/\text{s}$; $\dot{V}_2 = 339\,\text{mm}^3/\text{s}$

c) In the case of b), the calculation results in a larger volume flow rate because the viscosity is determined by the shear stress at the wall. The viscosity is the lowest at the wall due to the high shear rate. Consequently, the volume flow rate is greater. For this reason, the method of representative viscosity is presented in Chapter 10. If this representative point is used to calculate the viscosity, the results match.

18.9 Section 11.2.1.1, *Task*: Pressure Loss: Plate Geometry

Solution: The pressure required to fill the molded part is $\Delta p = 35.6$ bar.

18.10 Section 11.2.1.2, *Task*: Pressure Loss: Disk Geometry

Solution: The pressure loss in the bar sprue is $\Delta p = 263.3$ bar. The pressure required to fill the molded part is $\Delta p = 360.7$ bar. If the viscosities are calculated using the Carreau approach, the resulting pressure is slightly different. For the bar sprue, $\Delta p = 243.7$ bar is obtained, and for filling the DVD, a pressure requirement of $\Delta p = 386.1$ bar results. The difference is due to the slightly different material parameters.

18.11 Section 11.2.2, Influence of Material Properties on the Manufacturing Process

Solution 1: $\Delta p_{material\ 1} = 228$ bar and $\Delta p_{material\ 2} = 222$ bar.

Solution 2: $\Delta p_{material\ 1} = 186$ bar and $\Delta p_{material\ 2} = 192$ bar.

Solution 3: A closer look at the two flow curves reveals that they intersect twice (Figure 18.3). This is due to the different parameters A, B, and C.

Solution 4: $\eta_{material\ 1} = 122$ Pa·s and $\eta_{material\ 2} = 162$ Pa·s.

The standard temperature describes the temperature dependence of the plastic. The glass transition temperature of material 2 is much lower than that of material 1, which means that the viscosity of material 2 is much less temperature-dependent than that of material 1. The temperature dependence of the plastic and the glass transition temperature could be used to conclude that the plastic is amorphous (material 1) or semicrystalline (material 2).

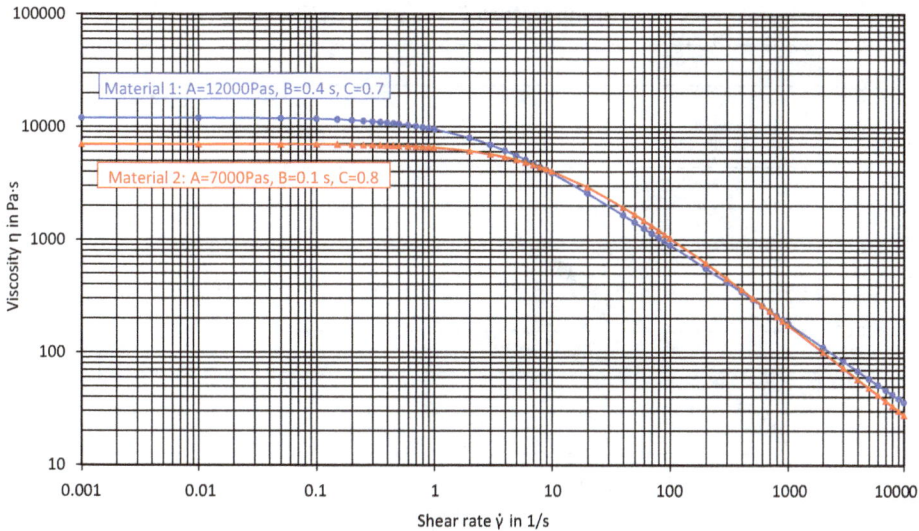

Figure 18.3 Graphical *solution*

18.12 Section 11.2.3, *Task*: Pressure Losses during Injection Molding and the Resulting Real Locking Force

Solution: The buoyant force is $F_A = 65$ kN and the locking force is 78 kN.

18.13 Section 11.2.4, *Task*: Consideration of Dissipation and Cooling Effects (Non-Isothermal Flow)

Solution

- The power approach initially results in a viscosity of $\eta = 229$ Pa·s and a filling pressure of $\Delta p = 76.4$ bar.

- The actual average temperature is $\vartheta = 5.22\,°C + 183.92\,°C = 189.14\,°C$ and, therefore, the actual viscosity is $\eta = 276$ Pa·s. The filling pressure thus increases to $\Delta p = 92.1$ bar.

- The solution of the partial differential equation can lead to Figure 18.4.

Temperature Curve Plate Half

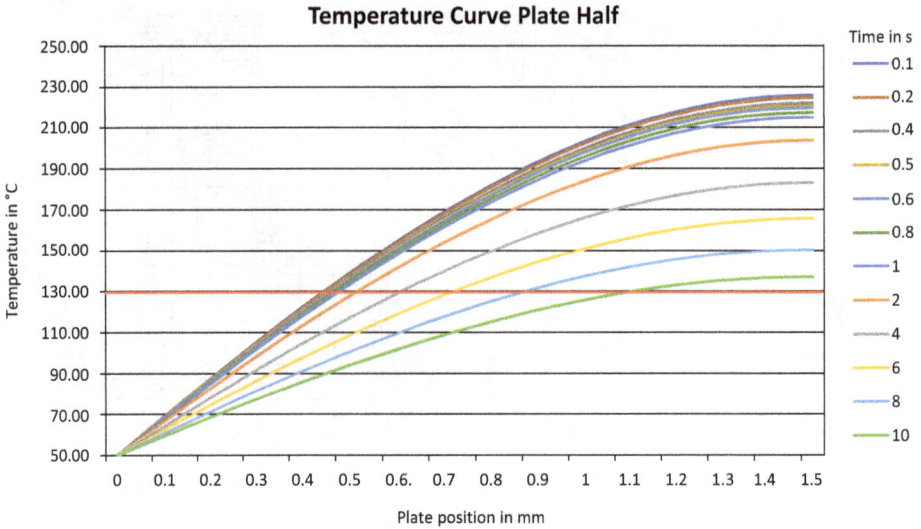

Figure 18.4 *Solution*: representation of the partial differential equation

18.14 Section 11.3, *Task*: Optimum Filling Time

Solution

- The optimum injection rate in this case is $v_m = 70.4$ mm/s.

- This results in a filling time of 8.5 s.

- If the cavity temperature is lowered to 40 °C, the injection rate increases to $v_m = 76.3$ mm/s.

18.15 Section 12.3.1, *Task*: Double Mold with Different Melt Distribution Systems

Solution

a) $D_{A1} = 1.67$ mm and $D_{A2} = 1.83$ mm.

b) $D_{A1} = 1.67$ mm and $D_{A2} = 2.98$ mm.

c) With the viscosity exponent $n = 0.4$, the given system can no longer be balanced. One possible solution would be to increase the length L_A. An extension to $L_A = 5$ mm leads to the diameters $D_{A1} = 1.67$ mm and $D_{A2} = 4.58$ mm.

18.16 Section 12.3.2, *Task*: Eight-Cavity Mold with Different Melt Distribution Systems

Solution

a) $D_{A2} = 1.649$ mm.

b) The maximum wall shear rate is, for Newtonian behavior, $\dot{\gamma}_N = 113.375$ s^{-1} and with the Weissenberg–Rabinowitsch correction, $\dot{\gamma}_S = 235.041$ s^{-1}. This is to be regarded as critical. According to the materials table in Chapter 17, the constants for the power approach no longer apply for this shear rate.

18.17 Section 12.3.3, *Task*: Six-Cavity Mold with Different Melt Distribution Systems

Solution

a) The diameter is $d_3 = 6$ mm.

b) The diameter is $d_1 = 4.98$ mm.

18.18 Section 12.3.4, *Task*: Family Mold with Two Different Cavities

Solution

1. In any case, the balancing must be carried out mathematically and rheologically, as the cavities are not identical. Natural rheological balancing makes no sense here.

2. The aim of balancing is to ensure that both cavities are filled at the same time!

3. $D_6 = 2.26$ mm.

18.19 Section 12.4.5, *Task*: Analytical Balancing of a Fish-Tail Distributor

Solution: The maximum distributor radius after the first calculation is $R_0 = 8.89$ mm and the maximum island length is $y_0 = 329$ mm. After the first iteration, $R_0 = 8.54$ mm.

18.20 Section 12.4.6, *Task*: Analytical Balancing of a Coat-Hanger Distributor

Solution: The operating-point-independent maximum distributor radius is $R_0 = 10.03$ mm, the maximum island length is $y_0 = 138.5$ mm, the average residence time is $t_V = 4.5$ s, and the pressure requirement is $\Delta p = 64.3$ bar. See Figure 18.5.

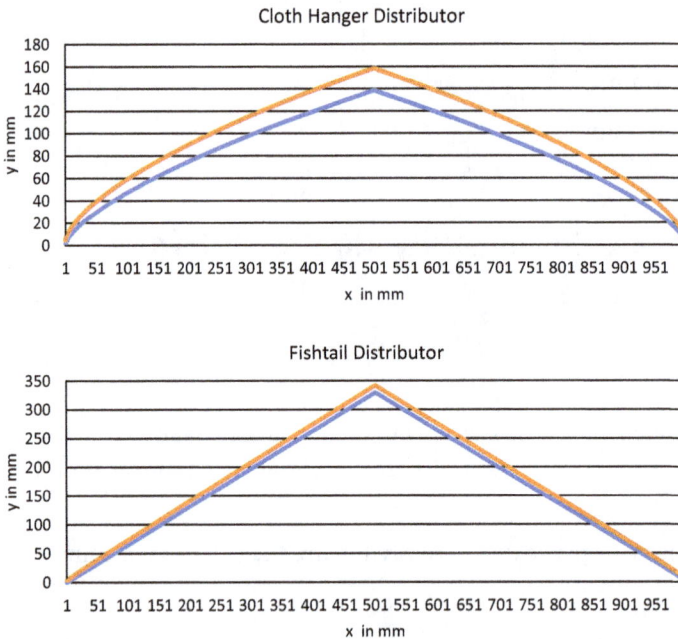

Figure 18.5 Graphical representation of the results

18.21 Section 12.4.7, *Task*: Numerical Balancing of a Wide Slot Nozzle with a Coat-Hanger Distributor with Segments

Solution: See Figure 18.6.

Coat Hanger Distributor (Numeric Solution)

Entry values

Type of calculation	⦿ operating point dependent design		
	○ operating point independent design		
Dimensions of nozzle	Height	H	10 mm
	Width	B	750 mm
	Max. island length	y_0	150 mm
Exit velocity		v_exit	25 m/min
Material constant of the Carreau-approach		P_1	3445 Pa*s
		P_2	0.03318 s
		P_3	0.8404 -
Temperature	Manufacturing temperature	T_ver	275 °C
	Reference temperature	T_bez	260 °C
	Standard temperature	T_st	210 °C

Start calculation

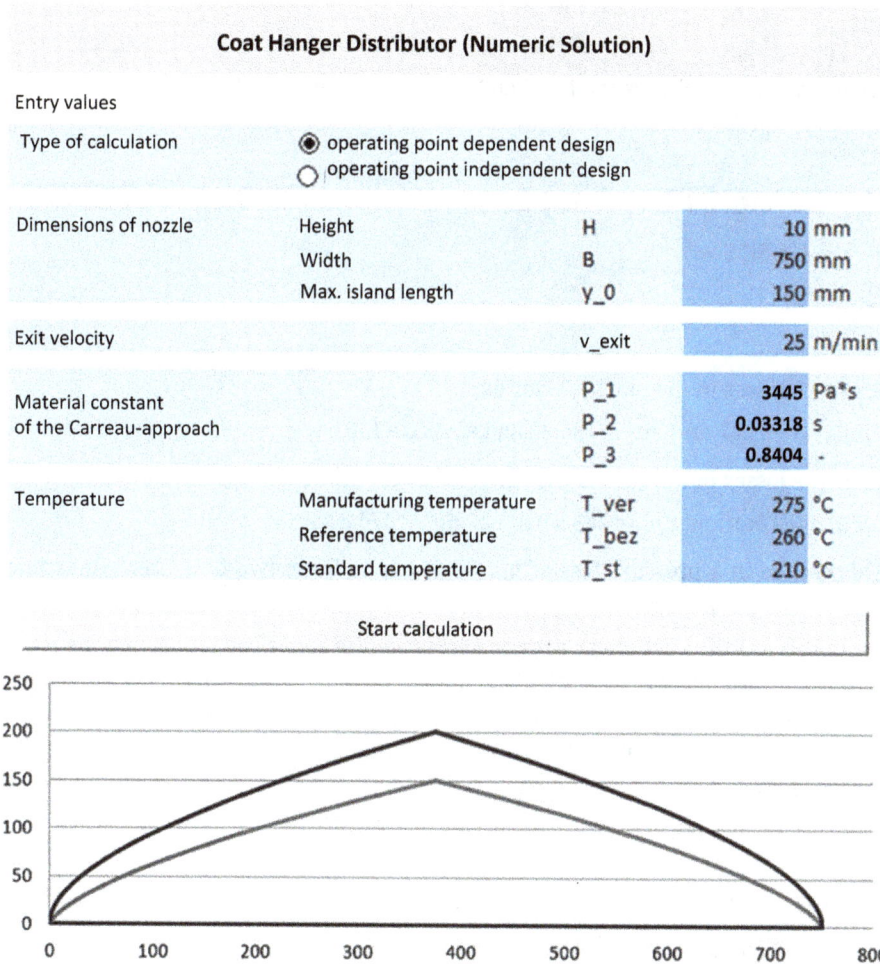

Figure 18.6 Example for the evaluation of the numerical calculation of a coat-hanger distributor

18.22 Section 12.4.8, *Task*: Calculation of the Output Rate of an Extruder

Solution: The output of the extrusion line is 261 kg/h.

18.23 Section 12.4.9, *Task*: Design of a Slot Die

Solution: The vertical forces are $F_V = 67{,}008$ N and the horizontal forces are $F_H = 670$ N.

18.24 Section 13.1, *Task*: Extensional and Shear Pressure Losses

Solution

- Pressure losses in the machine nozzle:

  $\Delta p_{shear} = 13.9$ bar and $\Delta p_{ext} = 136.3$ bar, i.e., $\Delta p_{tot} = 150$ bar

- Pressure losses in the gate of the "quarter circle disk":

  $\Delta p_{shear} = 6.2$ bar and $\Delta p_{ext} = 268.5$ bar, i.e., $\Delta p_{tot} = 274.8$ bar

- This results in a pressure loss of $\Delta p_{tot} = 425$ bar in these two flow areas (machine nozzle and gate)!

Added to this are the pure shear pressure losses in the sprue rod, in the sub-distributor, and in the molded part. The calculated value for the extensional pressure losses is slightly too high, as the extensional viscosity is determined in the calculation using the Trouton ratio. This assumption is not entirely correct, as the ratio only applies for low shear rates.

18.25 Section 14.5, *Task*: Proof of the Independence of the Filling Pattern Method from the Shear-Thinning Viscosity

Solution: If the specified steps are carried out, the solution to the problem shows that the filling pattern method does not depend on the material parameters of the plastic melt. This had to be proven.

18.26 Section 16.3.1, *Task*: Total Pressure Drop in an Extrusion Die

Solution

1. The total volume flow rate is $61.95 \, \text{cm}^3/\text{s}$. This results in a volume flow rate per hole of $0.30975 \, \text{cm}^3/\text{s}$. The apparent wall shear rate is $116.85 \, \text{s}^{-1}$. The viscosity exponent n is required to correct the shear rate. The calculation is made using two pairs of values from Figure 16.32 or Table 16.3 This results in $n = 0.26$. The correction is made using the Weissenberg–Rabinowitsch method. This results in a true wall shear rate of $200 \, \text{s}^{-1}$. For this shear rate, the viscous (shear stress $\tau = 1.41 \cdot 10^5 \, \text{Pa}$) and the elastic (first normal stress difference $N_1 = 4.93 \cdot 10^5 \, \text{Pa}$) components of the pressure loss can now be read from Table 16.3. Using Equation 16.9, this results in a total pressure loss of $\Delta p_{tot} = 18.8 \, \text{bar} + 17.26 \, \text{bar} = 36.06 \, \text{bar}$.

2. If the bore length is halved, the shear pressure loss is also halved and the total pressure loss is $\Delta p_{tot} = 26.66 \, \text{bar}$.

3. Melt fracture can be calculated using the Weissenberg number (Equation 16.8). This results in $We = 3.5 < 4.6$, which means that no melt fracture is to be expected!

Index

www.ingramcontent.com/pod-product-compliance
Lightning Source LLC
Chambersburg PA
CBHW081458190326
41458CB00015B/5278